METHODS IN MOLECULAR BIOLOGY

*Series Editor*
**John M. Walker**
**School of Life and Medical Sciences**
**University of Hertfordshire**
**Hatfield, Hertfordshire, AL10 9AB, UK**

For further volumes:
http://www.springer.com/series/7651

# Protein Amyloid Aggregation

## Methods and Protocols

Edited by

**David Eliezer**

*Weill Cornell Medical College, New York, NY, USA*

*Editor*
David Eliezer
Weill Cornell Medical College
New York, NY, USA

ISSN 1064-3745 ISSN 1940-6029 (electronic)
Methods in Molecular Biology
ISBN 978-1-4939-2977-1 ISBN 978-1-4939-2978-8 (eBook)
DOI 10.1007/978-1-4939-2978-8

Library of Congress Control Number: 2015949999

Springer New York Heidelberg Dordrecht London


Printed on acid-free paper

Humana Press is a brand of Springer
Springer Science+Business Media LLC New York is part of Springer Science+Business Media (www.springer.com)

# Preface

This volume is focused on methods for the characterization of aggregation processes that lead to the formation of amyloid fibrils and amyloid oligomers that feature in the etiology of a variety of human disorders collectively known as amyloidoses. The focus includes techniques for visualizing early steps on the amyloid formation pathway, methods for capturing and characterizing oligomeric, potentially toxic, intermediates, strategies for preparing and characterizing mature amyloid fibrils, and approaches for understanding templating and transmission of amyloid aggregates. The target audience includes biochemists and biophysicists with an interest in elucidating the mechanisms of protein amyloid formation, as well as chemists, pharmacologists, and clinicians with an interest in leveraging an understanding of such mechanisms for the purpose of therapeutic development.

Chapter 1 treats methods to prepare posttranslationally modified amyloid proteins with a focus on the production of phosphorylated forms of the Parkinson's disease-associated protein alpha-synuclein. Posttranslational modifications of synuclein and other amyloid proteins are often associated with disease pathology yet their role in disease etiology has remained unclear, in part because of the difficulty of producing homogeneously modified proteins for in vitro studies. Chapter 2 describes both chemical synthesis and native chemical ligation strategies for the production of isotopically labeled amyloid proteins for characterization by spectroscopic techniques such as two-dimensional infrared spectroscopy or NMR spectroscopy. Detailed procedures are provided for the production of the diabetes-linked amyloidogenic peptide amylin as well as for the amyloid-forming protein alpha-beta crystallin. Chapter 3 describes the use of paramagnetic relaxation enhancement NMR spectroscopy to detect and describe the earliest interactions between amyloid monomers, using alpha-synuclein as an example, while Chapter 4 describes the use of circular dichroism spectroscopy to detect the presence of helical intermediates during the formation of amyloid fibrils, in this case using amylin as an example. The formation of helical intermediates has been implicated as a potentially critical step in the formation of fibrillar aggregates by a number of amyloid proteins. Chapter 5 describes innovative applications of fluorescence correlation spectroscopy to measure oligomer formation both for purified amyloid proteins in vitro and also for fluorescently labeled amyloid proteins in intact cells, providing a unique approach to observing the amyloid formation process in vivo using huntingtin exon 1 polypeptides as an example. Chapter 6 describes the application of advanced Raman Spectroscopy methods for the characterization of the process by which amyloid fibrils form, allowing for the characterization of different fibril regions, such as the core or the surface, as well as for determining the order in which secondary structure is formed during fibril assembly. The method is illustrated using amyloid formation by lysozyme. Chapter 7 describes an innovative use of quantitative electron microscopy to determine the parameters that govern the fibrillization kinetics of the Alzheimer's protein tau.

Chapters 8, 9, and 10 focus on the characterization of oligomeric species formed on the pathway of amyloid fibril assembly. Chapter 8 describes the use of a powerful combination of mass spectrometry techniques: ion mobility spectrometry and electrospray ionization, in order to characterize the gas phase collision cross sections of different oligomeric species that are present simultaneously in samples undergoing amyloid fibril assembly. The heterogeneity of species in such samples has long been a major hurdle to characterizing the fibril formation process, and this technique is one of very few that is able to provide a simultaneous analysis and resolution of different species. An application to the aggregation of beta-2-microglobulin is described. Chapter 9 describes techniques to produce stable homogeneous preparations of oligomers formed by alpha-synuclein as well as a variety of methods to characterize these oligomers, including SDS-PAGE, circular dichroism, electron microscopy, atomic force microscopy, Fourier-transform infrared spectroscopy, and fluorescence assays of phospholipid vesicle permeabilization. Chapter 10 also treats the characterization of oligomers formed from the protein alpha-synuclein but describes the application of a novel single-molecule fluorescence photobleaching approach to characterizing the number density of the oligomers. Despite intense efforts, reliably determining the distributions of the number of molecules in amyloid oligomers has remained a frustrating challenge, and this method provides a reliable solution to this long-standing issue.

Chapters 11 through 14 describe methods for the characterization of mature amyloid fibrils. Chapter 11 describes protocols for the preparation of alpha-synuclein amyloid fibrils for characterization by solid-state NMR spectroscopy. Solid-state NMR has provided some of our most detailed insights into the structures of amyloid fibrils of a number of proteins and continues to be at the forefront of amyloid fibril structure determination. Key to the success of this method, however, is the production of high quality samples of fibrils, and this chapter provides an avenue for achieving this. Chapter 12 describes the characterization of amyloid fibrils formed from the protein tau using electron paramagnetic resonance spectroscopy, which can produce information on local environments within fibrils, provide powerful distance constraints on fibril conformation, and also distinguish different fibril populations. Chapter 13 describes the preparation of amyloid fibrils for structure determination by x-ray crystallography. This approach has provided the highest resolution structural views of the central spines of a variety of amyloid-forming sequences. In addition, this method has recently succeeded in resolving the atomic resolution structures of amyloid oligomers, and the preparation of such samples is also described. Chapter 14 describes methods for the analysis of amyloid fibril structure using amide proton hydrogen exchange monitored by NMR spectroscopy. This approach can provide information, at the single residue level, on solvent accessible regions and hydrogen bonding patterns within fibrils.

Chapters 15, 16, and 17 describe computational approaches towards understanding the structure and assembly of amyloid aggregates. This is a rapidly growing area that provides novel insights that are difficult or impossible to obtain via experimental methods. Chapter 15 describes a protocol for executing replica exchange molecular dynamics simulations of amyloid proteins, using a fragment of the protein tau as an example. Chapter 16 describes the use of molecular dynamics simulations to model the structures of amyloid ion channels, as well as to calculate their ion permeation properties, using the Alzheimer's amyloid-beta (A-beta) peptide as an example. Ion channel formation by amyloid oligomers is an important potential mechanism for the toxic effects of such species. Chapter 17 describes a protocol for a method that employs Bayesian statistics to leverage experimental data on amyloid proteins for the identification of ensembles of model structures that "best" represent the experimental observables, including statistical parameters to evaluate the significance of various properties of the resulting ensembles.

Chapters 18, 19, and 20 describe methods for evaluating processes that may influence the toxicity and pathology of amyloid proteins in vivo. Chapter 18 describes experimental methods for evaluating the ability of amyloid species to permeabilize phospholipid bilayers using amylin as an example. Indiscriminant membrane permeabilization by amyloid species has been proposed as a potentially general mechanism for their toxicity. Chapter 19 describes protocols to study the transmission of amyloid species between cells, using the example of alpha-synuclein. Cell-to-cell spread of amyloid species, potentially via a prion-like mechanism, has emerged as an area of tremendous interest and may explain observations of how amyloid pathology spreads through the body and brain in neurodegenerative disease. Chapter 20 describes the preparation of amyloid fibrils seeded using material obtained from diseased human or mouse brain tissues in a way that preserves the ultrastructure of the material in the original tissues, using Alzheimer's brain derived A-beta fibrils as an example. Such samples can then be characterized structurally, in this example using solid-state NMR, in order to delineate the structural basis for different disease-associated fibrillar states and to characterize amyloid strains. The possibility that different amyloid strains, corresponding to different molecular structures of amyloid fibrils, may be associated with different disease presentation and phenotype goes hand-in-hand with the idea that a single or a small number of nucleating aggregation events in the brain or body can lead, via cell-to-cell transmission, to a single or a few dominant fibril forms.

In summary, this volume presents modern methods and protocols for characterizing amyloid aggregation, amyloid aggregates, and amyloid spread and toxicity from the very earliest manifestations in the form of nucleating conformers or transiently interacting monomers, through the formation of helical intermediates, oligomeric species, membrane-bound oligomers or channels, and finally arriving at mature amyloid fibrils, which can be spread from cell to cell, and the molecular details of which may underlie the specific features of human disease presentation.

*New York, NY, USA* *David Eliezer*

# Contents

## Contributors

ANDISHEH ABEDINI • *Diabetes Research Program, Division of Endocrinology, Diabetes and Metabolism, NYU School of Medicine, New York, NY, USA*

ANDREI T. ALEXANDRESCU • *Department of Molecular and Cell Biology, University of Connecticut, Storrs, CT, USA*

FERNANDO TERAN ARCE • *Department of Bioengineering, Materials Science Program, University of California, San Diego, La Jolla, CA, USA; Department of Mechanical and Aerospace Engineering, Materials Science Program, University of California, San Diego, La Jolla, CA, USA*

ALISON E. ASHCROFT • *Astbury Centre for Structural Molecular Biology, School of Molecular and Cellular Biology, University of Leeds, Leeds, UK*

EUN-JIN BAE • *Neuroscience Research Institute and Department of Medicine, Seoul National University College of Medicine, Jongro-gu, South Korea*

ALEXANDER M. BARCLAY • *Center for Biophysics and Computational Biology, University of Illinois at Urbana-Champaign, Urbana, IL, USA*

JEAN BAUM • *Department of Chemistry and Chemical Biology, Rutgers University, Piscataway, NJ, USA; Center for Integrative Proteomics Research, Rutgers University, Piscataway, NJ, USA*

PING CAO • *Structural Biology Program, Kimmel Center for Biology and Medicine at the Skirball Institute, New York University School of Medicine, New York, NY, USA*

JOSEPH M. COURTNEY • *Department of Chemistry, University of Illinois at Urbana-Champaign, Urbana, IL, USA*

KENNETH W. DROMBOSKY • *Department of Structural Biology and Pittsburgh Institute for Neurodegenerative Diseases, University of Pittsburgh School of Medicine, Pittsburgh, PA, USA*

DAVID EISENBERG • *Department of Biological Chemistry, Howard Hughes Medical Institute (HHMI), University of California Los Angeles (UCLA), Los Angeles, CA, USA*

BRUNO FAUVET • *Laboratory of Molecular and Chemical Biology of Neurodegeneration, Brain Mind Institute, Ecole Polytechnique Fédérale de Lausanne, Lausanne, Switzerland*

CHARLES K. FISHER • *Physics Department, Boston University, Boston, MA, USA*

ALAN L. GILLMAN • *Department of Bioengineering, Materials Science Program, University of California, San Diego, La Jolla, CA, USA*

MAKSIM GRECHKO • *Department of Chemistry, University of Wisconsin-Madison, Madison, WI, USA*

THOMAS GURRY • *Computational and Systems Biology Initiative, Massachusetts Institute of Technology, Cambridge, MA, USA; Research Laboratory of Electronics, Massachusetts Institute of Technology, Cambridge, MA, USA*

JOSEPH D. HANDEN • *Department of Chemistry, University at Albany, SUNY, Albany, NY, USA*

CAROL J. HUSEBY • *Interdisciplinary Biophysics Graduate Program, Department of Molecular and Cellular Biochemistry, The Ohio State University College of Medicine, Columbus, OH, USA*

HYUNBUM JANG • *Cancer and Inflammation Program, Leidos Biomedical Research, Inc., Frederick National Laboratory for Cancer Research, National Cancer Institute at Frederick, Frederick, MD, USA*

MARIA K. JANOWSKA • *Department of Chemistry and Chemical Biology, Rutgers University, Piscataway, NJ, USA*

BRUCE L. KAGAN • *Department of Psychiatry, David Geffen School of Medicine, Semel Institute for Neuroscience and Human Behavior, University of California, Los Angeles, CA, USA*

JEFF KURET • *Department of Molecular and Cellular Biochemistry, The Ohio State University College of Medicine, Columbus, OH, USA*

RATNESH LAL • *Department of Bioengineering, Materials Science Program, University of California, San Diego, La Jolla, CA, USA; Department of Mechanical and Aerospace Engineering, Materials Science Program, University of California, San Diego, La Jolla, CA, USA*

MEYTAL LANDAU • *Department of Biology, Technion-Israel Institute of Technology, Haifa, Israel*

HILAL A. LASHUEL • *Laboratory of Molecular and Chemical Biology of Neurodegeneration, Brain Mind Institute, Ecole Polytechnique Fédérale de Lausanne, Lausanne, Switzerland; Qatar Biomedical Research Institute, Hamad Bin Khalifa University, Qatar Foundation, Doha, Qatar*

IGOR K. LEDNEV • *Department of Chemistry, University at Albany, SUNY, Albany, NY, USA*

HE-JIN LEE • *Institute of Biomedical Science and Technology, Konkuk University, Seoul, South Korea; Department of Anatomy, School of Medicine, Konkuk University, Seoul, South Korea*

JOON LEE • *Department of Mechanical and Aerospace Engineering, Materials Science Program, University of California, San Diego, La Jolla, CA, USA*

SEUNG-JAE LEE • *Neuroscience Research Institute and Department of Medicine, Seoul National University College of Medicine, Jongro-gu, South Korea*

ZACHARY A. LEVINE • *Department of Chemistry and Biochemistry, University of California Santa Barbara, Santa Barbara, CA, USA; Department of Physics, University of California Santa Barbara, Santa Barbara, CA, USA*

NIKOLAI LORENZEN • *Department of Molecular Biology, Center for Insoluble Protein Structures (inSPIN), Interdisciplinary Nanoscience Center (iNANO), Aarhus University, Aarhus C, Denmark; Department of Protein Biophysics and Formulation, Novo Nordisk A/S, Måløv, Denmark*

JUN-XIA LU • *Laboratory of Chemical Physics, NIDDK, National Institutes of Health, Bethesda, MD, USA*

MARTIN MARGITTAI • *Department of Chemistry and Biochemistry, University of Denver, Denver, CO, USA*

STEPHEN C. MEREDITH • *Department of Pathology, Biochemistry and Molecular Biology, The University of Chicago, Chicago, IL, USA*

VIRGINIA MEYER • *Department of Chemistry and Biochemistry, University of Denver, Denver, CO, USA*

SEAN D. MORAN • *Department of Chemistry, University of Wisconsin-Madison, Madison, WI, USA*

ASHER MOSHE • *Department of Biology, Technion-Israel Institute of Technology, Haifa, Israel*

RUTH NUSSINOV • *Cancer and Inflammation Program, Leidos Biomedical Research, Inc., Frederick National Laboratory for Cancer Research, National Cancer Institute at Frederick, Frederick, MD, USA; Department of Human Molecular Genetics and Biochemistry, Sackler School of Medicine, Tel Aviv University, Tel Aviv, Israel*

DANIEL E. OTZEN • *Department of Molecular Biology, Center for Insoluble Protein Structures (inSPIN), Interdisciplinary Nanoscience Center (iNANO), Aarhus University, Aarhus, Denmark*

WOJCIECH PASLAWSKI • *Department of Molecular Biology, Center for Insoluble Protein Structures (inSPIN), Interdisciplinary Nanoscience Center (iNANO), Aarhus University, Aarhus C, Denmark; Department of Clinical Neuroscience, Karolinska Institutet, Stockholm, Sweden*

SHEENA E. RADFORD • *Astbury Centre for Structural Molecular Biology, School of Molecular and Cellular Biology, University of Leeds, Leeds, UK*

DANIEL P. RALEIGH • *Department of Chemistry, Stony Brook University, Stony Brook, NY, USA; Graduate Program in Biochemistry and Structural Biology, Stony Brook University, Stony Brook, NY, USA*

SRINIVASAN RAMACHANDRAN • *Department of Bioengineering, Materials Science Program, University of California, San Diego, La Jolla, CA, USA; Department of Mechanical and Aerospace Engineering, Materials Science Program, University of California, San Diego, La Jolla, CA, USA*

CHAD M. RIENSTRA • *Department of Chemistry, University of Illinois at Urbana-Champaign, Urbana, IL, USA; Department of Biochemistry, University of Illinois at Urbana-Champaign, Urbana, IL, USA; Center for Biophysics and Computational Biology, University of Illinois at Urbana-Champaign, Urbana, IL, USA*

BANKANIDHI SAHOO • *Department of Structural Biology and Pittsburgh Institute for Neurodegenerative Diseases, University of Pittsburgh School of Medicine, Pittsburgh, PA, USA*

CHARLOTTE A. SCARFF • *Astbury Centre for Structural Molecular Biology, School of Molecular and Cellular Biology, University of Leeds, Leeds, UK*

KATHRYN P. SCHERPELZ • *Department of Biochemistry and Molecular Biology, The University of Chicago, Chicago, IL, USA*

NATHALIE SCHILDERINK • *Nanobiophysics, MESA+ Institute for Nanotechnology, Faculty of Science and Technology, University of Twente, Enschede, The Netherlands; Nanobiophysics, MIRA Institute for Biomedical Technology and Technical Medicine, University of Twente, Enschede, The Netherlands*

MOLLY SCHMIDT • *Research Laboratory of Electronics, Massachusetts Institute of Technology, Cambridge, MA, USA; Department of Electrical Engineering and Computer Science, Massachusetts Institute of Technology, Cambridge, MA, USA*

JOAN-EMMA SHEA • *Department of Chemistry and Biochemistry, University of California Santa Barbara, Santa Barbara, CA, USA; Department of Physics, University of California Santa Barbara, Santa Barbara, CA, USA*

COLLIN M. STULTZ • *Computational and Systems Biology Initiative, Massachusetts Institute of Technology, Cambridge, MA, USA; Research Laboratory of Electronics, Massachusetts Institute of Technology, Cambridge, MA, USA; Department of Electrical Engineering and Computer Science, Massachusetts Institute of Technology, Cambridge, MA, USA;*

*The Institute for Medical Engineering and Science, Massachusetts Institute of Technology, Cambridge, MA, USA*

VINOD SUBRAMANIAM • *FOM Institute AMOLF, Amsterdam, The Netherlands; Nanobiophysics, MESA+ Institute for Nanotechnology, Faculty of Science and Technology, University of Twente, Enschede, The Netherlands; Nanobiophysics, MIRA Institute for Biomedical Technology and Technical Medicine, University of Twente, Enschede, The Netherlands*

MARCUS D. TUTTLE • *Department of Chemistry, University of Illinois at Urbana-Champaign, Urbana, IL, USA*

ROBERT TYCKO • *Laboratory of Chemical Physics, NIDDK, National Institutes of Health, Bethesda, MD, USA*

RONALD WETZEL • *Department of Structural Biology and Pittsburgh Institute for Neurodegenerative Diseases, University of Pittsburgh School of Medicine, Pittsburgh, PA, USA*

MARTIN T. ZANNI • *Department of Chemistry, University of Wisconsin-Madison, Madison, WI, USA*

TIANQI O. ZHANG • *Department of Chemistry, University of Wisconsin-Madison, Madison, WI, USA*

NIELS ZIJLSTRA • *FOM Institute AMOLF, Amsterdam, The Netherlands; Nanobiophysics, MESA+ Institute for Nanotechnology, Faculty of Science and Technology, University of Twente, Enschede, The Netherlands*

# Part I

## Labeling Strategies

# Chapter 1

# Semisynthesis and Enzymatic Preparation of Post-translationally Modified α-Synuclein

**Bruno Fauvet and Hilal A. Lashuel**

## Abstract

Posttranslational modifications (PTMs) serve as molecular switches for regulating protein folding, function, and interactome and have been implicated in the misfolding and amyloid formation by several proteins linked to neurodegenerative diseases, including Alzheimer's and Parkinson's disease. Understanding the role of individual PTMs in protein misfolding and aggregation requires the preparation of site-specifically modified proteins, as well as the identification of the enzymes involved in regulating these PTMs. Recently, our group has pioneered the development of enzymatic, synthetic, and semisynthetic strategies that allow site-specific introduction of PTMs at single or multiple sites and generation of modified proteins in milligram quantities. In this chapter, we provide detailed description of enzymatic and semisynthetic strategies for the generation of the phosphorylated α-Synuclein (α-Syn) at S129, (pS129), which has been identified as a pathological hallmark of Parkinson's disease. The semisynthetic method described for generation of α-Syn-pS129 requires expertise with protein chemical ligation, but can be used to incorporate other PTMs (single or multiple) within the α-Syn C-terminus if desired. On the other hand, the in vitro kinase-mediated phosphorylation strategy does not require any special setup and is rather easy to apply, but its application is restricted to the generation of α-Syn_pS129. These methods have the potential to increase the availability of pure and homogenous modified α-Syn reagents, which may be used as standards in numerous applications, including the search for potential biomarkers of synucleinopathies.

**Key words** Parkinson's disease, Posttranslation modification, Alpha-synuclein, Amyloid, Phosphorylation, Semisynthesis, Native chemical ligation, Desulfurization

## 1 Introduction

The misfolding and aggregation of the normally soluble neuronal protein α-Synuclein (α-Syn) play key roles in the pathogenesis of Parkinson's disease (PD) and several associated neurodegenerative disorders colloquially known as synucleinopathies [1]. One of most widely recognized clinical hallmarks of PD consists of intracellular proteinaceous inclusions known as Lewy Bodies (LBs), composed mainly of insoluble α-Syn amyloid fibrils. This fibrillar form of α-Syn bears several covalent posttranslational modifications (PTMs) including phosphorylation, ubiquitination, nitrative

David Eliezer (ed.), *Protein Amyloid Aggregation: Methods and Protocols*, Methods in Molecular Biology, vol. 1345, DOI 10.1007/978-1-4939-2978-8_1, 

oxidation, and proteolysis [2], suggesting that these modifications play a role in α-Syn aggregation, LB formation, and/or clearance [2, 3]. The most abundant PTM in α-Syn is serine 129 phosphorylation; its distinctive upregulation under pathological conditions [4–6] and in the brain of transgenic mice of synucleinopathies has triggered great interest in this PTM as a potential biomarker for early diagnosis of synucleinopathies [6–8] and a potential target for therapeutic intervention. In addition to its probable role in the pathogenesis of PD, several recent reports suggest that phosphorylation at S129 plays important roles in regulating the α-Syn degradation via autophagy or the proteasome, its subcellular localization [9, 10], and putative physiological function(s) [11]. Therefore, a better understanding of how PTMs may influence α-Syn's behavior in health and disease is crucial for understanding the normal function(s) of α-Syn and developing novel diagnostic and therapeutic strategies for early intervention and treatment strategies of PD and related disorders.

Elucidating the effect of PTMs on the structure, aggregation, and toxicity of α-Syn requires homogenous preparations of chemically well-defined α-Syn PTMs. A limited number of α-Syn PTMs, such as serine 129 phosphorylation, are amenable to preparation with great site specificity using enzymatic methods [2, 9], while others (especially ubiquitinated variants) can only be obtained as heterogeneous mixtures of products. For example, our group recently reported the in vitro and in vivo identification of Y39 phosphorylation in α-Syn, with c-Abl as the principal kinase phosphorylating at this site [12]. Although c-Abl mainly phosphorylates α-Syn at Y39, it also targets Y125 [12], especially under preparative in vitro phosphorylation conditions. Thus c-Abl phosphorylation is not suitable for studying the effects of α-Syn Y39 phosphorylation in isolation. Likewise, Y125-phosphorylated α-Syn cannot be efficiently prepared enzymatically, due to the lack of either efficiency or specificity of kinases phosphorylating the C-terminal tyrosine residues of α-Syn [13, 14]. Similar problems were reported in studies of α-Syn nitration, since chemically induced nitration generally shows very little site specificity [15, 16]. This limitation has been addressed by the introduction of mutations to allow chemically induced nitration or enzyme-mediated phosphorylation at a single site. However, this approach results in the introduction of up to three mutations, which could dramatically alter the conformational and aggregation properties of the protein. In order to overcome these limitations, our group has developed semisynthetic approaches based on expressed protein ligation (EPL [17, 18]) that provide access to milligram-scale preparations of site-specifically modified α-Syn that has been used to introduce N-terminal acetylation [19], mono-ubiquitination at K6 [20] and K12 [21], poly-ubiquitination [21, 22], phosphorylation at Y125 and S129 [23], and FRET probe pairs [24]. In addition, we developed a total chemical synthesis approach [25] that allows greater flexibility in the introduction of single or multiple PTMs or unnatural amino acids (such as fluorescent probes) at single or multiple sites within any region of the protein sequence. In particular, these tools greatly facilitate the study of cross-talks between different PTMs, which have been demonstrated before between pY125 and pS129 [26], as well as between K12 poly-ubiquitination and pS129 [22].

In this chapter, we focus on the S129 phosphorylation (α-Syn_pS129) of α-Syn, one of the most actively studied α-Syn PTM, and present two methods for preparing and purifying site-specifically S129-phosphorylated α-Syn. The first method makes use of a native chemical ligation (NCL) between a recombinant fragment consisting of α-Syn residues 1–106 and a synthetic peptide containing the residues 107–140 C-terminal domain and bearing the phosphorylated residue. This method could easily be adapted to accommodate other α-Syn PTMs within the C-terminal domain.

We also present an alternative approach to prepare α-Syn_pS129 using in vitro phosphorylation, which is significantly faster to perform than the semisynthetic method and offers the possibility to perform homogeneous isotopic labeling of the whole protein for NMR studies, whereas the semisynthesis-based method is suitable to introduce isotopic labeling only within the first 106 residues (i.e., the recombinantly expressed fragment), due to the fact that isotope-labeled amino acid building blocks for peptide synthesis are prohibitively expensive. However, a significant limitation of the in vitro phosphorylation approach is that it is restricted to a single PTM (pS129), as no other enzymes have been identified that both efficiently and site specifically modify α-Syn at the C-terminus so far.

## 2 Materials

All chromatography buffer solutions should be filtered (0.65 μm pore size or smaller) and degassed before use.

### 2.1 Instrumentation

1. Incubator (static, 37 °C).
2. Agitator for 1.5 mL tubes with heating capability (such as Eppendorf Thermo-Mixers).
3. pH microelectrode able to take measurements in 1.5 mL plastic tubes (e.g., Hanna Instruments HI2212 pH meter with model HI1083B microelectrode).
4. Standard instrumentation for casting polyacrylamide gels, for electrophoresis, gel staining, and imaging.
5. Chromatography system (FPLC) for protein purification under aqueous conditions, such as Äkta systems (GE Healthcare). The FPLC system should be placed in a cold room.

6. Chromatography system for reversed-phase HPLC, for example Waters 2535 pump and Waters 2489 UV/VIS detector.
7. Mass spectrometer or LC-ESI-MS system, for example Thermo Scientific LTQ ion trap. An appropriate deconvolution software such as ProMass (http://www.enovatia.com/products/promass/) should be installed in order to obtain the zero-charge mass spectra from the ESI charge envelopes.

### 2.2 Native Chemical Ligation (NCL)

1. Purified recombinant α-Syn fragment comprising residues 1–106 with a C-terminal thioester functionality, α-Syn(1–106)SR. Expression and purification of α-Syn(1–106)SR has been described in detail previously, *see* ref. 27.
2. Purified synthetic α-Syn fragment comprising residues 107–140 with the temporary point mutation A107C and bearing a phosphoserine residue at position 129. The synthesis and purification of α-Syn(A107C-140)_pS129 has been described by Hejjaoui and colleagues [27]. If the peptide is to be synthesized in-house, the following residues should be double-coupled: all residues following a proline residue, all glutamine (Fmoc-Gln(Trt)-OH) and asparagine (Fmoc-Asn(Trt)-OH), the phosphoserine residue, and the N-terminal cysteine residue (Fmoc-Cys(Trt)-OH). Alternatively, the synthetic α-Syn C-terminal peptides may be purchased from specialized vendors. In the latter case, the purity of purchased peptides should always be verified again using both mass spectrometry and analytical HPLC. Common impurities in such peptides (which are avoided if proper synthesis, peptide cleavage, and purification procedures are followed, as described in ref. 27) are incomplete deprotection of the *t*-butyl side chain protecting groups (+56 Da adduct) and methionine oxidation (as methionine sulfoxide, +16 Da).
3. Guanidine hydrochloride solution: 6 M guanidine hydrochloride, 0.2 M sodium phosphate, pH 7.2 (*see* **Note 1**). To make 100 mL of this solution, dissolve 57.3 g guanidine hydrochloride and 2.4 g sodium phosphate monobasic into an initial volume of 40 mL of water. Slowly add water while vigorously stirring until complete dissolution; adjust the pH to 7.2 and finally bring up to 100 mL. Store at room temperature.
4. NCL buffer: phosphate-buffered guanidine hydrochloride solution containing 20 mM tris(2-carboxyethyl)phosphine (TCEP). For 5 mL of NCL buffer, weigh 29 mg of TCEP.HCl and dissolve into guanidinium hydrochloride solution and mix thoroughly until complete dissolution. Addition of TCEP.HCl will significantly decrease the pH of the solution, which must be adjusted back to 7.2 using small aliquots of 1 M aqueous NaOH. After adjusting the pH, start degassing the buffer by bubbling inert gas (preferably argon) inside it. The buffer must be prepared freshly (*see* **Note 2**) and degassed (sparged with nitrogen

or preferably argon) for 10 min before use in order to minimize oxidative side reactions (*see* **Note 3**).

5. Sample dilution solution: water with 0.1 % v/v trifluoroacetic acid (TFA)
6. Desalting column: PD-10 columns (manual, gravity-based flow) or HiPrep 26/10 Desalting (for automated operation using a FPLC system).

### 2.3 Desulfurization

1. Guanidine hydrochloride solution: 6 M guanidine hydrochloride, 0.2 M sodium phosphate, pH 7.2 (same as in Subheading 2.2, **item 3**).
2. Radical initiator stock solution: 100 mM 2,2′-Azobis[2-(2-imidazolin-2-yl)propane]dihydrochloride (VA-044, Wako Chemicals GmbH, Neuss, Germany). To prepare a 1 mL solution, weigh 32.3 mg VA-044 (MW: 323.3 g/mol) into a 1.5 mL plastic tube, add 1 mL guanidinium hydrochloride solution (see above) and vortex until the solution becomes clear. Keep on ice and use within 30 min.
3. 2-methylpropane-2-thiol (*t*-butyl mercaptan, Sigma).

### 2.4 In Vitro Phosphorylation

1. Purified recombinant WT human α-Syn. We recommend a purification protocol that includes a reversed-phase HPLC purification step, in order to ensure complete removal of any enzymes (especially proteases) that may cause problems during the phosphorylation reactions. The expression and purification protocol is described in detail in [28].
2. Purified recombinant PLK3 (Life Technologies, cat. # PV3812). Make 1 μL aliquots upon receiving the stock solution from the vendor and immediately flash-freeze them and store at −80 °C.
3. 0.5 M HEPES, pH 7.4. To prepare a 100 mL solution, dissolve 11.9 g HEPES (free acid) in 75 mL water, adjust the pH to 7.4 using sodium hydroxide, and then bring up to 100 mL. Store at room temperature.
4. 0.1 M dithiothreitol (DTT). Prepare this solution freshly each time since DTT is unstable in aqueous solutions (*see* **Note 4**). Dissolve 15.4 mg DTT in 1 mL of water; keep on ice.
5. 0.5 M EGTA solution. Obtained from Boston Bioproducts, cat. # BM-151
6. 0.2 M $MgCl_2$. To prepare a 100 mL solution, weigh 4.07 g of magnesium chloride hexahydrate and dissolve to a final volume of 100 mL with water. Store at room temperature.
7. 0.1 M Mg-ATP. To prepare a 5 mL solution, weigh 254 mg of adenosine 5′ triphosphate, magnesium salt and dissolve in 5 mL of water. Adjust the pH to ~7.0; be extremely careful not to overshoot when adjusting the pH, in order to avoid causing

significant hydrolysis of the ATP. Make 100 μL aliquots and store at −20 °C.

8. Phosphorylation reaction buffer: 50 mM HEPES, 1 mM $MgCl_2$, 1 mM EGTA, 1 mM DTT. This solution is to be freshly prepared each time. For 1 mL, mix 100 μL 0.5 M HEPES, 10 μL 0.1 M DTT, 2 μL 0.5 M EGTA, 5 μL 0.2 M $MgCl_2$, and 883 μL water. Keep on ice.

### 2.5 LC-ESI-MS Mass Spectrometry

1. LC-ESI-MS solvent A: nano-pure water containing 0.1 % v/v formic acid (FA).
2. LC-ESI-MS solvent B: gradient-grade (HPLC) acetonitrile containing 0.1 % v/v FA.
3. Sample dilution solvent: water containing 0.1 % v/v TFA.
4. LC-MS column: Agilent Poroshell 300SB C3 column (1.0 mm ID, 75 mm length), cat. # 661750-909.

### 2.6 Protein Purification (Semisynthesis Protocol)

1. Strong cation-exchange (SCX) column: HiTrap SP HP 5 mL (GE Healthcare, cat. # 17-1152-01).
2. SCX buffer A1: 20 mM sodium citrate, pH 4.0. To prepare 500 mL of solution, dissolve 2.14 g of sodium citrate monobasic (MW: 214.1 g/mol) into 475 mL of water, adjust the pH to 4.0, bring up to 500 mL. Store at room temperature.
3. SCX buffer B1: 20 mM sodium citrate, 500 mM NaCl, pH 4.0. To prepare 500 mL of solution, dissolve 2.14 g of sodium citrate monobasic (MW: 214.1 g/mol) and 14.6 g NaCl (MW: 58.4 g/mol) into 475 mL of water, adjust the pH to 4.0, bring up to 500 mL. Store at room temperature.
4. SCX buffer A2: 20 mM sodium citrate, pH 5.0. To prepare a 500 mL solution, dissolve 2.14 g of sodium citrate monobasic into 475 mL of water, adjust the pH to 5.0, and bring up to 500 mL. Store at room temperature.
5. SCX buffer B2: 20 mM sodium citrate, 250 mM NaCl, pH 5.0. To prepare a solution of 500 mL, dissolve 2.14 g of sodium citrate monobasic and 7.31 g NaCl into 475 mL of water, adjust the pH to 4.0, and bring up to 500 mL. Store at room temperature.

### 2.7 Protein Purification (In Vitro Phosphorylation Protocol)

1. Inertsil WP300-C8 semiprep column, 7.6 mm ID × 250 mm, 5 μm particles with 300 Å pores (GL Sciences, cat. # 5020-05968). In order to extend column life, an appropriate guard column is recommended (GL Sciences, cat. # 5020-05969 (guard column cartridge) and 5020-06920 (guard column holder)).
2. Reversed-phase HPLC solvent A: nano-pure water containing 0.1 % v/v TFA.

3. Reversed-phase HPLC solvent B: HPLC-grade acetonitrile containing 0.1 % v/v TFA.
4. HPLC loading buffer: 95:5 water:acetonitrile containing 0.1 % v/v TFA.

## 3 Methods

### 3.1 Semisynthesis of α-Syn Containing C-Terminal Modifications

The following protocol describes the semisynthesis of the disease-relevant α-Syn phosphorylated on serine 129 (α-Syn_pS129), modified from the semisynthesis of pY125 α-Syn previously described by Hejjaoui and colleagues [27]. The reaction scheme is described in Fig. 1a.

1. Weigh 15 mg (1.4 μmol) of lyophilized α-Syn(1–106)SR (Fig. 1, *protein 1*) recombinant thioester protein into a 1.5 mL plastic tube at room temperature using a microbalance.
2. Similarly, weigh two molar equivalents (2.8 μmol, 11 mg) of synthetic phosphorylated peptide α-Syn(A107C-140)_pS129 (Fig. 1, *peptide 2*) into a 1.5 mL plastic tube (Eppendorf).

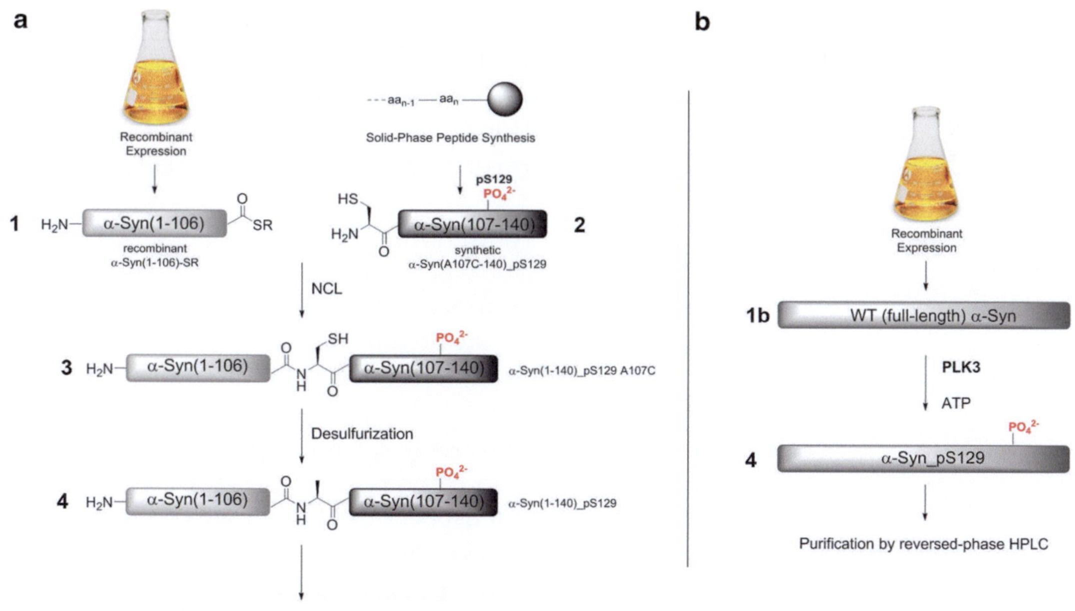


**Fig. 1** Schematic depiction of the methods used to prepare α-Syn_pS129. (**a**) Protein semisynthesis method: the first step involves a native chemical ligation between the α-Syn(1–106)SR recombinant thioester (*protein 1*) and the synthetic α-Syn(A107C-140)_pS129 peptide (*peptide 2*) to obtain full-length α-Syn_pS129 with the temporary A107C mutation (*protein 3*). In the second step, a cysteine-specific desulfurization reaction restores the native alanine residue at position 107 to produce the final α-Syn_pS129 (*protein 4*). (**b**) α-Syn_pS129 preparation using in vitro phosphorylation with PLK3. In this scheme, purified recombinant WT (full-length) α-Syn (*protein 1b*) is incubated with ATP and active recombinant PLK3 which will site specifically phosphorylate α-Syn at S129, thereby producing *protein 4*

3. Dissolve α-Syn(A107C-140) first using 1 mL of degassed NCL buffer (*see* **Note 5**). Once the peptide is fully dissolved, add this solution to the lyophilized α-Syn(1–106)SR powder and vortex until complete dissolution.
4. Immediately take 4 μL of solution at the initial time point and quench the reaction by diluting the aliquot with 36 μL of sample dilution solution (water + 0.1 % TFA) and keep on ice or store at −20 °C until use.
5. Blanket the ligation solution with argon and incubate at 37 °C without agitation.
6. Take and quench additional aliquots several times at later time points, for example 30 min, 1 h, and 2 h (as described in Subheading 3.1, **step 5**).
7. Dilute 10 μL of the quenched reaction (*see* Subheading 3.1, **step 5**) into 90 μL of water + 0.1 % TFA; then briefly spin down to remove any aggregates or dust particles before injecting into the LC system. Inject 5 μL (approximately 750 ng) into the LC-ESI-MS system. In order to obtain a good signal with the Poroshell 300SB C3 column (1.0 mm ID, 75 mm length), run a gradient from 10 to 90 % LC-ESI-MS solvent B over 10 min at 0.3 mL/min, and acquire the MS data in positive ionization mode. The desired product (Fig. 1, *protein 3*) has an expected mass of 14,572 Da. The ligation can be considered as completed when the recombinant thioester protein α-Syn(1–106SR) (expected mass: 10,742 Da) is entirely consumed. The hydrolyzed form of the α-Syn(1–106) fragment is observed at 10,619 Da.
8. SDS-PAGE: Mix 10 μL of the quenched reaction to 10 μL of 2× Laemmli (SDS-PAGE sample buffer), centrifuge at 20,000 × *g* for 5 min to remove precipitated guanidinium salts (which precipitate in the presence of SDS). Analyze the supernatant by electrophoresis on a 15 % polyacrylamide SDS gel (load 10 μL per lane).
9. Note that analysis by reversed-phase HPLC is not recommended due to the co-elution of α-Syn(1–106) with the full-length protein, which happens using all of the commonly used stationary phases (C4, C8, and C18).
10. Figure 2 shows a typical example of NCL reaction monitored by mass spectrometry and SDS-PAGE. Even small changes in the initial pH of the reaction may slow its kinetics; however if properly adjusted, SDS-PAGE analysis shows the ligation appears complete after 30–45 min (Fig. 2b). Once the ligation reaction is confirmed to be completed by both mass spectrometry and SDS-PAGE, the resulting full-length A107C/pS129 α-Syn must be desulfurized in order to restore the native alanine residue at position 107. Desulfurization is achieved by

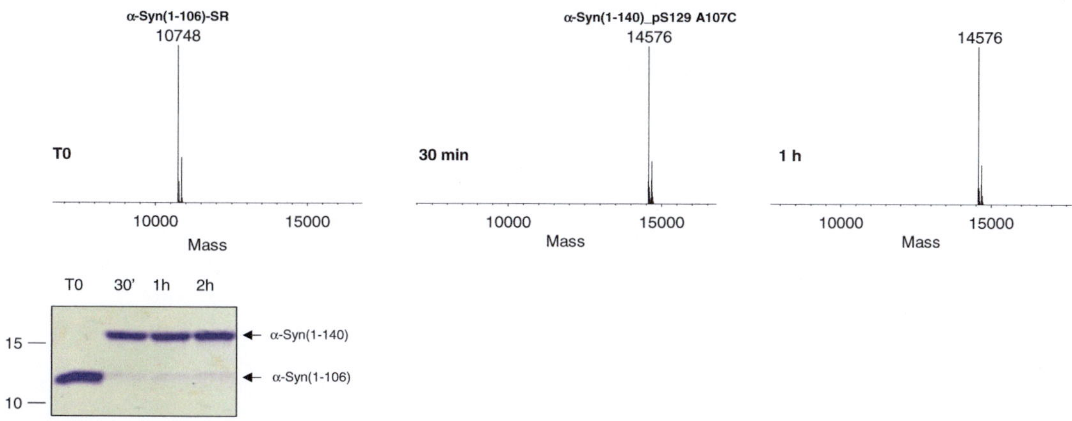


**Fig. 2** Monitoring of the NCL between α-Syn(1–106)SR and the α-Syn(A107C-140)_pS129 peptide by mass spectrometry (**a**) and SDS-PAGE/Coomassie staining (**b**). Note that the 4 kDa α-Syn(A107C-140)_pS129 peptide is not detectable by SDS-PAGE. Calculated mass for α-Syn(1–106)SR: 10,742 Da (observed: 10,748 Da); calculated mass for the ligation product α-Syn(1–140)_pS129 A107C: 14,572 Da (observed: 14,572 Da)

means of a radical-based reaction previously described by Wan and colleagues [29].

11. Add TCEP to a final concentration of 100 mM to the ligation reaction. This can be done by directly adding TCEP.HCl powder, or from a freshly prepared, pH-adjusted 1 M aqueous TCEP solution. Note that adding TCEP.HCl powder will require the pH to be adjusted again to 7.2.
12. Add 64 μL of radical initiator stock solution to achieve a final VA-044 concentration of 6 mM.
13. Add 50 μL of *t*-butyl mercaptan (final concentration: 400 mM). From this point onwards, all steps should be performed under a properly ventilated fume hood, due to the volatility and extremely unpleasant odor of *t*-butyl mercaptan (*see* **Note 6**). The reaction is then blanketed with inert gas and incubated at 37 °C with orbital shaking at 600–800 rpm.
14. Monitor the progress of the reaction by mass spectrometry analysis. Withdraw aliquots and perform dilutions and mass spec analysis as described under Subheading 3.1, **steps 4–7**, with the exception that the tubes should not be opened outside of the fume hood, and the LC-ESI-MS sample vials should be appropriately sealed under the fume hood before being transferred to the mass spectrometer. The desulfurized product (Fig. 1, *protein 4*) is identified by a 32 Da mass loss compared to the starting material, corresponding to one sulfur atom, i.e., the expected mass of the desired product is 14,540 Da. Once the starting material is no longer detected by mass spectrometry (typically after 3–4 h), incubate the reaction for an additional 30 min before performing the next step in order to ensure completion of the reaction.

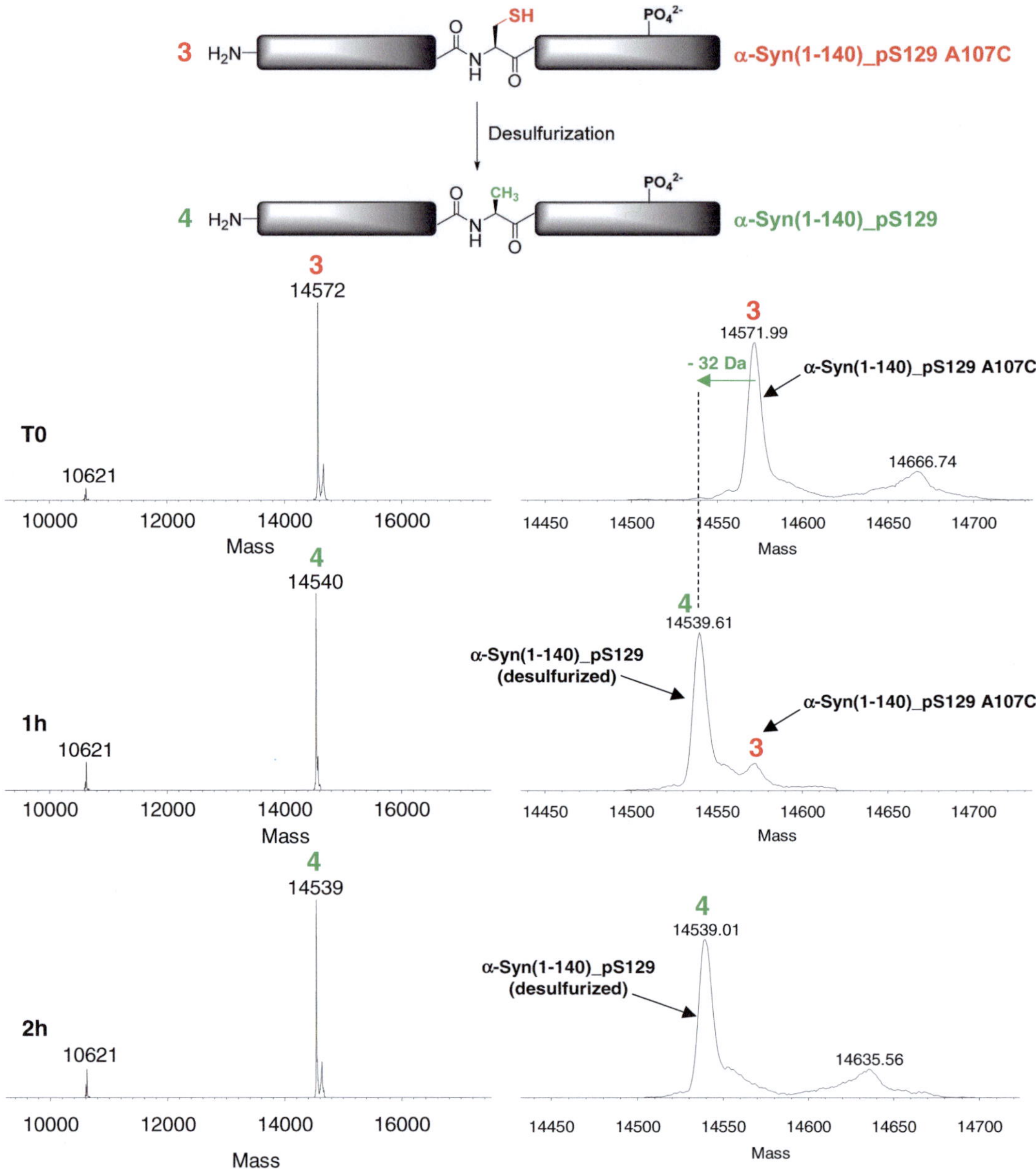


**Fig. 3** Desulfurization of α-Syn(1–140)_pS129 A107C (calculated mass: 14,572 Da; observed: 14,572 Da at *t*= 0) to obtain the final product α-Syn(1–140)_pS129 (calculated mass: 14,540 Da; observed: 14,539 Da)

15. Analysis by mass spectrometry of the desulfurization reaction is shown in Fig. 3. Care must be taken to analyze a narrow mass range around the desired product (calculated mass for α-Syn(1–140)_pS129: 14,540 Da) since the mass difference between the desulfurized product and the starting material is only 32 Da. The desulfurized protein is purified using a two-step cation-exchange chromatography method. Due to the co-elution of hydrolyzed α-Syn(1–106) fragment (and eventually

remaining α-Syn(1–106)SR thioester) with the full-length protein (desulfurized and non-desulfurized) on reversed-phase HPLC columns, RP-HPLC purification cannot be performed. The next steps describe how to separate the desired full-length desulfurized pS129 α-Syn protein from the main contaminants, namely the excess of α-Syn(107–140)_pS129 synthetic peptide and the α-Syn(1–106) fragment.

16. Wash the PD-10 desalting column with 25 mL of water, then equilibrate it with 25 mL of SCX buffer A1. Transfer the column under the fume hood and desalt the desulfurized protein as described in the manufacturer's protocol [30].
17. *SCX Purification Step 1* (*see* Fig. 4a): Pool the desalted protein fractions and manually inject them into the HiTrap SP HP strong cation-exchange column at an approximate flow rate of 1 mL/min. During this step, all protein components from the desulfurization reaction will bind onto the column and the excess of α-Syn(107–140)_pS129 peptide should be found in the flow-through (Fig. 4a, b). Manual loading is preferred to minimize sample loss; however the desalted protein fractions can also be further diluted using SCX buffer A1 and loaded on the column using a chromatography system by the means of a sample loop or equivalent loading mechanism. In all cases, save the flow-through and analyze by mass spectrometry. The absence of protein in this fraction ensures complete binding of the protein onto the column.
18. While the full-length protein remains bound on the column, elute any bound α-Syn(107–140)_pS129 peptide by executing the following program on the FPLC system: set the flow rate to 2 mL/min and collect 5 mL fractions. Wash the column with 70 mL of SCX buffer A1, then perform a short linear gradient from 0 to 15 % of SCX buffer B1 over 20 mL to ensure complete removal of the α-Syn(107–140) peptide. Set the buffer composition back to 100 % of SCX buffer A1 and wash for an additional 30 mL.
19. *SCX Purification Step 2*: This step is aimed at separating the full-length protein from the α-Syn(1–106) fragment based on their charge difference at pH 5.0 (Fig. 4a). Prime the FPLC system's buffer lines with SCX buffers A2 and B2, respectively, then execute the following program on the FPLC system: set the flow rate to 1 mL/min and start collecting 2.5 mL fractions. After 5 mL with 100 % SCX buffer A2, perform a 150 mL-long linear gradient from 0 to 70 % SCX buffer B2 (Fig. 4a, c). Hold this composition for 5 mL, then switch to 100 % SCX buffer B2 and wash for 10 mL. Stop collecting fractions and re-equilibrate the column with either SCX buffer A1 if another purification is planned, or with water followed by 20 % ethanol if the column will be stored. During the gradient,

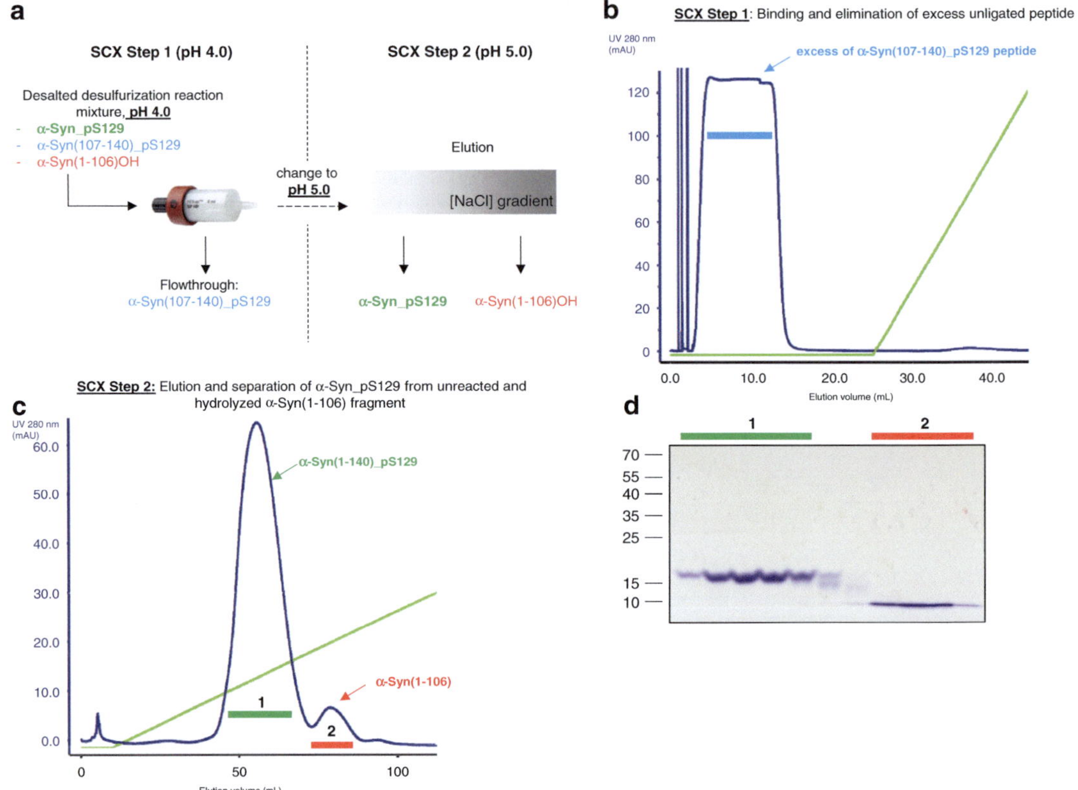


**Fig. 4** Purification of semisynthetic α-Syn(1–140)_pS129 by cation-exchange chromatography. (**a**) Scheme depicting the main steps of the purification protocol. Adapted from Hejjaoui et al. [23]. (**b**) SCX Step 1: Binding step at pH 4.0 where the full-length ligation product as well as the unreacted, hydrolyzed α-Syn(1–106) fragment bind onto the cation-exchange column and the excess of the (now desulfurized) synthetic α-Syn(107–140)_pS129 peptide is eliminated. (**c**) SCX Step 2: Elution step at pH 5.0 where the desired product α-Syn(1–140)_pS129 is separated from the α-Syn(1–106) fragment. The full-length phosphorylated protein (peak 1, *blue bar*) elutes before the α-Syn(1–106) fragment (peak 2, *red bar*). (**d**) SDS-PAGE/Coomassie analysis of fractions collected during SCX Step 2 (*see panel* **c**) showing the complete separation between the full-length α-Syn(1–140)_pS129 and the α-Syn(1–106) fragment

the desired full-length α-Syn elutes first, followed by the α-Syn(1–106) fragment (*see* Fig. 4a, c). Analyze all the fractions corresponding to the observed peaks on the chromatogram by SDS-PAGE and pool them according to purity (Fig. 4d).

20. Desalt or dialyze the protein against water, then quantify using UV absorbance to measure the yield. Concentration determination may be performed on a nanodrop UV spectrophotometer (or any cuvette-based UV spectrophotometer) using the same water solution as that used for dialysis or desalting for the blank reading. The concentration is calculated by measuring the absorbance of the dialyzed or desalted protein solution at 275 nm, where the extinction coefficient of α-Syn_pS129 is 5974 $M^{-1}$ $cm^{-1}$ (or equivalently, 0.4109 $g^{-1}$ L $cm^{-1}$).

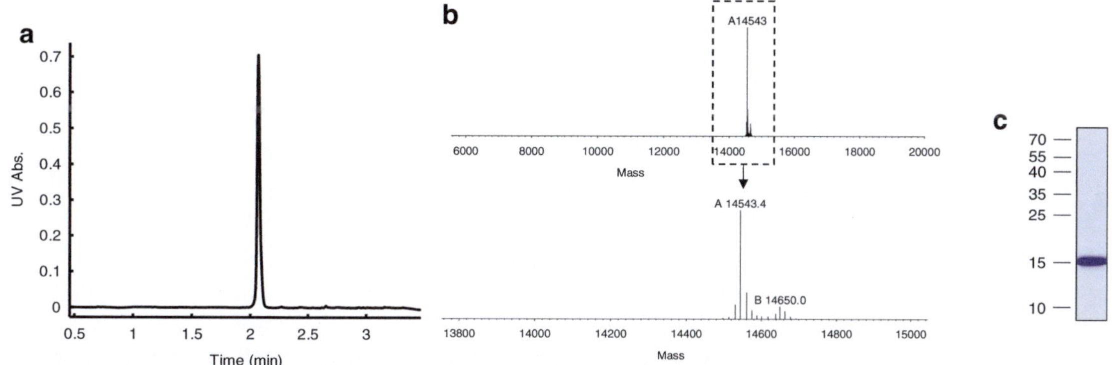


**Fig. 5** Purity analyses of α-Syn(1–140)_pS129 after cation-exchange chromatography and dialysis by analytical RP-UHPLC (**a**); mass spectrometry (**b**, calculated mass: 14,540 Da; observed: 14,543 Da; the *bottom panel* shows an expanded view of the mass range shown in the *dotted box* from the *top panel*); and SDS-PAGE/Coomassie staining (**c**)

21. Perform final purity analyses using mass spectrometry, SDS-PAGE, and analytical HPLC (*see* Fig. 5); then lyophilize. Keep the lyophilized protein at −20 °C until use.

### 3.2 Enzyme-Based Preparation of S129-Phosphorylated α-Syn

Previous studies from our group and others have shown that the members of the polo like family of kinases PLK2 and PLK3 phosphorylate α-Syn efficiently and specifically at S129 [3, 10]. Because of the high specificity of α-Syn S129 phosphorylation by PLK2 and PLK3, pS129 α-Syn may also be obtained by co-incubation of recombinant full-length WT α-Syn with purified PLK3 (Fig. 1b). The following protocol describes how to perform preparative-scale in vitro α-Syn phosphorylation using commercially available recombinant PLK3.

1. Weigh 500 μg of purified, lyophilized WT α-Syn (Fig. 1b, *protein 1b*) using an analytical microbalance.
2. Freshly prepare the phosphorylation reaction buffer. During this time, thaw one aliquot of PLK3 stock solution and one aliquot of 100 mM Mg-ATP on ice.
3. Dissolve the lyophilized WT α-Syn in 195 μL of phosphorylation buffer.
4. Add 4 μL of 100 mM Mg-ATP (final concentration: 2 mM) and 0.42 μg (1 μL) of PLK3. Mix by pipetting up and down (do not vortex) and incubate for 12 h at 30 °C without agitation. Depending on the quantity of phosphorylated α-Syn material desired, several of these reactions may be performed in parallel. We have observed that scaling up of a single reaction produced reduced the phosphorylation efficiency; thus parallel, smaller-scale reactions are preferred.

5. Verify the extent of the phosphorylation reaction by mass spectrometry: take a 1 μL aliquot from the reaction tube and add 49 μL of water containing 0.1 % v/v TFA to quench the reaction. Briefly spin down and inject 10 μL for analysis by LC-ESI-MS (positive ionization mode). α-Syn phosphorylated at S129 has an expected mass (M+H) of 14,541 Da, while unphosphorylated (WT) α-Syn is expected at 14,461 Da (M+H).
6. It is expected that 500 μg of WT α-Syn should be completely phosphorylated after 12 h of incubation; however we observed this is not always the case, presumably due to factors such as PLK3 storage time and possible batch-to-batch variability in PLK3 activity. If incomplete phosphorylation is observed after 12 h of incubation, add again 0.45 μg of PLK3 and 2 mM of fresh Mg-ATP, then incubate for another 12 h, in order to ensure complete phosphorylation. This is particularly important since unphosphorylated α-Syn and α-Syn_pS129 are difficult to separate; co-eluting unphosphorylated α-Syn contaminates phosphorylated α-Syn fractions during purification, thus decreasing the yields.
7. Once unphosphorylated α-Syn is confirmed to be undetectable by mass spectrometry, proceed to purify α-Syn_pS129 (Fig. 1b, *protein 4*) by reversed-phase HPLC, using a semipreparative (7.8 mm ID × 250 mm, 5 μm, 300 Å) C8 column. Before loading any sample, the column should be well equilibrated, by flowing 5 % solvent B at 3 mL/min for at least 20 min.
8. Dilute the contents of up to two reaction tubes (400 μL, 1 μg of protein) into 2 mL of HPLC loading buffer. Filter (0.22 μm) or centrifuge before loading on the HPLC column. After loading the sample onto the loop, execute the following program on the HPLC system, with a constant flow rate of 3 mL/min: run isocratically at 5 % solvent B for 10 min, then increase to 20 % B over 3 min; then perform the separating linear gradient from 20 % B to 70 % B over 30 min. Then, wash the column by increasing the proportion of solvent B to 95 % over 3 min, leave at 95 % B for 5 min, then gradually switch back to 5 % B over 3 min, and finally re-equilibrate (5 % B) for 20 min.
9. We recommend collecting fractions manually during the gradient elution in order to best discriminate between closely eluting peaks. Figure 6 shows a typical semipreparative chromatogram. Note that the actual retention time will vary depending on the specific configuration of the HPLC system (pre- and post-column volumes, presence and type of precolumn) and the specific column model (*see* **Note 7**). Fractions should be analyzed by mass spectrometry to assess purity. Typically, α-Syn_pS129 elutes immediately after a shoulder containing methionine-oxidized α-Syn_pS129 (expected

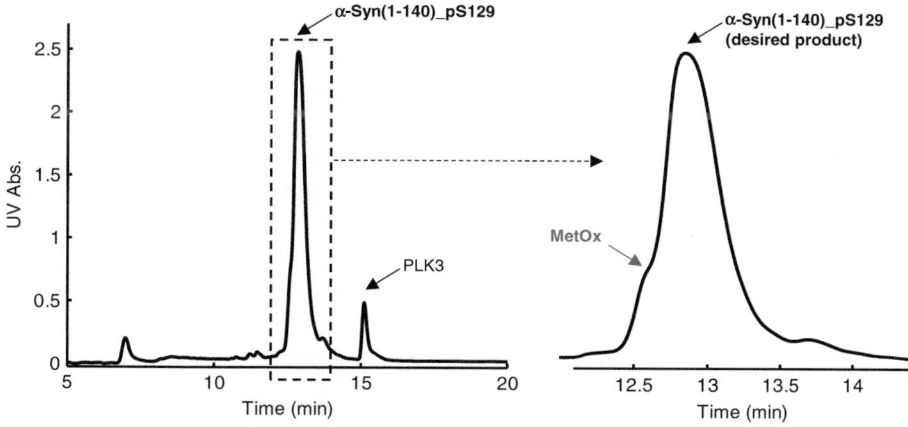


**Fig. 6** Semipreparative RP-HPLC purification of α-Syn(1–140)_pS129 prepared by in vitro phosphorylation of recombinant WT α-Syn by PLK3; using a 7.6 mm ID × 250 mm column. The chromatograms show the absorbance at 214 nm. The *right panel* shows an expanded view of the area shown in the *dotted box* on the *left panel*, highlighting the shoulder eluting just before the desired protein and containing one methionine sulfoxide residue (labeled "MetOx" on the chromatogram)

mass: 14,556 Da) which can be minimized by using proper buffer degassing of the phosphorylation buffer (*see* **Note 3**).

10. Pool fractions according to purity (as assessed by mass spectrometry), then perform a final purity analysis using SDS-PAGE, analytical reversed-phase HPLC, and mass spectrometry (similarly as in Fig. 5); determine the yield by UV absorbance, and finally lyophilize. Keep the lyophilized protein at −20 °C until use.

## 4 Notes

1. The NCL reaction is optimally performed at near-physiological pH. It is critically important that the pH of the reaction buffer is verified before each reaction. This requires a pH microelectrode (*see* Subheading 2.1, **item 3**) with a diameter small enough to fit into a 1.5 mL Eppendorf tube. Moreover, the pH meter should be re-calibrated immediately before the measurements. Significant deviations from the optimal pH reduce the yields of the reaction: a more basic pH will increase the rate of hydrolysis of the thioester-containing fragment, and high reaction pH also leads to nonspecific ligation events [31]. On the other hand, lowering the pH will result in drastically lower reaction rates due to decreased transthioesterification reaction rates under these conditions [31].
2. It is recommended to add the TCEP to the NCL buffer shortly before performing the reaction, since TCEP is somewhat prone to oxidation in phosphate-containing buffers [32],

although complete oxidation requires ~72 h in a pH 7.0 phosphate buffer [32].

3. Degassing of the NCL buffer is useful to further minimize disulfide formation by the N-terminal cysteine and thus keeping it available for reacting with the thioester-containing fragment. Furthermore, removing dissolved oxygen helps keeping methionine residues in a reduced state. If the formation of methionine sulfoxides during NCL (as seen by a +16 Da shift by mass spectroscopy) is not minimized by simply degassing the NCL buffer, adding 5 mM to 20 mM l-methionine to the reaction has proven effective in our hands.
4. DTT oxidizes relatively quickly when in solution, especially at room temperature (the half-life of DTT is only a few hours at 20 °C and pH 7.5 [33]). DTT solutions should thus be freshly made before each use and stored on ice during the working day.
5. It is desirable to dissolve the fragment containing the N-terminal cysteine first, and then add the peptide thioester, in order to minimize the risks of hydrolyzing the thioester which might happen (albeit at a slow rate) if it is alone in solution at neutral pH.
6. In order to minimize the spreading of *t*-butyl mercaptan vapors, bleach traps (200–300 mL of ~5 % aqueous sodium hypochlorite in plastic or glass containers) should be set up inside the hood near the reactions [34]. A bleach-containing waste container should also be used to discard any disposable plastics that have been in contact with solutions containing *t*-butyl mercaptan. Thiol-contaminated wastes should be left in a bleach solution for ~24 h before final disposal [34]. Moreover, we advise to dedicate a set of lab coats for work involving thiols. These lab coats should remain in the room where thiols are stored and used and should not be worn outside of that room.
7. We have observed significant differences (up to ±5 min) between columns of the same dimensions, particle size, and phase but from different manufacturers. Thus it is generally preferable to monitor the entire run and collect fractions manually, especially during the first purification.

## Acknowledgements

This work has been possible thanks to the tremendous efforts of all the members of the Lashuel group that have contributed to developing and optimizing the protocols described here. We wish to specially thank Mr. John Perrin for the enzymatic preparation of α-Syn_pS129; as well as Mr. Anass Chiki, Dr. Sean Deguire, and Dr. John Warner for helpful comments on the manuscript. This work was supported by grants from the European Research Council (ERC grant n° 243182), the Michael J. Fox Foundation for Parkinson's Research (grant n° 531107), and Ecole Polytechnique Fédérale de Lausanne Swiss National Science Foundation (grant n (31003A_120653).

## References


1. Lashuel HA, Overk CR, Oueslati A, Masliah E (2013) The many faces of alpha-synuclein: from structure and toxicity to therapeutic target. Nat Rev Neurosci 14:38–48
2. Oueslati A, Fournier M, Lashuel HA (2010) Role of post-translational modifications in modulating the structure, function and toxicity of alpha-synuclein implications for Parkinson's disease pathogenesis and therapies. Prog Brain Res 183C:115–145
3. Oueslati A, Schneider BL, Aebischer P, Lashuel HA (2013) Polo-like kinase 2 regulates selective autophagic alpha-synuclein clearance and suppresses its toxicity in vivo. Proc Natl Acad Sci U S A 110(41):E3945–E3954
4. Anderson JP, Walker DE, Goldstein JM, de Laat R, Banducci K, Caccavello RJ, Barbour R, Huang J, Kling K, Lee M, Diep L, Keim PS, Shen X, Chataway T, Schlossmacher MG, Seubert P, Schenk D, Sinha S, Gai WP, Chilcote TJ (2006) Phosphorylation of Ser-129 is the dominant pathological modification of alpha-synuclein in familial and sporadic Lewy body disease. J Biol Chem 281:29739–29752
5. Fujiwara H, Hasegawa M, Dohmae N, Kawashima A, Masliah E, Goldberg MS, Shen J, Takio K, Iwatsubo T (2002) alpha-Synuclein is phosphorylated in synucleinopathy lesions. Nat Cell Biol 4:160–164
6. Wang Y, Shi M, Chung KA, Zabetian CP, Leverenz JB, Berg D, Srulijes K, Trojanowski JQ, Lee VM-Y, Siderowf AD, Hurtig H, Litvan I, Schiess MC, Peskind ER, Masuda M, Hasegawa M, Lin X, Pan C, Galasko D, Goldstein DS, Jensen PH, Yang H, Cain KC, Zhang J (2012) Phosphorylated α-Synuclein in Parkinson's disease. Sci Transl Med 4:121ra120
7. Foulds PG, Mitchell JD, Parker A, Turner R, Green G, Diggle P, Hasegawa M, Taylor M, Mann D, Allsop D (2011) Phosphorylated alpha-synuclein can be detected in blood plasma and is potentially a useful biomarker for Parkinson's disease. FASEB J 25:4127–4137
8. Schmid AW, Fauvet B, Moniatte M, Lashuel HA (2013) Alpha-synuclein post-translational modifications as potential biomarkers for Parkinson disease and other synucleinopathies. Mol Cell Proteomics 12:3543–3558
9. Paleologou KE, Schmid AW, Rospigliosi CC, Kim HY, Lamberto GR, Fredenburg RA, Lansbury PT Jr, Fernandez CO, Eliezer D, Zweckstetter M, Lashuel HA (2008) Phosphorylation at Ser-129 but not the phosphomimics S129E/D inhibits the fibrillation of alpha-synuclein. J Biol Chem 283:16895–16905
10. Mbefo MK, Paleologou KE, Boucharaba A, Oueslati A, Schell H, Fournier M, Olschewski D, Yin G, Zweckstetter M, Masliah E, Kahle PJ, Hirling H, Lashuel HA (2010) Phosphorylation of synucleins by members of the Polo-like kinase family. J Biol Chem 285: 2807–2822
11. Burre J, Sharma M, Tsetsenis T, Buchman V, Etherton MR, Sudhof TC (2010) Alpha-synuclein promotes SNARE-complex assembly in vivo and in vitro. Science 329:1663–1667
12. Mahul-Mellier AL, Fauvet B, Gysbers A, Dikiy I, Oueslati A, Georgeon S, Lamontanara AJ, Bisquertt A, Eliezer D, Masliah E, Halliday G, Hantschel O, Lashuel HA (2014) c-Abl phosphorylates alpha-synuclein and regulates its degradation: implication for alpha-synuclein clearance and contribution to the pathogenesis of Parkinson's disease. Hum Mol Genet 23(11):2858–2879
13. Negro A, Brunati AM, Donella-Deana A, Massimino ML, Pinna LA (2002) Multiple phosphorylation of alpha-synuclein by protein tyrosine kinase Syk prevents eosin-induced aggregation. FASEB J 16:210–212
14. Ellis CE, Schwartzberg PL, Grider TL, Fink DW, Nussbaum RL (2001) alpha-synuclein is phosphorylated by members of the Src family of protein-tyrosine kinases. J Biol Chem 276: 3879–3884
15. Hodara R, Norris EH, Giasson BI, Mishizen-Eberz AJ, Lynch DR, Lee VM, Ischiropoulos H (2004) Functional consequences of alpha-synuclein tyrosine nitration: diminished binding to lipid vesicles and increased fibril formation. J Biol Chem 279:47746–47753
16. Uversky VN, Yamin G, Munishkina LA, Karymov MA, Millett IS, Doniach S, Lyubchenko YL, Fink AL (2005) Effects of nitration on the structure and aggregation of alpha-synuclein. Brain Res Mol Brain Res 134:84–102
17. Muir TW, Sondhi D, Cole PA (1998) Expressed protein ligation: a general method for protein engineering. Proc Natl Acad Sci U S A 95: 6705–6710

18. Evans TC Jr, Benner J, Xu MQ (1998) Semisynthesis of cytotoxic proteins using a modified protein splicing element. Protein Sci 7:2256–2264
19. Fauvet B, Fares MB, Samuel F, Dikiy I, Tandon A, Eliezer D, Lashuel HA (2012) Characterization of semisynthetic and naturally Nalpha-acetylated alpha-synuclein in vitro and in intact cells: implications for aggregation and cellular properties of alpha-synuclein. J Biol Chem 287:28243–28262
20. Hejjaoui M, Haj-Yahya M, Kumar KS, Brik A, Lashuel HA (2011) Towards elucidation of the role of ubiquitination in the pathogenesis of Parkinson's disease with semisynthetic ubiquitinated alpha-synuclein. Angew Chem Int Ed Engl 50:405–409
21. Shabek N, Herman-Bachinsky Y, Buchsbaum S, Lewinson O, Haj-Yahya M, Hejjaoui M, Lashuel HA, Sommer T, Brik A, Ciechanover A (2012) The size of the proteasomal substrate determines whether its degradation will be mediated by mono- or polyubiquitylation. Mol Cell 48:87–97
22. Haj-Yahya M, Fauvet B, Herman-Bachinsky Y, Hejjaoui M, Bavikar SN, Karthikeyan SV, Ciechanover A, Lashuel HA, Brik A (2013) Synthetic polyubiquitinated alpha-Synuclein reveals important insights into the roles of the ubiquitin chain in regulating its pathophysiology. Proc Natl Acad Sci U S A 110:17726–17731
23. Hejjaoui M, Butterfield SM, Fauvet B, Vercruysse F, Cui J, Dikiy I, Prudent M, Olschewski D, Zhang Y, Eliezer D, Lashuel HA (2012) Elucidating the role of C-terminal post-translational modifications using protein semisynthesis strategies: alpha-synuclein phosphorylation at tyrosine 125. J Am Chem Soc 134(11):5196–5210
24. Wissner RF, Wagner AM, Warner JB, Petersson EJ (2013) Efficient, traceless semi-synthesis of alpha-synuclein labeled with a fluorophore/thioamide FRET pair. Synlett 24:2454–2458
25. Fauvet B, Butterfield SM, Fuks J, Brik A, Lashuel HA (2013) One-pot total chemical synthesis of human alpha-synuclein. Chem Commun (Camb) 49:9254–9256
26. Chen L, Periquet M, Wang X, Negro A, McLean PJ, Hyman BT, Feany MB (2009) Tyrosine and serine phosphorylation of alpha-synuclein have opposing effects on neurotoxicity and soluble oligomer formation. J Clin Invest 119:3257–3265
27. Hejjaoui M (2012) Elucidating the role of post-translational modifications of alpha-synuclein using semisynthesis - phosphorylation at Tyrosine 125 and monoubiquitination at Lysine 6, EPFL
28. Khalaf O, Fauvet B, Oueslati A, Dikiy I, Mahul-Mellier AL, Ruggeri FS, Mbefo MK, Vercruysse F, Dietler G, Lee SJ, Eliezer D, Lashuel HA (2014) The H50Q mutation enhances alpha-synuclein aggregation, secretion, and toxicity. J Biol Chem 289:21856–21876
29. Wan Q, Danishefsky SJ (2007) Free-radical-based, specific desulfurization of cysteine: a powerful advance in the synthesis of polypeptides and glycopolypeptides. Angew Chem Int Ed Engl 46:9248–9252
30. PD-10 gravity-flow column user manual: https://www.gelifesciences.com/gehcls_images/GELS/Related%20Content/Files/1314723116657/litdoc52130800BB_20110830191706.pdf
31. Hackenberger CP, Schwarzer D (2008) Chemoselective ligation and modification strategies for peptides and proteins. Angew Chem Int Ed Engl 47:10030–10074
32. Han JC, Han GY (1994) A procedure for quantitative determination of tris(2-carboxyethyl)phosphine, an odorless reducing agent more stable and effective than dithiothreitol. Anal Biochem 220:5–10
33. Stevens R, Stevens L, Price NC (1983) The stabilities of various thiol compounds used in protein purifications. Biochem Educ 11:70
34. Singletary AM (1997) Hazardous laboratory chemicals disposal guide margaret-Ann Armour, CRC Press, Inc., Boca Raton, FL, (1996), 546 Pages, [ISBN No.: 1–56670–108–2] U.S. List Price: 7$79.95. Environ Prog 16:S5

# Chapter 2

# Isotope-Labeled Amyloids via Synthesis, Expression, and Chemical Ligation for Use in FTIR, 2D IR, and NMR Studies

Tianqi O. Zhang, Maksim Grechko, Sean D. Moran, and Martin T. Zanni

## Abstract

This chapter provides protocols for isotope-labeling the human islet amyloid polypeptide (hIAPP or amylin) involved in type II diabetes and γD-crystallin involved in cataract formation. Because isotope labeling improves the structural resolution, these protocols are useful for experiments using Fourier transform infrared (FTIR), two-dimensional infrared (2D IR), and NMR spectroscopies. Our research group specializes in using 2D IR spectroscopy and isotope labeling. 2D IR spectroscopy provides structural information by measuring solvation from 2D diagonal lineshapes and vibrational couplings from cross peaks. Infrared spectroscopy can be used to study kinetics, membrane proteins, and aggregated proteins. Isotope labeling provides greater certainty in the spectral assignment, which enables new structural insights that are difficult to obtain with other methods. For amylin, we provide a protocol for $^{13}C/^{18}O$ labeling backbone carbonyls at one or more desired amino acids in order to obtain residue-specific structural resolution. We also provide a protocol for expressing and purifying amylin from *E. coli*, which enables uniform $^{13}C$ or $^{13}C/^{15}N$ labeling. Uniform labeling is useful for measuring the monomer infrared spectrum in an amyloid oligomer or fiber as well as amyloid protein bound to another polypeptide or protein, such as a chaperone or an inhibitor. In addition, our expression protocol results in 2–2.5 mg of amylin peptide per 1 L cell culture, which is a high enough yield to straightforwardly obtain the 2–10 mg needed for high resolution and solid-state NMR experiments. Finally, we provide a protocol to isotope-label either of the two domains of γD-crystallin using expressed protein ligation. Domain labeling makes it possible to resolve the structures of the two halves of the protein in FTIR and 2D IR spectra. With modifications, these strategies and protocols for isotope labeling can be applied to other amyloid polypeptides and proteins.

**Key words** Two-dimensional infrared spectroscopy, 2D IR, FTIR, hIAPP, Amylin, Amyloid, Isotope labeling, Expressed protein ligation, Native chemical ligation, γD-crystallin, NMR spectroscopy

## 1 Introduction

Infrared spectroscopy is one of the most commonly used techniques for assessing if a peptide or protein has formed amyloid fibers. The amide I mode of proteins is created by the stretching motions of the backbone carbonyl groups (with a little CN stretch). Amyloid fibers exhibit a very sharp and characteristic peak at

David Eliezer (ed.), *Protein Amyloid Aggregation: Methods and Protocols*, Methods in Molecular Biology, vol. 1345, DOI 10.1007/978-1-4939-2978-8_2, 

1620 $cm^{-1}$ due to strong vibrational coupling resulting from the carbonyl groups vibrating in unison across the strands [1]. While useful for identifying amyloid fibers and other secondary structures, infrared spectra are too congested to assign structure to specific residues in any but the smallest sequences. α-Helices or β-sheets can be identified and their relative abundance quantified [2, 3], but the residues that contribute to the structure cannot be identified. Isotope labeling overcomes this limitation. Individual residues can be resolved with $^{13}C$ and/or $^{13}C/^{18}O$ isotopes of the backbone carbonyl atoms incorporated into the sequence via Fmoc synthesis of the polypeptide [4–6]. $^{13}C$ labeling produces a 40 $cm^{-1}$ shift [7] while $^{13}C/^{18}O$ produces a 66 $cm^{-1}$ shift [8]. 66 $cm^{-1}$ is far outside the spectral width of all natural amide I bands and lies in a region of the spectrum largely absent of side-chain absorbance [9]. Using 2D IR spectroscopy, the secondary structure and solvation of the labeled residue can be deduced from its frequency, cross peaks between labeled and unlabeled modes, and 2D lineshape. The kinetics of amyloid formation can also be followed, residue by residue, either in neat solution or catalyzed by membranes [10–12]. By doing so, we recently identified an on-pathway, β-sheet intermediate in the FGAIL region of amylin that is ultimately disrupted to form the loop in the final fiber [13]. This intermediate may explain why aggregation is so sensitive to mutations in this region.

Another isotope labeling strategy is the expression of amylin in *E. coli*. Expression allows all of the residues to be $^{13}C$ or $^{13}C/^{15}N$ labeled simultaneously by carrying out expression in isotope-enriched growth media. This approach has uses in FTIR, 2D IR, and NMR spectroscopies. For 2D IR spectroscopy, it has two uses. First, it enables the monomer structure of amylin to be studied even when aggregated with many other amylin molecules [7, 14]. As stated above, amide I vibrations become delocalized across multiple polypeptides when the coupling is strong enough. As a result, the infrared spectra become insensitive to the structure of the individual monomers. Due to the frequency difference, isotope labeling prevents delocalization. Thus, by mixing in a small amount of isotope-labeled protein with a larger portion of unlabeled protein, the labeled portion of the spectrum will be dominated by the structure and couplings inherent to the monomer. We have used this fact to determine the number of strands that each monomer contributes to amyloid fibers made from γD-crystallin and to determine that $K_2Q_{24}K_2$ adopts an antiparallel hairpin rather than a beta-turn in its fibers [7, 14]. Second, uniform labeling allows mixtures of different proteins to be studied. In unpublished work, we have mixed isotope-labeled amylin with unlabeled αB-crystallin, which is a chaperone protein that is known to bind to amyloid fibers. Amylin is well resolved from the crystallin, which is enabling us to study its structure and binding to αB-crystallin.

A third strategy is to label pieces of proteins, such as domains, using native chemical ligation (a variant of expressed protein ligation) [15–17]. One domain can then be resolved from the other, enabling independent structural kinetics. γD-crystallin has two domains, each formed from very similar Greek key motifs and connected by a flexible linker (*see* Fig. 1). By $^{13}C$ labeling the C-terminal domain, we discovered that it formed the β-sheet core

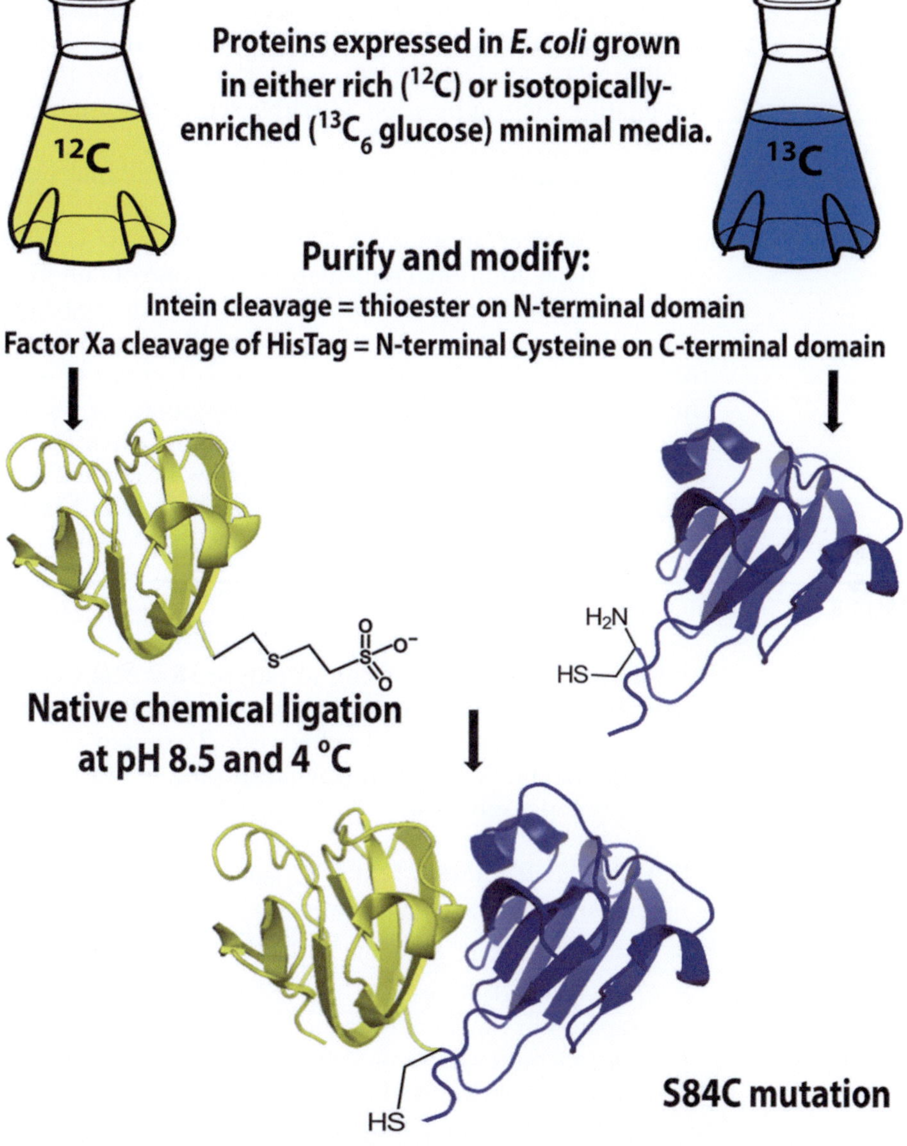


**Fig. 1** Isotope labeling and native chemical ligation of γD-crystallin. The N-terminal domain is expressed in $^{12}C$ media and purified with intein-mediated cleavage. The C-terminal domain is expressed in $^{13}C$ media and purified with Ni affinity column. The His tag is cleaved with Factor Xa before ligation. The ligated protein contains a mutation S84C which does not change the structural and chemical property of the protein

of the amyloid fibers, not the N-terminal domain as originally thought from fluorescence and other studies [7, 18]. Mixtures can also be studied as described above, but now with domain-specific resolution.

In this chapter, we provide very specific protocols that have been developed, tested, and used over the course of several years. First, we describe a protocol for $^{18}O$ exchanging Fmoc-protected amino acids, although other methods also exist [19, 20]. Second, a protocol for amylin expression is given. Amylin has been expressed before [21–24], but our protocol produces C-terminally amidated amylin at a higher yield. Third, we provide a protocol for domain labeling γD-crystallin by expressing the two domains separately and ligating them at position 84 with a serine-to-cysteine mutation (S84C).

How exactly does one obtain precise structural information from these three labeling schemes? When should one use one labeling strategy over another? Is 2D IR required or is FTIR good enough? What additional information does one obtain from 2D IR spectroscopy? These questions and others are addressed in a recent review about vibrational couplings, infrared spectroscopy, and isotope labeling [25]. 2D IR spectroscopy is coming of age. In just the last few years the theoretical underpinnings of the technique have become well enough understood and the experimental methods well enough established that it can now be applied to sophisticated problems in biophysics and structural biology [26–30].

## 2 Materials

All solutions are prepared using ultrapure 18.2 MΩ water, HPLC-grade reagents, and analytical grade chemicals without further purification, unless specified otherwise. Restricted waste disposal protocols should be followed when disposing of biochemical material and organic solvents.

### 2.1 Synthesis of $^{13}C/^{18}O$-Labeled hIAPP Using Fmoc Chemistry

#### 2.1.1 $^{18}O/^{16}O$ Exchange of Fmoc-Labeled Amino Acids

1. 1-$^{13}C$-labeled amino acid ($^{13}C$ isotope label on the backbone carbonyl) with or without Fmoc-protecting groups (*see* **Note 1**).
2. Reaction solvent: Dioxane and $^{18}O$-water in 1 g aliquots.
3. Oven-dried glasswares (assembled as shown in Fig. 2) including a separation funnel, a branched adapter, a small round-bottom flask, a condenser, round-bottom flask with one sidearm, a stopcock as shown in the middle of the figure, and a long needle. Before starting the reaction, leave the needle out of the round-bottom flask with the amino acids.
4. A lyophilizer.
5. Schlenk line (optional) or balloons that are filled with $N_2$ gas to put on top of the condenser and the separation funnel.

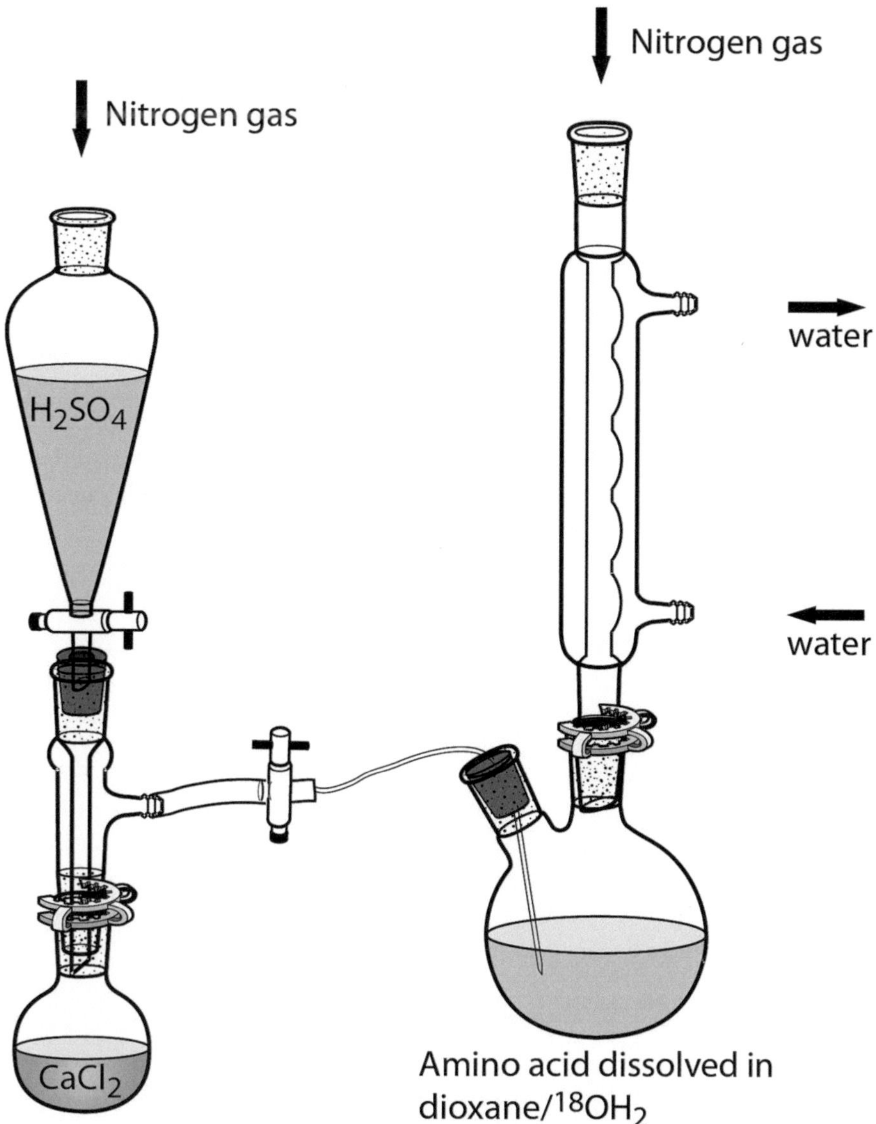


**Fig. 2** Experimental setup for $^{18}O$ exchange of amino acids. All parts should be dried in oven before use and assembled in a fume hood. The unit on the *left* should be assembled first. Allow the reaction between $H_2SO_4$ and $CaCl_2$ proceed for at least 2 min before inserting the syringe into the side-armed flask on the *right*. The nitrogen gas source can be a Schlenk line or simply balloons filled with $N_2$ gas

6. Hydrogen chloride gas-generating reagents: Anhydrous $CaCl_2$, concentrated $H_2SO_4$.
7. Centrifuge capable of 5000 × *g* RCF.

*2.1.2 Solid-Phase Synthesis for 0.1 mM Scale*

1. Solid-phase synthesizer (*see* **Note 2**).
2. Solid-phase synthesis resin: Fmoc-protected PAL-PEG-PS (*see* **Note 3**) with loading capacity 0.16 mmol/g (0.16 free amine

for coupling per 1 g of beads). For 0.1 mM-scale synthesis, weigh out 0.625 g of resin.

3. Deprotection solution: 600 mL of 20 % piperidine in dimethylformamide (DMF) with 0.1 M hydroxybenzotriazole (HOBt). Dissolve 103.3 g of HOBt in 480 mL of DMF and mix well with 120 mL piperidine (*see* **Note 4**).
4. Activation solution: 0.5 M HBTU in DMF. Dissolve 15.2 g of HBTU in 80 mL DMF.
5. Activator base solution: Mix 17 mL of *N, N-diisopropylethylamine* (DIEA) with 30 mL of *N*-methyl-2-pyrrolidone (NMP) to make total 47 mL of reagent.
6. Amino acids: 0.2 M in DMF. Weigh out proper amount of each amino acid and dissolve in DMF in individual tubes. Five times excess of amino acid is used for each coupling reaction (*see* **Note 5**).
7. Kaiser test reagent A: Dissolve 0.5 g ninhydrin in 10 mL ethanol.
8. Kaiser test reagent B: 0.005 mM KCN in pyridine. Dissolve 65.12 mg of potassium cyanide (KCN) in 10 mL of $H_2O$ to make 0.1 M stock solution. Add 4 μL of KCN stock solution to 20 mL of pyridine.

#### 2.1.3 Cleavage of Peptide Off the Resin

1. Cleavage cocktail: 90 % trifluoroacetic acid (TFA), 5 % ethanedithiol (EDT), 2.5 % thioanisole, and 2.5 % anisole. In a centrifuge tube, add 9 mL of TFA, 500 μL thioanisole, 300 μL of EDT, and 200 μL of anisole (*see* **Note 6**).
2. Ether: Chill on ice for at least 30 min.

#### 2.1.4 HPLC Purification

1. HPLC solvent A: 0.046 % HCl in $H_2O$. Dissolve 5 mL of 37 % HCl in 4 L of $H_2O$ in a fume hood. Mix well.
2. HPLC solvent B: 80 % Acetonitrile in $H_2O$ with 0.046 % HCl. Mix 3.2 L of acetonitrile with 0.8 L of $H_2O$. Add 5 mL of 37 % HCl and mix well.

### 2.2 Expression and Purification of Uniform $^{13}C$ hIAPP from E. coli

1. BL21 DE3 *E. coli* cells transfected with a PTXB1 plasmid coding for hIAPP fused to chitin-binding domain.
2. Ampicillin plate: Dissolve 5.0 g of tryptone, 2.5 g of yeast extract, and 5.0 g of NaCl in 0.5 L $H_2O$. Autoclave with liquid cycle at 121 °C for 30 min. Prepare a 60 °C water bath to cool the solution after autoclaving. When the solution is cooled to 60 °C, add 0.5 mL of 100 mg/mL ampicillin solution. Mix the solution well and pour it into sterile disposable petri dishes. Cool at room temperature for 2 h and then store at 4 °C.
3. Minimal media buffer: Dissolve 1.25 g of $(NH_4)_2SO_4$, 4.5 g of $KH_2PO_4$, 3.0 g of $K_2HPO_4$, 0.25 g of citric acid monohydrate

in 0.5 L $H_2O$. Adjust the pH of solution to 6.6 using KOH. Autoclave buffer at 121 °C for 30 min. Store at room temperature if not used immediately.

4. Trace element solution: Dissolve 16.2 g of $FeCl_3{\cdot}6H_2O$, 2.4 g of $ZnSO_4{\cdot}7H_2O$, 4.2 g of $CoCl_2{\cdot}6H_2O$, 4.2 g of $Na_2MoO_4{\cdot}2H_2O$, 4.8 g of $CuSO_4{\cdot}5H_2O$, 1.2 g of $H_3BO_3$, 3.0 g of $MnSO_4$, and 30 mL of 37 % HCl into 570 mL of $H_2O$. Autoclave at 121 °C for 30 min [31].
5. Vitamin solution: Dissolve 0.4 g of each of the following ingredients: pantothenic acid (calcium salt), choline chloride, folic acid, nicotinamide, pyridoxal hydrochloride, and thiamine hydrochloride in 800 mL of $H_2O$. In the same solution, dissolve 0.8 g of myoinositol and 0.04 g of riboflavin. Adjust solution to pH 7.2 and filter with sterile filter. Aliquot the solution, cover the outsides of the tubes with aluminum foil, and place in −80 °C freezer for long-term storage.
6. Stock $MgSO_4$ solution: Weigh 10 g of anhydrous $MgSO_4$ and dissolve in 200 mL of $H_2O$. Autoclave this solution.
7. Minimal media: In 0.5 L of autoclaved minimal media buffer, add 1 g of $^{13}C_6$-glucose, 10 mL of 0.05 g/mL $MgSO_4$ (autoclaved separately), 500 μL of trace element solution, 325 μL of vitamin solution, 35 mg of thiamine HCl, and 500 μL of 100 mg/mL ampicillin solution (*see* **Note 7**).
8. French pressure cell press (used in this protocol) or sonicator for cell lysis.
9. hIAPP column buffer: 20 mM HEPES, 0.1 mM EDTA, 50 mM NaCl, 2 M urea, pH 8.0. Dissolve 3.182 g of HEPES, 30.4 mg of EDTA, 2.337 g of NaCl, and 96.1 g of urea in 800 mL $H_2O$. Adjust pH to 8.0 at 4 °C with HCl. Store at 4 °C.
10. hIAPP cleavage buffer: 100 mM DTT, 2 M ammonium bicarbonate. In 50 mL hIAPP column buffer, dissolve 0.73 g of dithiothreitol (DTT) and 6.4 g of ammonium bicarbonate (*see* **Note 8**). The cleavage buffer should be prepared fresh every time.
11. Chitin resin column.

### 2.3 γD-Crystallin

#### 2.3.1 Expression of γD-Crystallin Domains for Native Chemical Ligation

1. The expression of γD-crystallin N-terminal domain uses the same material as hIAPP expression described in Subheading 2.2 except different column and cleavage buffer.
2. BL21 DE3 *E. coli* cells transfected with a PTXB1 plasmid coding for the γD-crystallin N-terminal domain fused to chitin-binding domain.
3. γD-crystallin column buffer: 20 mM HEPES, 200 mM NaCl, pH 8.5. Dissolve 3.81 g of HEPES and 9.36 g NaCl in 800 mL of $H_2O$. Cool to 4 °C and adjust pH to 8.5 with NaOH.

4. γD-crystallin cleavage buffer: Column buffer with 50 mM MESNA. Dissolve 0.32 g of MESNA in 40 mL γD-crystallin column buffer. Adjust pH to 8.5 at 4 °C.
5. γD-crystallin storage buffer: 5 mM Bis-Tris, pH 6.5, 250 mM NaCl.
6. BL21 DE3 *E. coli* cells transfected with a modified pet16b plasmid coding for the C-terminal domain of γD-crystallin fused to a His tag and a factor Xa cleavage site IEGR. The actual protein sequence is MHHHHHHXXXIEGRCYYYY (XXX stands for the sequence of ligation site, YYYY stands for the actual protein sequence starting from position 85). Factor Xa cleaves after IEGR and leaves a cysteine at the N-terminus of the protein.
7. Ni column resuspension buffer (NiRB): 50 mM $Na_2HPO_4$, 500 mM NaCl, 50 mM imidazole, pH 7.5. Dissolve 5.68 g of $Na_2HPO_4$, 23.38 g of NaCl, and 2.72 g of imidazole in 800 mL of $H_2O$. Adjust pH to 7.5 with HCl.
8. Ni column elution buffers (NiEB): 50 mM $NaHPO_4$, 500 mM NaCl, 500 mM imidazole, pH 7.5. Dissolve 5.68 g of $Na_2HPO_4$, 23.38 g of NaCl, and 27.23 g of imidazole in 800 mL of $H_2O$. Adjust pH to 7.5 with HCl.
9. Factor Xa enzyme: Purchase before protein expression. Factor Xa does not have a very long shelf life time. Store at −80 °C before use.
10. Factor Xa cleavage buffer: 10 mM Tris, 20 mM NaCl. Dissolve 0.78 g of Tris base and 5.8 g of NaCl in 500 mL of $H_2O$. Adjust pH to 7.5 with HCl. To make 5 L buffer, dissolve 7.8 g of Tris base and 58.4 g of NaCl in 5 L of $H_2O$. Adjust pH to 7.5 with HCl.
11. $CaCl_2$ stock solution, 1.0 M: Dissolve 2.22 g anhydrous $CaCl_2$ in 20 mL of $H_2O$.

### 2.3.2 Native Chemical Ligation of Segmentally Labeled γD-Crystallin and Purification

1. Ion-exchange chromatography (IEC) column: A column with 16 mm inner diameter and 200 mm bed height is packed with Q Sepharose Fast Flow preswollen beads (prepacked columns are also available through several venders).
2. IEC buffer A: 20 mM Tris, pH 8.5 at 4 °C. Dissolve 2.42 g of Tris base in 1 L of $H_2O$. Adjust pH to 8.5 at 4 °C using HCl.
3. IEC buffer B: 20 mM Tris, 1.0 M NaCl, pH 6 at 4 °C. Dissolve 2.42 g of Tris base and 58.44 g of NaCl in $H_2O$. Adjust pH to 6 at 4 °C using HCl.
4. Size-exclusion chromatography (SEC) column: A column with 16 mm inner diameter and 600 mm bed height is packed with Superdex 75 prep grade gel filtration medium.
5. SEC buffer: 20 mM $Na_2HPO_4$, pH 5.5 at 4 °C. Dissolve 0.14 g of $Na_2HPO_4$ in $H_2O$. Adjust pH to 5.5 using HCl 4 °C.

## 3 Methods

### *3.1 Synthesis and Purification of hIAPP*

#### *3.1.1 Synthesis of Specific Amino Acid-Labeled hIAPP Using Fmoc Chemistry*

1. Routine maintenance of the solid-phase synthesizer is necessary before any synthesis. Personal protective equipment, such as lab coats, gloves, and goggles, are required for Fmoc peptide synthesis. The solvents used in solid-phase synthesis could dissolve plastic, so it is recommended to use glass vessels, pipets, and flasks. If plastic is inevitable, make the contact time as short as possible. If an isotope-labeled amino acid is used in the synthesis, a small-scale cleavage is recommended to check the synthesis efficiency before adding the isotope-labeled amino acid. Note that in Fmoc solid-phase synthesis, the peptide is synthesized from the C-terminus to the N-terminus, which is opposite from the direction of peptide expression.
2. The sequence of hIAPP is KCNTA TCATQ RLANF LVHSS NNFGA ILSST NVGSN TY (from N-terminus to C-terminus). The underlined amino acids can be replaced by pseudoproline derivatives to facilitate the solid-phase synthesis [32].
3. Swell resin in 7 mL of DMF for 1–2 h at room temperature before synthesis (this step is usually programmed into the synthesizer).
4. Use double coupling for the first amino acid because it is hard to add onto the resin. The following procedure is programmed in the synthesizer. However, the same procedure can be followed for manual synthesis (*see* **Note 9**).
5. Wash resin with 7 mL of DMF. Drain. Add 7 mL of deprotection solution. Incubate with 40 W microwave power at 75 °C for 3 min. Drain.
6. Repeat **step 5**.
7. Add 2.5 mL of amino acid solution (0.2 M in DMF). Add 1 mL of activation solution and 0.5 mL of activator base solution. Microwave coupling method at 75 °C for 5 min. Drain. Wash with 7 mL DMF.
8. Repeat **step** 7. Wash with additional 7 mL DMF.
9. Use single coupling at 75 °C for 5 min for most of the amino acids. Use 20 W microwave power. Two amino acids, cysteine and arginine, required special treatments. Cysteine is added at 50 °C and incubated for 2 min without microwave power, followed by 4-min incubation with 25 W microwave power. The different temperature is used to prevent the degradation of cysteine. Arginine is usually added with double coupling at 75 °C because its side chain can decrease coupling efficiency. Incubate arginine for 25 min without microwave power and another 5 min with 25 W microwave power. Same washing, deprotecting, and activating method is used as described in **steps 5–8**.

10. Use Kaiser test to monitor the coupling reaction for isotope-labeled amino acids. When synthesizing a peptide by hand, perform Kaiser test after coupling every amino acids. If a solid-phase synthesizer is used, a "pause" should be programmed in the sequence to allow for sampling of the resin.
11. From the reaction vessel, take out a few resin beads and wash with ethanol in a glass test tube. Rinse well and dry the resin until it becomes powder.
12. Add 100 μL each of Kaiser test reagent A and B. Mix well.
13. Heat this solution on a heating block at 100–115 °C for 5 min. A dark blue color suggests an unfinished coupling reaction. If the reaction is complete, the color of the solution should be pink.
14. After all the amino acids are added to the sequence, the last step should be deprotection (as described in **steps 5** and **6**) to produce a free $NH_3$ at the N-terminus. Resin beads can be rinsed with dichloromethane, dried, and stored at −20 °C before cleavage (*see* **Note 10**).

### *3.1.2 Cleavage of Peptide from the Resin*

1. Measure out about 300 mg of resin beads with peptides into a glass test tube (*see* **Note 10**).
2. In a fume hood, add 6 mL of cleavage cocktail to the resin. Let cleavage reaction proceed at room temperature for 4 h and stir the solution every 30 min. The cleavage solution color should change from yellow to brown.
3. Prepare a Teflon filter and fit it tightly inside of a syringe or an empty column. The material of the syringe or column should be tested for chemical compatibility before use (*see* **Note 11**). A vacuum can be applied to help remove the solvent by assembling the syringe as shown in Fig. 3. A needle is attached to the bottom of the syringe. The needle goes through a septum that is on a clean side-arm Erlenmeyer flask.
4. Carefully transfer the cleavage cocktail into the syringe. Pull a vacuum to withdraw all the solution from the syringe. Add the rest of the peptide cleavage cocktail to rinse the resin again. Pull a vacuum once more to obtain all the solution.
5. Disassemble the setup and dry the solution in the Erlenmeyer flask with $N_2$ gas (*see* **Note 12**).
6. Add 5–10 mL of cold ethyl ether to the flask. Break down and scrape the precipitate in the flask with a spatula to recover as much peptide as possible.
7. Transfer the ethyl ether and the precipitate into a 50 mL centrifuge tube (with known weight to calculate the yield). Rinse the flask with another 5 mL of ethyl ether. Vortex for 30 s and leave on ice for 45 min.

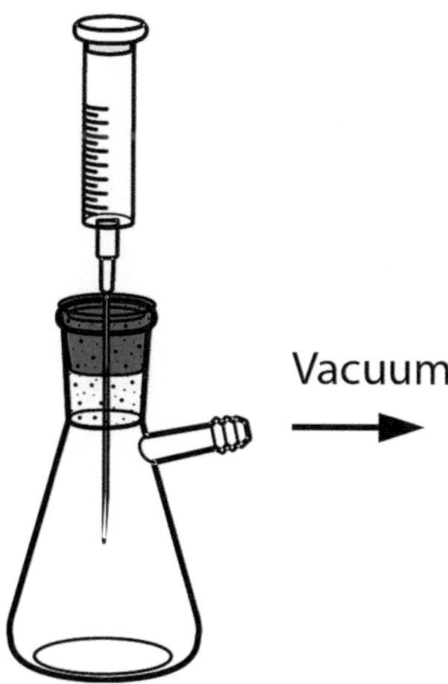


**Fig. 3** Experimental setup for peptide cleavage from resin. A Teflon filter should be fitted tightly at the bottom of the syringe

8. Spin down in a centrifuge at 5000 × *g* for 10–20 min. Decant the ether.
9. Repeat **steps 6–8**. Dry the precipitate under gentle $N_2$ gas until it forms a white powder.
10. Dissolve precipitate in 15–25 mL of acetic acid (use 25 mL acetic acid for every 100 mg of peptide) and sonicate for 30 min. Lyophilize this solution overnight. Dissolve the lyophilized power in dimethyl sulfoxide (DMSO) (1 mL DMSO per 2 mg of power) and let it sit at room temperature for 48 h.

#### 3.1.3 Purification of the Peptide with HPLC

1. Mix 10 μL of hIAPP in DMSO and 10 μL 20 % acetic acid in $H_2O$. Load this solution onto a C18 analytical HPLC reverse-phase column. Use a gradient of 10–60 % HPLC solvent B for 50 min at 1 mL/min and collect all peaks separately and lyophilize each peak. Identify the peak that corresponds to hIAPP by MALDI. The correct peak should be eluted around 40 % solvent B.
2. For large-quantity purification, mix 2 mL of hIAPP in DMSO and 2 mL 20 % acetic acid in $H_2O$ right before HPLC injection. Load this solution onto a C18 preparation HPLC reverse-phase column (*see* **Note 13**). The peak containing hIAPP should be eluted around the same percent of solvent B as on the analytical reverse-phase column.
3. Lyophilize the solutions that contain hIAPP. Dried hIAPP powder can be dissolved in hexafluoroisopropanol (HFIP) and sonicated for 4 h. This HFIP solution can be stored at −20 °C.

4. A second round of purification is recommended if the resolution of the peaks is not ideal. Mix 1 mL of peptide dissolved in HFIP into 3 mL of $H_2O$ and purify with a C18 preparation column using the same gradient as in **step 2**.
5. Redissolve hIAPP powder into HFIP or HFIP-d to make a 1 mM stock and sonicate for 4 h before use. Exact concentration of the stock solution should be measured by nano-drop UV–Vis spectrometry. Aliquot 2 μL of hIAPP in HFIP or HFIP-d, dry under $N_2$ gas, and redissolve in $H_2O$ before the concentration is measured. The final concentration is determined as the mean of three measurements.

### 3.2 Expression of $^{13}C$ Uniformly Labeled hIAPP

#### 3.2.1 First Day

1. Streak the stock cells on an ampicillin plate and incubate at 37 °C overnight (*see* **Note 14**).

#### 3.2.2 Second Day

1. Pick a round-shaped colony that is well separated from the others. Grow the colony in 10 mL of minimal media or LB broth overnight at 37 °C (*see* **Note 15**).

#### 3.2.3 Third Day

1. Add 5 μL of the cell culture from second day to another 10 mL of minimal media and incubate at 37 °C overnight (*see* **Note 16**).

#### 3.2.4 Fourth Day

1. Add 10 mL of the overnight cell culture from **step 3** to 0.5 L of minimal media and incubate at 37 °C until $OD_{600\ nm} = 0.6–0.8$. Induce with 0.5 mM isopropyl-beta-D-thiogalactopyranoside (IPTG) and incubate for 6–8 h (*see* **Note 17**).
2. Harvest cells by centrifugation for 15 min at 5000 × *g*. Resuspend the cell pellets in hIAPP column buffer on ice and immediately crush with French pressure cell press at 4 °C. Centrifuge the homogeneous solution at 50,000 × *g* for 45 min to obtain cell lysate as the supernatant. Load the cell lysate onto a 10 mL chitin column at 0.5 mL/min and then wash with hIAPP column buffer for at least 15 column volumes at 1 mL/min.
3. Flush column with 3 volumes of hIAPP cleavage buffer and incubate for 12–24 h at 4 °C (*see* **Note 18**). The use of ammonium bicarbonate in the hIAPP cleavage buffer leads to the production of a C-terminally amidated cleavage product.

#### 3.2.5 Fifth Day

1. Elute the column with 2.5 volumes of hIAPP column buffer to harvest MhIAPP (*see* **Note 19**).
2. Dialyze the MhIAPP with 1000 MWCO membrane in 5 L baskets. Dialyze MhIAPP against deionized water in the first round and 0.3 % HCl in the second and third rounds each for 2–4 h (*see* **Note 20**).

3. Add 150 mg of cyanogen bromide to 30 mL MhIAPP in 0.3 % HCl solution (5 mg cyanogen bromide per mL solution). Allow cleavage of methionine residue at room temperature for 24 h in the dark (*see* **Note 21**).
4. The cleavage process can be monitored by MALDI. Samples can be mixed with sinapic acid matrix directly. The MhIAPP peak shows up around 4206.3 *m/z*, while the hIAPP peak shows up around 4070.1 *m/z*. Multiple other MALDI peaks might be detected since the chitin bead affinity purification is not perfect.

#### 3.2.6 Sixth Day

1. Freeze the hIAPP solution (methionine residue is now cleaved) and lyophilize. Lyophilization time can be significantly reduced when the solution is partitioned into several aliquots (*see* **Note 22**).

#### 3.2.7 Seventh Day

1. Dissolve lyophilized residues in 2.5 mL of DMSO (*see* **Note 23**). Purify the peptide with C18 reverse-phase columns as described in Subheading 3.1.3.

#### 3.2.8 Eighth Day

1. Dissolve hIAPP powder in $D_2O$. Measure the concentration using nano-drop UV–Vis spectrometry. Aliquot and lyophilize the solutions (*see* **Note 24**).
2. To redissolve and use hIAPP powder, follow the same steps as described in Subheading 3.1.3, **step 5**.

### 3.3 Expression of γD-Crystallin N-Terminal Domain

1. The schematic flowchart is shown in Fig. 1, in which the C-terminal domain is $^{13}C$ isotope labeled. The reversely labeled protein ($^{13}C$ N-terminal domain and $^{12}C$ C-terminal domain) can be made by switching the growth conditions of *E. coli*.
2. To grow $^{12}C$ γD-crystallin N-terminal domain, $^{12}C_6$-glucose is used instead of $^{13}C_6$-glucose in the minimal media.
3. The growth of γD-crystallin N-terminal domain follows Subheadings 3.2.1–3.2.4 using γD-crystallin cleavage buffer instead of hIAPP cleavage buffer. In **step 3** of Subheading 3.2.4, the incubation time is longer for the γD-crystallin N-terminal domain because MESNA is used instead of DTT, which cleaves much faster. Take an aliquot every 12 h for up to 3 days to examine the cleavage progress using MALDI.
4. After at least 48 h of incubation, elute the N-terminal domain protein with 3 column volumes of γD-crystallin cleavage buffer. Use a spin column (5000 Da MWCO) to obtain 1 mL of solution (protein concentration greater than 1 mM) with high protein concentration.
5. Use freshly cleaved protein for ligation to maximize the yield. Protein can be dialyzed into γD-crystallin storage buffer and

stored at −80 °C if it will not be used right away. Frozen samples should be dialyzed back into γD-crystallin cleavage buffer prior to ligation.

### 3.4 Expression of γD-Crystallin C-Terminal Domain

#### 3.4.1 Days 1–3

The growth of γD-crystallin C-terminal domain follows **steps 1–3** in Subheading 3.2 for the first 3 days (*see* **Note 25**).

#### 3.4.2 Fourth Day

1. Add 10 mL of the overnight cell culture to 0.5 L of minimal media and incubate at 37 °C until $OD_{600\ nm}=0.6–0.8$. Induce with 0.5 mM IPTG and incubate for 6–8 h.
2. Harvest cells by centrifugation for 15 min at 5000×*g*. Resuspend the cell pellets in NiRB on ice. Crush cells with French pressure cell press at 4 °C and centrifuge at 50,000×*g* for 45 min to obtain cell lysate. Load the cell lysate onto a 10 mL Ni column immediately with 0.3 mL/min and wash with NiRB for at least 15 column volumes at 1 mL/min.
3. Elute with 3 column volumes of NiEB. To obtain purer proteins, use a gradient elution starting from 100 mM imidazole and increase the imidazole concentration with steps of 50 mM. Purest protein is eluted around 200 mM imidazole (*see* **Note 26**).
4. Dialyze protein-containing eluate against 5 L of Factor Xa cleavage buffer three times (3, 3, and 4 h).

#### 3.4.3 Fifth Day

1. Use spin column (5000 Da MWCO) to spin down the protein solution to 4 mL (to obtain protein concentration greater than 1 mM) and transfer to a centrifuge tube. Add 20 μL of Factor Xa (in glycerol) and 5 μL $CaCl_2$ stock solution (1.0 M) slowly while shaking to make 1.0 mM in the final concentration.
2. Incubate the solution in the dark overnight (*see* **Note 27**).

#### 3.4.4 Sixth Day

1. Load the protein solution through a Ni column. Collect the solution that does not bind to the column (the flow through).
2. Dialyze the flow through (γD-crystallin C-terminal domain with N-terminal cysteine) against γD-crystallin column buffer.

### 3.5 Native Chemical Ligation and Purification of Segmentally Labeled γD-Crystallin

1. Add both N-terminal domain (with MESNA) and C-terminal domain (with N-terminal cysteine). When mixing the two domains with 1:4 volume ratio, the final buffer will have 10 mM MESNA, which is the condition suitable for ligation. Incubate the solution at 4 °C overnight.
2. Take 5 μL of solution for MALDI every 4 h after the first 8 h if possible.

3. The ligation reaction should be tested without $^{13}C$ isotope labels first.
4. Purify the ligated protein with IEC using IEC buffer A and B with a gradient of 0–50 % of buffer B in 2 h at 1 mL/min flow rate (the retention time depends on the column). Collect all peaks and identify ligated protein peak using SDS-PAGE gel.
5. Purify the IEC peak of ligated protein with SEC using SEC buffer. The retention time is around 30 min on our column.
6. Use MALDI to verify the mass of the ligated protein.
7. Aliquot the ligated protein and store at −80 °C. Only thaw the sample before use. Frequent freeze-and-thaw cycles can cause precipitation of the protein.

### 3.6 Infrared Spectroscopy Measurement of hIAPP

1. Dry 2 μL hIAPP (1 mM) in HFIP-d under $N_2$ in a fume hood. A speed vacuum (concentration centrifuge) is also recommended.
2. Aggregation is initiated by dissolving hIAPP in 2 μL of 20 mM Tris $D_2O$ buffers, pH 7.4 to make 1 mM final concentration of peptides (*see* **Note 28**).
3. Place 2 μL of peptide solution immediately between two calcium fluoride ($CaF_2$) windows after dissolving in buffer. Use a 56 μm spacer to fix the path length (*see* **Note 29**).
4. For FTIR, the kinetics of stationary-phase measurements can be set up with the instrument settings. It is important to collect a background spectrum without solution and a $D_2O$ buffer background spectrum before measuring the peptide sample. A HeNe beam can be used to ensure that light is going through the sample.
5. For 2D IR spectroscopy, the same sample cell and assembly can be used in both our FTIR and 2D IR setups. Before measuring the sample, the laser setup is optimized using a standard calibration molecule, *N*-acetyl-proline (NAP). The peptide sample is placed where the pump and probe pulses are overlapped and focused. We collect frequency domain data on the probe axis with a mercury cadmium telluride (MCT) detector. With a pulse shaper, time *t* between the two pump pulses is scanned (time domain data) and a Fourier transformation gives us the pump frequencies. 2D IR spectroscopy with a pulse shaper is suitable for kinetics measurements of protein aggregation because a single spectrum can be collected in less than 1 min.

## 4 Notes

1. This protocol mainly applies to labeling the amino acids with hydrophobic side chains (Ala, Gly, Ile, Leu, Phe, and Val). Many side chains require protection for Fmoc synthesis.

The side chain-protecting groups are acid labile (removable by acid) and so will be removed by the protocol given here for $^{18}O$ labeling the amino acids, which is catalyzed by acid. The Fmoc group is base labile, so it will not be affected during $^{18}O$ labeling. If amino acids without the Fmoc group are purchased, the Fmoc group can be added before $^{18}O$ labeling.

2. The following recipes are used for a particular hIAPP synthesis. The final numbers presented in this protocol are chosen to have 5 % excess over what is actually needed. These numbers are given for a 0.1 mM-scale synthesis. However, the actual amount needed in any particular synthesis will depend on the solid-phase synthesizer being used; use the software provided by the synthesizer company to calculate the right amount of reagents for your instrument. Alternatively, the correct amount of reagents can be calculated based on the concentration and volume required for each step. The concentrations given here are provided to aid manual synthesis of shorter peptides.
3. We typically use Fmoc-PAL-PEG-PS resin for hIAPP synthesis because it produces an amidated C-terminus. The resulting peptide contains the C-terminal amide and free $NH_3$, which gives positive charge. The loading capacity of this resin can vary depending on the supplier.
4. HOBt is classified as an explosive and must be stored following special requirements. We recommend that you contact the safety department of your institute for proper chemical storage procedures. Special permission is needed to purchase piperidine because it is a restricted substance. It is recommended to start the approval process early to avoid unnecessary delays in research.
5. If a solid-phase synthesis is used, the amount needed should be calculated. A small amount of excess can be used to ensure a successful synthesis.
6. Other cleavage cocktails are also available. The one presented here is the one that works the best for hIAPP according to our results. All cleavage or synthesis should be performed in a fume hood. This particular cleavage cocktail uses thioanisole, ethanedithiol, and anisole as scavengers; therefore a strong thiol smell will be generated. Gloves and all other wastes should be sprayed with bleach and left in the fume hood overnight before disposal. Another commonly used recipe uses 9 mL TFA, 800 μL ethanedithiol, and 200 μL $H_2O$ for a total volume of 10 mL. This amount of cleavage cocktail is used to cleave about 300 mg of resin (20–25 % of the total resin used in a 0.1 mM scale). We do not recommend cleaving all the resin at once.
7. Another option is to make a nutrient stock solution by dissolving 1 g of $^{13}C_6$-glucose into 10 mL of 0.05 g/mL $MgSO_4$

(autoclaved separately) and add 500 μL of trace element solution, 325 μL of vitamin solution, and 35 mg of thiamine HCl. This can then be added directly to 0.5 L of minimal media buffer, followed by addition of 500 μL ampicillin solution (100 mg/mL).

8. It takes time for the ammonium bicarbonate to dissolve. Add ammonium bicarbonate first and then add DTT. Prepare the solution on ice in the fume hood. Constantly release pressure if the solution is prepared in a closed tube or flask because ammonium bicarbonate will release gas as it dissolves.
9. The first amino acid is usually added with double coupling because it is hard to add onto the resin due to steric hindrance. If peptide is synthesized manually, the microwave-coupling step can be replaced by bubbling with $N_2$ gas for at least 1 h. The deprotection step is performed twice for every amino acid to increase yield.
10. The amount of resin beads is measured out to obtain the best yield. When the peptides are still attached to the beads, they are stable and can be stored for many months. However, when the peptides are cleaved from the resin, they are susceptible to deamidation [33] even when stored at −20 °C. We typically cleave 300 mg (about 25 % of the total resin) at a time, which yields 50 mg of hIAPP peptide that is sufficient for dozens of 2D IR measurements.
11. We use a 30 mL syringe with chemical resistance to the cleavage cocktail. The Teflon filter is placed inside of the syringe without the plunger.
12. It takes 5–12 h to completely dry the solvent. It is important that the solvent is completely dried to obtain good peptide with high yield. You would expect to see pale yellow precipitate if the synthesis is successful. Use a gentle flow of $N_2$ gas.
13. DMSO is used to form disulfide bonds for synthesized hIAPP. Mixing DMSO and acetic acid can cause breakdown of the disulfide bonds. Thus, it is important that the two solutions are mixed immediately prior to injection. DMSO goes through the reverse-phase column with little interaction. A large saturation peak of DMSO will show up after the dead time of the column. You can start with a larger range of the gradient, for example 10–60 % solvent B in 40 min, and narrow down the range to obtain a better peak resolution.
14. You can grow the overnight cell culture from a frozen stock of cells that has been stored in −80 °C in glycerol. However, cells that grow from a single colony are healthier than those from a frozen stock. By screening the cell stock on a plate, you can also select for the antibiotic resistance and the preferred morphology of the colonies.

15. The minimal media can be stored at 4 °C overnight. You can also make a small amount of minimal media fresh using nutrient stock as described in **Note 7**. Add 200 μL of nutrient stock to 10 mL of minimal media buffer to make a small amount of minimal media in a Falcon tube. 10 μL of ampicillin (100 mg/mL) should be added to the solution before adding the cell colony.
16. This step is optional. If $^{12}C$ peptide is desired, this step should be skipped. When making $^{13}C$ peptide, this step is recommended for a higher labeling efficiency.
17. The doubling time for cell concentration is longer in minimal media than in LB broth. For Bl21 DE3 cells in minimal media, the doubling time is about 1–1.5 h.
18. There will be bubbles generated in the process. Open columns are recommended. Leave the column open to air to release pressure during overnight incubation in cleavage buffer.
19. The start codon codes for methionine in most protein expressions. Matured IAPP in humans does not contain methionine at the N terminus because it is cut during posttranslational modification. There is no other methionine in the hIAPP sequence, so cyanogen bromide can be used to cleave the N terminus methionine. When designing the expression of other peptides that might contain methionine in other positions of the sequence, different approaches are needed.
20. Since ammonia bicarbonate buffer was used, dialyzing against HCl during the first round of dialysis will generate $CO_2$ gas and potentially break the dialysis tubing. The smell of DTT might still be detected after three rounds of dialysis. This does not seem to affect the following steps.
21. Cyanogen bromide is highly volatile, light sensitive, and toxic. Cyanogen bromide readily hydrolyzes into hydrogen cyanide, a powerful hemotoxin. Cyanogen bromide is also skin permeable. Work involving cyanogen bromide should be conducted in a fume hood and should not be conducted alone.
22. The solution takes around 12–24 h to lyophilize. The remaining residue is usually yellow and gooey which dissolves completely in DMSO.
23. When working with hIAPP that is synthesized on a solid-phase synthesizer, the peptides that are cleaved from resin beads need to sit in DMSO for 48 h before purification with HPLC to form a disulfide bond between Cys2 and Cys7. Expressed hIAPP contains the disulfide bond, so the 2-day incubation period can be skipped.
24. You can also dissolve the hIAPP powder directly into d-HFIP. Dissolving hIAPP in $D_2O$ makes it easier to aliquot and improves the H/D exchange efficiency.

25. The C-terminal domain (with an N-terminal cysteine) is relatively stable because it does not have a leaving group like MESNA. It should be made before making the N-terminal domain protein.
26. Because some impurities bind to Ni column at low imidazole concentration, purer target protein can be obtained by a gradient elution. You can load the 10 μL of the elution from each concentration step onto an SDS-PAGE gel to determine which fraction of protein is the purest.
27. Different condition of Factor Xa can be used. Incubation at 37 °C is the optimal working condition for Factor Xa when working with proteins that are not stable. Incubation overnight at room temperature works well for the γD-crystallin C-terminal domain.
28. Other $D_2O$ buffers can also be used. Different salt content, concentration, and pH value will affect the kinetics of the aggregation, as well as the final fiber stability [34].
29. Different thicknesses of spacers are available: 12, 25, 56, 75, and 100 μm. A 56 μm spacer works best for our experiments in $D_2O$ buffer. However, a thinner spacer can be used to reduce scattering when working with lipids and micelles.

## Acknowledgments

We thank NIH NIDDK DK79895 for supporting this work.

*Statement*: Martin Zanni is a cofounder of PhaseTech Spectroscopy, Inc. which manufactures 2D IR spectrometers.

### References

1. Hahn S, Kim SS, Lee C, Cho M (2005) Characteristic two-dimensional IR spectroscopic features of antiparallel and parallel beta-sheet polypeptides: simulation studies. J Chem Phys 123:084905
2. Baiz CR, Peng CS, Reppert ME, Jones KC, Tokmakoff A (2012) Coherent two-dimensional infrared spectroscopy: quantitative analysis of protein secondary structure in solution. Analyst (Cambridge, U K) 137:1793–1799
3. Strasfeld DB, Ling YL, Shim S-H, Zanni MT (2008) Tracking fiber formation in human islet amyloid polypeptide with automated 2D-IR spectroscopy. J Am Chem Soc 130:6698–6699
4. Cheng-Yen Huang ZG, Wang T, DeGrado WF, Gai F (2001) Time-Resolved Infrared Study of the helix-coil transition using $^{13}C$-labeled helical peptides. J Am Chem Soc 123:12111–12112
5. Brewer SH, Song B, Raleigh DP, Dyer RB (2007) Residue specific resolution of protein folding dynamics using isotope-edited infrared temperature jump spectroscopy†. Biochemistry 46:3279–3285
6. Decatur SM, Antonic J (1999) Isotope-edited infrared spectroscopy of helical peptides. J Am Chem Soc 121:11914–11915
7. Sean D, Moran AMW, Buchanan LE, Bixby E, Zanni MT (2012) Two-dimensional IR spectroscopy and segmental $^{13}C$ labeling reveals the domain structure of human γD-crystallin amyloid fibrils. Proc Natl Acad Sci U S A 109: 3329–3334

8. Woys AM, Almeida AM, Wang L, Chiu CC, McGovern M et al (2012) Parallel beta-sheet vibrational couplings revealed by 2D IR spectroscopy of an isotopically labeled macrocycle: quantitative benchmark for the interpretation of amyloid and protein infrared spectra. J Am Chem Soc 134:19118–19128
9. Hamm P, Zanni M (2011) Concepts and methods of 2D infrared spectroscopy. Cambridge University Press, Cambridge
10. Shim SH, Gupta R, Ling YL, Strasfeld DB, Raleigh DP, Zanni MT (2009) Two-dimensional IR spectroscopy and isotope labeling defines the pathway of amyloid formation with residue-specific resolution. Proc Natl Acad Sci U S A 106:6614–6619
11. Falvo C, Zhuang W, Kim YS, Axelsen PH, Hochstrasser RM, Mukamel S (2012) Frequency distribution of the amide-I vibration sorted by residues in amyloid fibrils revealed by 2D-IR measurements and simulations. J Phys Chem B 116:3322–3330
12. Remorino A, Korendovych IV, Wu Y, DeGrado WF, Hochstrasser RM (2011) Residue-specific vibrational echoes yield 3D structures of a transmembrane helix dimer. Science 332: 1206–1209
13. Buchanan LE, Dunkelberger EB, Tran HQ, Cheng PN, Chiu CC et al (2013) Mechanism of IAPP amyloid fibril formation involves an intermediate with a transient beta-sheet. Proc Natl Acad Sci U S A 110:19285–19290
14. Buchanan LE, Carr JK, Fluitt AM, Hoganson AJ, Moran SD et al (2014) Structural motif of polyglutamine amyloid fibrils discerned with mixed-isotope infrared spectroscopy. Proc Natl Acad Sci U S A 111:5796–5801
15. Pentelute BL, Kent SBH (2006) Selective desulfurization of cysteine in the presence of Cys(Acm) in polypeptides obtained by native chemical ligation. Org Lett 9:687–690
16. Devaraneni PK, Komarov AG, Costantino CA, Devereaux JJ, Matulef K, Valiyaveetil FI (2013) Semisynthetic K+ channels show that the constricted conformation of the selectivity filter is not the C-type inactivated state. Proc Natl Acad Sci U S A 110:15698–15703
17. Flavell RR, Muir TW (2009) Expressed protein ligation (EPL) in the study of signal transduction, ion conduction, and chromatin biology. Acc Chem Res 42:107–116
18. Moran SD, Decatur SM, Zanni MT (2012) Structural and sequence analysis of the human gammaD-crystallin amyloid fibril core using 2D IR spectroscopy, segmental 13C labeling, and mass spectrometry. J Am Chem Soc 134: 18410–18416
19. Seyfried MS, Lauber BS, Luedtke NW (2010) MultipleTurnover isotopic labeling of Fmoc- and Boc-protected amino acids with oxygen isotopes. Org Lett 12:104–106
20. Middleton CT, Woys AM, Mukherjee SS, Zanni MT (2010) Residue-specific structural kinetics of proteins through the union of isotope labeling, mid-IR pulse shaping, and coherent 2D IR spectroscopy. Methods 52:12–22
21. Lopes DH, Colin C, Degaki TL, de Sousa AC, Vieira MN et al (2004) Amyloidogenicity and cytotoxicity of recombinant mature human islet amyloid polypeptide (rhIAPP). J Biol Chem 279:42803–42810
22. Mazor Y, Gilead S, Benhar I, Gazit E (2002) Identification and characterization of a novel molecular-recognition and self-assembly domain within the islet amyloid polypeptide. J Mol Biol 322:1013–1024
23. Kosicka I, Kristensen T, Bjerring M, Thomsen K, Scavenius C et al (2014) Preparation of uniformly (13)C, (15)N-labeled recombinant human amylin for solid-state NMR investigation. Protein Expr Purif 99:119–130
24. Williamson JA, Miranker AD (2007) Direct detection of transient alpha-helical states in islet amyloid polypeptide. Protein Sci 16:110–117
25. Moran SD, Zanni MT (2014) How to get insight into amyloid structure and formation from infrared spectroscopy. J Phys Chem Lett 5:1984–1993
26. King JT, Kubarych KJ (2012) Site-specific coupling of hydration water and protein flexibility studied in solution with ultrafast 2D-IR spectroscopy. J Am Chem Soc 134:18705–18712
27. Chung HS, Khalil M, Smith AW, Tokmakoff A (2007) Transient two-dimensional IR spectrometer for probing nanosecond temperature-jump kinetics. Rev Sci Instrum 78:063101
28. Ihalainen JA, Bredenbeck J, Pfister R, Helbing J, Chi L et al (2007) Folding and unfolding of a photoswitchable peptide from picoseconds to microseconds. Proc Natl Acad Sci U S A 104: 5383–5388
29. Ghosh A, Wang J, Moroz YS, Korendovych IV, Zanni M et al (2014) 2D IR spectroscopy reveals the role of water in the binding of channel-blocking drugs to the influenza M2 channel. J Chem Phys 140:235105
30. Manor J, Mukherjee P, Lin YS, Leonov H, Skinner JL et al (2009) Gating mechanism of the influenza A M2 channel revealed by 1D and 2D IR spectroscopies. Structure 17:247–254
31. Jansson M, Li YC, Anderson S, Montelione G, Nilsson B (1996) High-level production of uniformly $^{15}$N-and $^{13}$C-enriched fusion proteins in *Escherichia coli*. J Biomol NMR 7:131–141

32. Abedini A, Raleigh DP (2005) Incorporation of pseudoproline derivatives allows the facile synthesis of human IAPP, a highly amyloidogenic and aggregation-prone polypeptide. Org Lett 7:693–696
33. Dunkelberger EB, Buchanan LE, Marek P, Cao P, Raleigh DP, Zanni MT (2012) Deamidation accelerates amyloid formation and alters amylin fiber structure. J Am Chem Soc 134:12658–12667
34. Marek PJ, Patsalo V, Green DF, Raleigh DP (2012) Ionic strength effects on amyloid formation by amylin are a complicated interplay among Debye screening, ion selectivity, and Hofmeister effects. Biochemistry 51: 8478–8490

# Part II

## Kinetics/Mechanism

# Chapter 3

# Intermolecular Paramagnetic Relaxation Enhancement (PRE) Studies of Transient Complexes in Intrinsically Disordered Proteins

**Maria K. Janowska and Jean Baum**


## Abstract

NMR interchain paramagnetic relaxation enhancement (PRE) techniques are a very powerful approach for detecting transient interchain interactions between intrinsically disordered proteins. These experiments, requiring a mixed sample containing a 1:1 ratio of isotope-labeled $^{15}$N protein and natural abundance $^{14}$N protein with a paramagnetic spin label, provide data that is limited to interchain interactions only. Application of these experiments to weakly associated transient species such as those that are present in the very early stages of self-assembly processes will aid our understanding of protein aggregation or fibril formation processes.


**Key words** Aggregation, Intermolecular interactions, NMR, Paramagnetic relaxation enhancement, PRE

## 1 Introduction

Many devastating neurodegenerative diseases are associated with proteins that convert from their normal soluble forms to amyloid fibrils that accumulate in the brain, and the mechanism by which this occurs remains poorly understood. It is critically important to characterize the species formed during the very early stages of aggregation, as increasing evidence suggests that small protein oligomers may be more toxic than the final fibrillar aggregates. Atomic characterization of domain-domain interactions or interchain interactions at the earliest times is therefore key to understanding the structural transformation from monomer to fibril. Describing the dimer encounter complex is extremely challenging as these self-associated species are transient and exist at very low populations. In addition, proteins involved in neurodegenerative disease are often intrinsically disordered proteins (IDPs) such as

David Eliezer (ed.), *Protein Amyloid Aggregation: Methods and Protocols*, Methods in Molecular Biology, vol. 1345,
DOI 10.1007/978-1-4939-2978-8_3, © Springer Science+Business Media New York 2016

α-synuclein, the primary protein in the Lewy bodies of patients with Parkinson's, or Aβ, the main component of amyloid plaques in Alzheimer's disease.

Interchain NMR paramagnetic relaxation enhancement (PRE) experiments allow the direct visualization and characterization of lowly populated transient encounter complexes in IDPs and establish the nature of the interchain interactions that may be present in the self-assembly process. $^{1}$H interchain NMR PRE experiments are performed by making 1:1 mixtures of $^{15}$N-labeled protein and $^{14}$N paramagnetic singly spin-labeled protein and detecting broadened resonances on the $^{15}$N-labeled NMR visible sample that arise from the paramagnetic spin label on the $^{14}$N chain. This experiment limits observation of PREs to interchain interactions only as the $^{14}$N protein that contains the paramagnetic spin label is NMR blind, thereby making detection of intrachain PREs impossible. The observed interchain transverse PRE rate on the $^{15}$N-labeled sample arises from the interaction of the paramagnetic center and the nucleus of interest is proportional to $<r^{-6}>$, and can provide distance information up to approximately 25 Å with an MTSL spin label. Detection is very sensitive to lowly populated states and the transient dimer interactions can be detected under equilibrium monomer conditions established in the 1:1 $^{15}$N-labeled and paramagnetic spin-labeled protein. The protocol for performing the interchain PRE experiments consists of six stages, including (a) preparation of NMR $^{15}$N-labeled protein, (b) preparation of $^{14}$N protein with single-cysteine mutants, (c) preparation of paramagnetic NMR sample, (d) preparation of diamagnetic sample, (e) experimental acquisition, and (f) data analysis (Fig. 1).

## 2 Materials

All solutions should be prepared using ultrapure water (prepared by purifying deionized water purifier with sensitivity of 18 MΩ cm). Filter all the solutions through a 22 μM filter. All the buffers that will be used for HPLC/FPLC have to be filtered and degassed. Follow closely all the regulations for waste disposal.

### 2.1 Equipment

1. NMR spectrometer (field suitable for 2D experiments).
2. FPLC or HPLC instrument.
3. Desalting column.
4. NMR tubes.
5. Protein preparation setup.
6. Buffer exchange setup.

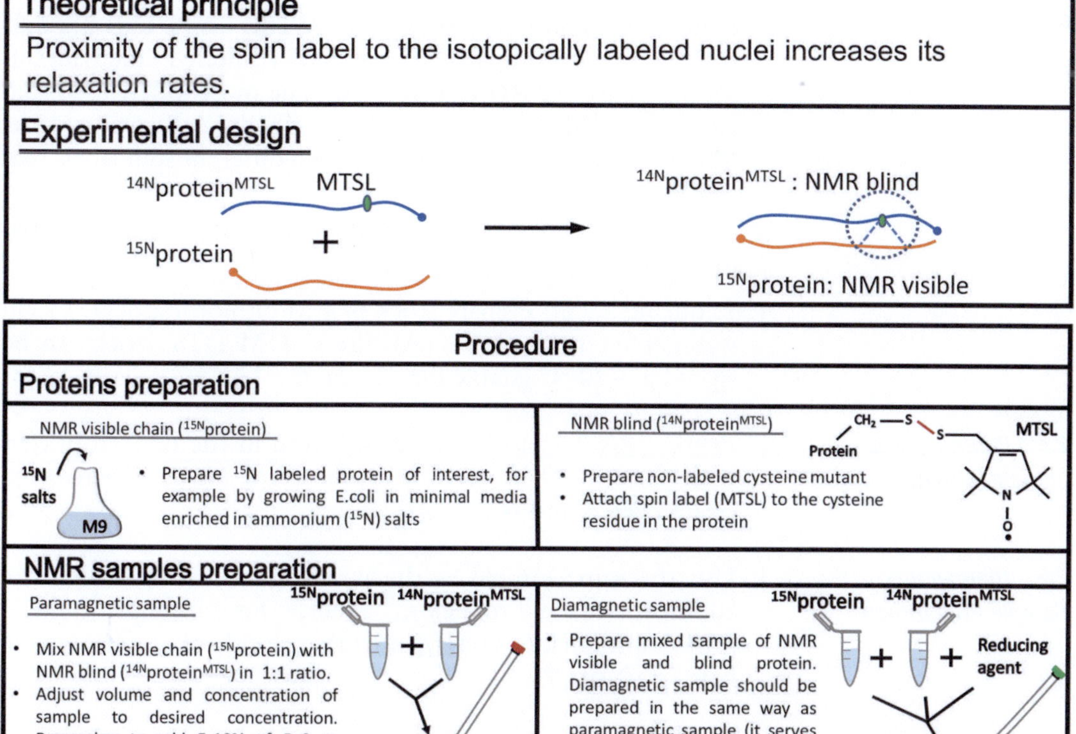


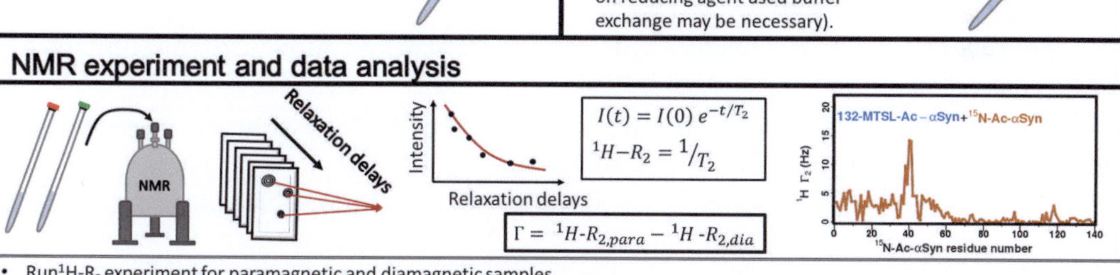


- Run $^1H$-$R_2$ experiment for paramagnetic and diamagnetic samples.
- Process $^1H$-$R_2$ experiment, get residue specific relaxation rates by fitting exponential decay with different relaxation delays.
- Paramagnetic relaxation enhancement (PRE value) is the difference in the relaxation rates of paramagnetic and diamagnetic samples. Diamagnetic relaxation rates are attributed to intrinsic relaxation of the protein, therefore by subtracting it we will obtain enhancement of relaxation rates (PRE values) caused by proximity of spin label.

**Fig. 1** Procedure and basic theoretical principle behind the interchain paramagnetic relaxation enhancement experiments

### 2.2 Protein Labeling Scheme and Purification

1. $^{15}N$-labeled protein (NMR visible chain): Grow cells, expressing the protein of interest in M9 minimal media with addition of $^{15}N$ ammonium salt to ensure uniform $^{15}N$ labeling. Follow standard purification procedure for the protein of interest.
2. Unlabeled ($^{14}N$, NMR blind)—cysteine mutant: Grow single-cysteine mutant in Luria Broth (LB) medium. Follow standard purification procedure for the protein of interest (*see* **Notes 1–3**).

#### 2.3 Spin Labeling of Cysteine Mutants

1. Unlabeled ($^{14}$N, NMR blind)—cysteine mutant.
2. Spin label solution: 10 mg of MTSL (*S*-(1-oxyl-2,2,5,5-tetramethyl-2,5-dihydro-1H-pyrrol-3-yl)methyl methanesulfonothioate) in 400 μL of acetone. The most widely used spin label is MTSL. The protocol described here assumes that MTSL will be used, but there are also different spin labels that can be used (*see* **Note 4**).
3. Standard buffer: Suggested buffers are PBS pH 7.4 or Tris pH 7.7; pH should be greater than 7.
4. Reducing agent: Prepare stock of 1 M dithiothreitol (DTT) in water, filter through 22 μM filters. 1 M DTT stock can be stored at −20 °C (stable for ~1 year).

#### 2.4 Paramagnetic Sample Preparation

1. Protein (as described above) solution in the desired buffer, with 10 % $D_2O$ in the final volume for NMR experiments (*see* **Notes 5** and **6**).

#### 2.5 Diamagnetic Sample Preparation

1. Use the same sample as the paramagnetic sample with the addition of a reducing agent, for example: DTT, β-mercaptoethanol (BME), ascorbate acid, or sodium ascorbate (*see* **Note 7**).

## 3 Methods

#### 3.1 Spin-Labeled Protein Preparation (MTSL-$^{14}$N-Labeled Cysteine Mutant)

1. Dissolve 5–10 mg of $^{14}$N-labeled single-cysteine mutant in a standard buffer.
2. Add 20 times molar ratio of DTT to solution using reducing agent stock solution.
3. Incubate for 4–6 h in the cold room to remove cysteine disulfide bonds.
4. Inject sample into desalting column according to the manufacturer's specifications. Our laboratory uses GE™ Healthcare HiPrep™ 26/10, but other desalting columns can be used (*see* **Note 8**).
5. Immediately add fivefold molar excess of freshly prepared MTSL using spin label solution.
6. Incubate in the dark overnight (4 °C) on a shaking platform; the sample is light sensitive.
7. Remove excess spin label either by dialysis or buffer exchange (*see* **Note 9**).
8. Lyophilize the protein or concentrate it for immediate NMR sample preparation.

### 3.2 Paramagnetic Sample Preparation

1. Mix $^{15}N$-labeled protein with $^{14}N$ MTSL-labeled protein in a 1:1 ratio to the desired final concentration. Buffer should contain 10 % $D_2O$ for NMR experiments. NMR experiments require sample concentrations of at least 0.1 mM for a small, unfolded protein. There are also upper limits to the concentration for the interchain PRE experiments (*see* **Note 6**).

### 3.3 Diamagnetic Sample Preparation

1. Reduce the paramagnetic samples with 10× excess of chosen reducing agent. Depending on the selection of reducing agent the sample may require buffer exchange (*see* **Note 7**).

### 3.4 NMR Experiment Acquisition: $^{1}H–R_2$ Experiments of Para- and Diamagnetic Samples

1. Two identical experiments will be performed, one with the paramagnetic sample and the second with the diamagnetic sample.
2. Contributions of the PRE effect to the relaxation rates are measured by detecting line broadening on the $^{15}N$-labeled NMR visible chain via standard $^{1}H_N$ transverse relaxation experiments ($^{1}H–R_2$) [1]. $^{1}H_N$ transverse relaxation experiments require acquisition of spectra with multiple time points (relaxation delays) (Eq. 1). For unfolded proteins optimal relaxation delay times are from 10 ms to at least 160 ms. T2 is obtained by fitting data at multiple relaxation delays to Eq. 1:

$$I(t) = I(0)e^{-t/T_2}; \quad \text{where}\,{}^{1}H - R_2 = 1/T_2 \tag{1}$$

For the calculation of error we measure duplicate time points (at least two) and use a standard error propagation routine.

### 3.5 PRE Data Analysis: Obtaining Paramagnetic Relaxation Enhancement Rates (PRE Rates, Γ)

1. Analyze the relaxation experiments using a standard processing procedure for relaxation experiments to obtain paramagnetic and diamagnetic relaxation rates ($^{1}H–R_{2,para}$ and $^{1}H–R_{2,dia}$). Diamagnetic relaxation rates ($^{1}H–R_{2,dia}$) are attributed to the intrinsic relaxation of the nuclei, while the paramagnetic relaxation rates ($^{1}H–R_{2,para}$) are the sums of the intrinsic relaxation rates and the enhancement of relaxation caused by the proximity of the spin label. Thus, the paramagnetic relaxation enhancement rate (PRE rate—Γ) is the difference of the relaxation rates of the paramagnetic and diamagnetic samples (Eq. 2):

$$\Gamma = {}^{1}H - R_{2,para} - {}^{1}H - R_{2,dia} \tag{2}$$

Direct correlation of PRE rates to distances is complicated due to the fact that the residue is experiencing a weighted average of all possible populations of the complex (*see* **Notes 10** and **11**). Due to the $<r^6>$ dependence, the populations that have closer distances are more heavily weighted.

2. To obtain a detailed analysis of transient interactions it is necessary to incorporate spin labels at many different positions. It is suggested that spin labels be placed at approximately every 10–30 residues for intra-chain PRE experiments, and inter-chain PRE require similar or even more extensive spacing of spin labels (*see* **Note 12**).

## 4 Notes

1. *Cysteine mutation requirements for PRE experiments.* Many of the spin labels that are used in the PRE experiments are thiol specific, which means that they interact specifically with cysteines to form disulfide bonds. For the interchain-PRE scheme to work successfully a single cysteine has to be present in the protein. Therefore site-directed mutagenesis schemes may have to be applied to either remove intrinsic cysteines or introduce single-cysteine mutations into the protein [2–5].
2. *Testing protein functionality upon mutation and spin labeling.* The PRE approach using site-directed mutagenesis has many advantages, but introducing mutations and MTSL modifications could cause changes in the protein function and structure. Thus it is recommended that a functionality test be performed on the mutated and/or MTSL spin-labeled proteins.
3. *Detection of distances in PRE experiments.* The positions of the spin labels should be chosen with care both to minimize the effect of the mutation on protein structure or function and to optimize detection of the interchain PRE effect. Typically the spin label is able to enhance relaxation rates of the nuclei for distances up to approximately 25 Å. Trial and error may be required for optimal selection of spin label positions. We recommend starting with spin labels near the termini as well as central regions of the protein to obtain preliminary results and then fine-tune around the interactive positions.
4. *Selection of the spin label.* Commonly used are cysteine-specific and nitroxide derivatives (for example: MTSL, TEMPO—((2,2,6,6-tetramethylpiperidin-1-yl)oxy), PROXYL—(3-(2-iodoacetamido)-2,2,5,5-tetramethyl-1-pyrrolidinyloxy)), or metal-chelating groups (*S*-(2-pyridylthio)-cysteaminyl-EDTA, which in the paramagnetic form chelates $Mn^{2+}$, and the diamagnetic form chelates $Ca^{2+}$) [6–8]. We use MTSL because it is small and generally stable and the reaction is highly cysteine specific and efficient.
5. *Optimize solution conditions to obtain maximum PRE effect.* PRE experiments are able to detect lowly populated interactions, even as low as 0.5–5 % [9]. However, for weakly interactive species in IDPs it is extremely important to optimize buffer

conditions and the experimental setup. For example many of the weak and transient interactions are stabilized through electrostatic interactions, so optimizing the ionic strength of the experiment will be important. Optimization includes selection of buffer concentration and type, ionic strength, ligand, and temperature [10, 11].

6. *Selection of sample concentration.* Sample concentration is another important variable in interchain PRE experiments. For weakly associating proteins, increasing the concentration of the spin-labeled protein may lead to an increase in nonspecific interactions driven by diffusion. We recommend using a low concentration of spin-labeled sample on the order of 0.5 mM or less to avoid collisional nonspecific interactions [1]. We mix 0.25 mM NMR visible chain with 0.25 mM NMR blind-spin-labeled chain to be able to detect PRE and avoid nonspecific interactions arising from collisional diffusion.
7. *Selection of paramagnetic sample reducing agent.* There are many reducing agents that can be used to reduce the paramagnetic form of the spin label to the diamagnetic form. Options include BME, DTT, and ascorbate ions (either as ascorbic acid or sodium ascorbate). BME and DTT break disulfide bonds and thus they are able to cleave the MTSL spin label attached to the cysteine. Ascorbate reduces the nitroxides to hydroxylamine with no cleavage of the MTSL [12]. Another option is to obtain the diamagnetic analogue of MTSL, MTS ((1-acetyl-2, 2,5,5-tetramethyl-3-pyrroline-3-methyl)-methanethiosulfonate), and attach this compound to the cysteine using the protocol described above. The drawbacks of BME and DTT are that elimination of the spin label results in different para- and diamagnetic samples. Additionally, DTT is pH sensitive and the reaction needs to be performed at pH higher than 7. We recommend using ascorbate ions to obtain the diamagnetic form of the protein as the MTSL spin label will remain and the paramagnetic and diamagnetic samples will thus be more identical. Care needs to be taken as ascorbic acid changes the pH of the sample (pH changes vary depending on the buffer), while sodium ascorbate changes the ionic strength of the sample (by ~5 mM). In order to readjust the pH or ionic strength buffer exchange may be necessary. Buffer exchange may change the sample concentration and para- and diamagnetic sample concentrations may not be identical.
8. *Preparation of the spin-labeled sample—usage of desalting column.* To prepare for the paramagnetic spin labeling reaction there are two important steps. (1) First all disulfide bonds that may have been formed between the cysteine-containing monomers need to be removed. This is achieved by incubating the sample with the reducing agent DTT for a few hours.

(2) Second, after the reduction of disulfide bonds it is critical that all DTT be removed from solution before the spin label reaction is performed. Thus, for fast and complete removal of DTT from the protein solution a desalting column should be used. The spin labeling reaction will not work in the presence of DTT (*see* **Note 7** for preparation of diamagnetic sample), and there is a danger that the reaction will not go to 100 % completion and the reaction product will be diamagnetic, not paramagnetic as desired.

9. *Completion of spin labeling reaction.* To test if the MTSL spin labeling is complete MALDI spectra of the sample can be acquired. Incomplete spin labeling, even at small percentages, will diminish the PRE values [13].
10. *Comparison of $^{1}H–R_2$ values of diamagnetic sample (with reduced spin label) and non-labeled sample.* Comparison of the diamagnetic sample with the wild-type unlabeled $^{14}N$ sample is a further check that the diamagnetic sample has maintained its integrity and that the conformational ensemble of the diamagnetic protein is similar to the unlabeled protein as sampled by $^{1}H–R_2$ values [11]. If there are big differences in the $^{1}H–R_2$ values of the diamagnetic sample and the non-spin-labeled control, it could mean that the sample is degrading or aggregating and should not be used.
11. *Interpretation of the PRE rates—protein and spin label flexibility.* Issues regarding the flexibility of spin labels and the effect on the PRE rates are thoroughly described in a highly recommended review by Iwahara and Clore [6]. If the protein belongs to the class of IDPs, or if the protein exists in more than one form, then the PRE rates are weighted averages over the interactions. PRE distances scale between the unpaired electron and the nucleus as $<r^6>$ and thus the fragments that have shorter distances will dominate the PRE rates.
12. *Spin label sampling for mapping of hetero-interactions.* To obtain a good sampling of protein contacts it is important to have an appropriate number of spin labels across the protein. Papers describing the density of spin labels for intra-chain PRE experiments suggest placing a spin label every 10–30 residues [14–18]. For interchain PRE experiments we suggest at least the same spacing of spin labels per chain.

## References


1. Donaldson LW, Skrynnikov NR, Choy WY, Muhandiram DR, Sarkar B et al (2001) Structural characterization of proteins with an attached ATCUN motif by paramagnetic relaxation enhancement NMR spectroscopy. J Am Chem Soc 123:9843–9847
2. Altenbach C, Marti T, Khorana HG, Hubbell WL (1990) Transmembrane protein-structure - spin labeling of bacteriorhodopsin mutants. Science 248:1088–1092
3. Poluektov OG, Utschig LM, Dalosto S, Thurnauer MC (2003) Probing local dynamics

of the photosynthetic bacterial reaction center with a cysteine specific spin label. J Phys Chem B 107:6239–6244

4. Berliner LJ, Grunwald J, Hankovszky HO, Hideg K (1982) A novel reversible thiol-specific spin label - papain active-site labeling and inhibition. Anal Biochem 119:450–455
5. Battiste JL, Wagner G (2000) Utilization of site-directed spin labeling and high-resolution heteronuclear nuclear magnetic resonance for global fold determination of large proteins with limited nuclear overhauser effect data. Biochemistry 39:5355–5365
6. Clore GM, Iwahara J (2009) Theory, practice, and applications of paramagnetic relaxation enhancement for the characterization of transient low-population states of biological macromolecules and their complexes. Chem Rev 109:4108–4139
7. Eliezer D (2012) Distance information for disordered proteins from NMR and ESR measurements using paramagnetic spin labels. Methods Mol Biol 895:127–138
8. Koehler J, Meiler J (2011) Expanding the utility of NMR restraints with paramagnetic compounds: background and practical aspects. Prog Nucl Magn Reson Spectrosc 59:360–389
9. Tang C, Ghirlando R, Clore GM (2008) Visualization of transient ultra-weak protein self-association in solution using paramagnetic relaxation enhancement. J Am Chem Soc 130:4048–4056
10. Anthis NJ, Clore M (2011) Intrinsic dynamics prime calmodulin for peptide binding: characterizing lowly populated states by paramagnetic relaxation enhancement. Biophys J 100:604
11. Wu KP, Baum J (2010) Detection of Transient Interchain Interactions in the Intrinsically Disordered Protein alpha-Synuclein by NMR Paramagnetic Relaxation Enhancement. J Am Chem Soc 132:5546, PMCID: PMC3064441
12. Bobko AA, Kirilyuk IA, Grigor'ev IA, Zweier JL, Khramtsov VV (2007) Reversible reduction of nitroxides to hydroxylamines: roles for ascorbate and glutathione. Free Radic Biol Med 42:404–412
13. Iwahara J, Tang C, Clore GM (2007) Practical aspects of (1)H transverse paramagnetic relaxation enhancement measurements on macromolecules. J Magn Reson 184:185–195
14. Silvestre-Ryan J, Bertoncini CW, Fenwick RB, Esteban-Martin S, Salvatella X (2013) Average conformations determined from PRE data provide high-resolution maps of transient tertiary interactions in disordered proteins. Biophys J 104:1740–1751
15. Marsh JA, Forman-Kay JD (2012) Ensemble modeling of protein disordered states: experimental restraint contributions and validation. Proteins 80:556–572
16. Ganguly D, Chen JH (2009) Structural interpretation of paramagnetic relaxation enhancement-derived distances for disordered protein states. J Mol Biol 390:467–477
17. Salmon L, Nodet G, Ozenne V, Yin G, Jensen MR et al (2010) NMR characterization of long-range order in intrinsically disordered proteins. J Am Chem Soc 132:8407–8418
18. Gillespie JR, Shortle D (1997) Characterization of long-range structure in the denatured state of staphylococcal nuclease.2. Distance restraints from paramagnetic relaxation and calculation of an ensemble of structures. J Mol Biol 268: 170–184

# Chapter 4

# Detection of Helical Intermediates During Amyloid Formation by Intrinsically Disordered Polypeptides and Proteins

**Andisheh Abedini, Ping Cao, and Daniel P. Raleigh**

## Abstract

Amyloid formation and aberrant protein aggregation are hallmarks of more than 30 different human diseases. The proteins that form amyloid can be divided into two structural classes: those that form compact, well-ordered, globular structures in their unaggregated state and those that are intrinsically disordered in their unaggregated states. The latter include the Aβ peptide of Alzheimer's disease, islet amyloid polypeptide (IAPP, amylin) implicated in type 2 diabetes and α-synuclein, which is linked to Parkinson's disease. Work in the last 10 years has highlighted the potential role of pre-amyloid intermediates in cytotoxicity and has focused attention on their properties. A number of intrinsically disordered proteins appear to form helical intermediates during amyloid formation. We discuss the spectroscopic methods employed to detect and characterize helical intermediates in homogenous solution and in membrane-catalyzed amyloid formation, with the emphasis on the application of circular dichroism (CD). IAPP is used as an example, but the methods are generally applicable.

**Key words:** Amyloid, Aβ, Islet amyloid polypeptide, Amylin, Helical intermediate, Oligomer, CD

## 1 Introduction

Amyloidoses are protein aggregation diseases in which normally soluble, functional polypeptides and proteins self-assemble into partially ordered insoluble amyloid fibrils that deposit in tissues and organs. More than 30 different amyloidogenic polypeptides or proteins are associated with human disorders, including systematic amyloidosis, neurodegenerative disorders such as Alzheimer's disease and Parkinson's disease, and type 2 diabetes (T2D) [1–4]. Amyloid fibrils share common structural features even though there is no sequence homology among the proteins that form amyloid in vivo; they are long, unbranched, and rich in β-sheet structure. The β-strands run perpendicular to the long axis of the fibril with the hydrogen bonds aligned along the fibril axis. This cross-β conformation is common to all amyloids characterized to date.

David Eliezer (ed.), *Protein Amyloid Aggregation: Methods and Protocols*, Methods in Molecular Biology, vol. 1345,
DOI 10.1007/978-1-4939-2978-8_4, © Springer Science+Business Media New York 2016

Amyloidogenic proteins can be divided into two structural classes: those that form compact globular structures in their unaggregated state and those that are intrinsically disordered in their unaggregated states. Examples of intrinsically disordered proteins (IDPs) that form amyloid include islet amyloid polypeptide (IAPP, amylin), responsible for pancreatic islet amyloidosis in T2D; the Aβ peptide of Alzheimer's disease; and α-synuclein, which is involved in Parkinson's disease. There is considerable interest in characterizing pre-amyloid intermediates as they are now considered to be the most toxic species in a variety of amyloid deposition diseases. A number of IDPs that form amyloid appear to do so via the formation of an α-helical intermediate [5].

A range of spectroscopic methods are available for detecting and characterizing helical intermediates. Far-UV circular dichroism (CD) and NMR are the most widely applied to amyloidogenic proteins. In the context of protein studies, far-UV CD is normally used to refer to data recorded from 260 nm to below. Far-UV CD spectra are sensitive to protein secondary structure. The term near-UV CD is typically used to refer to measurements above 260 nm and is sensitive to the environment of Trp and Tyr residues in proteins. Near-UV CD measurements are the relevant ones for detecting helical intermediates. Intrinsic fluorescent probes that rely on the use of fluorescent amino acids together with a residue that quenches fluorescence upon formation of an α-helical conformation have been applied to protein folding studies, and should be applicable to studies of amyloid formation [6]. Infrared spectroscopy is generally less sensitive to α-helical structure and often requires concentrated samples, although two-dimensional IR (2DIR) methods are being developed that extend the range of traditional IR studies and allow site-specific detection of secondary structure during amyloid formation [7]. These advanced techniques, which have largely been used to study the formation of β-sheet structure, now offer the prospect of defining the location of helices in proteins when used in conjunction with site-specific isotopically labeled samples. This is achieved by exploiting characteristic patterns of couplings between vibrational modes that arise because of α-helical structure. NMR is the method of choice for defining the secondary structure of soluble proteins in solution. The advantage of NMR is that it provides site-specific information through analysis of secondary chemical shifts, J-couplings, nuclear Overhauser effects, and residual dipolar couplings. The use of NMR to characterize structured and partially structured states of proteins is extensively discussed in the literature. The disadvantage of NMR is that detection requires that the species be trapped for a relatively long time at a suitable concentration in order to facilitate the collection of multidimensional NMR spectra. Modern NMR relies on isotopic labeling and this can be expensive for some amyloidogenic polypeptides that contain certain posttranslational modifications and/or are difficult to recombinantly express. For example, IAPP contains an amidated C-terminus that is required for full biological activity [4]. Both NMR and IR require more effort and higher protein concentrations than CD, and the instrumentation needed for real-time 2DIR measurements is currently expensive and limited to a small number of laboratories. Here we focus on CD methodology as it is probably the most generally accessible technique and does not require isotopic labeling and because the instrumentation is widely available.

Human IAPP is used as a model system for example, but the approaches are general and can be applied to the study of other proteins. The choice of buffer, pH, added salts, or other additives depends on the stability and solubility of the particular protein of interest.

In its physiologically functional native state, IAPP, a 37-residue posttranslationally modified polypeptide hormone, is secreted from pancreatic β-cells in concert with insulin, and plays a role in regulating metabolism, including adiposity and glucose homeostasis [8–10]. In T2D, the polypeptide converts from its functional form to amyloid by a process that is toxic to pancreatic islet β-cells, depositing as plaques in the islets of Langerhans [4, 11–14]. IAPP has been proposed to form amyloid via an α-helical intermediate in solution [5].

IAPP is a hydrophobic polypeptide and is cationic at physiological pH; as expected, it interacts with anionic membranes and other negatively charged surfaces. Interactions of IAPP with model membranes containing a significant portion of anionic lipids have been widely studied. Anionic lipid vesicles, monolayers, and supported bilayers all accelerate amyloid formation by IAPP and there is good evidence that IAPP forms α-helical intermediates when it interacts with model membranes [15]. The relationship between reductionist in vitro studies with model membranes and the situation in vivo is ambiguous; nonetheless, considerable attention is being paid to membrane-catalyzed amyloid formation [16]. We also describe CD studies of IAPP/membrane interactions in this chapter. The anionic lipid composition in the most common model membrane systems typically used ranges from 20 to 50 mol%. In the protocol described here, we use a model membrane with 25 mol% anionic lipids. The methods are not limited to a specific lipid composition and can be applied to other symmetric vesicles.

## 2 Materials

General: Deionized water and the highest grade reagents should be used. All MSDS data sheets should be carefully studied before using any reagents or solvents and appropriate waste disposal regulations should be followed. Appropriate personal protective equipment (including goggles) should be worn.

1. IAPP is usually prepared by solid-phase peptide synthesis (SPPS) since the peptide is toxic to many cell lines and contains posttranslational modification (C-terminally amidation) that are not generated by most expression systems. IAPP can be synthesized via 9-fluorenylmethoxycarbonyl (Fmoc) chemistry or *tert*-Butyl carbamate (t-Boc) methods (*see* **Note 1**). The Alzheimer's Aβ peptide is also typically prepared by SPPS, usually by t-Boc approaches. IAPP is purified by reverse-phase high-pressure liquid chromatography (HPLC) using a C18 preparative column (*see* **Note 2**). The identity of the purified peptide should be confirmed by mass spectrometry, even for commercially purchased samples. Chemically synthesized polypeptides can contain residual "scavengers"; these are small molecules added during the cleavage of the peptide from the polymeric resin during the synthesis, which can influence the properties of polypeptides. Thus, it is important to use samples of the highest possible purity. IAPP contains multiple Asn residues and can undergo spontaneous deamidation in which Asn residues are converted into mixtures of L-Asp, D-Asp, L-iso-Asp, and D-iso-Asp [17]. It is important to check the integrity of the polypeptide before commencing experiments. The peptide should be stored as a dry powder at −20 °C or −80 °C in a vacuum-sealed container with desiccant. Care should be used when removing samples from the freezer. The peptide, still in its container, should be placed in a desiccator and allowed to warm to room temperature before the container is opened. This will help to minimize undesired absorption of water upon thawing.
2. Lipid stocks: 1,2-Dioleoyl-sn-glycero-3-phosphocholine (DOPC) and 1,2-dioleoyl-sn-glycero-3-phospho-(1′-rac-glycerol) (DOPG) are used in the example described here and were obtained from Avanti Polar Lipids. Lipid stock solutions were prepared in chloroform and stored at −80 °C (*see* **Notes 3** and **4**).
3. General considerations for the choice of buffers for CD studies: The highest grade analytical reagents should be used. A key concern with CD studies is the necessity to avoid significant background absorbance. CD is a difference technique with the signal representing the difference in absorbance of right versus left circular polarized light. The differential absorbance is small, typically $10^{-4}$ to $10^{-6}$ absorbance units (A.U.) for a biomolecule that contributes an optical density of 1.0. This means that less than 0.1 % of the total absorbance signal must be measured accurately and precisely. The practical consequence is that a small difference needs to be measured precisely and a significant background absorbance from buffer or salts can be a complication. For example, a 10 mm solution of NaCl will contribute to an absorbance of greater than 0.5

A.U. at 190 nm in a 0.1 cm cell. Protein CD spectra are often recorded using NaF instead of NaCl since the fluoride salt has a smaller absorbance. However, the rate of IAPP amyloid formation is sensitive to the choice of the anion in solution, even at moderate salt concentrations, likely due to ion binding [18]. Control experiments should be conducted if one changes buffers or salts to ensure that the rate of amyloid formation is not significantly impacted. The same considerations hold for other amyloidogenic polypeptides and proteins. Some common buffers also have significant background absorbance in the far UV and are best avoided, if possible. Tricine, Hepes, and to a lesser extent, Pipes, all absorb in the far UV. Phosphate, borate, Tris, Mes, and cacodylate absorb less. It is important to use a buffer system whose pH is insensitive to temperature if temperature-dependent studies are to be conducted.

## 3 Methods

All sample preparation procedures can be carried out at room temperature unless otherwise specified.

### *3.1 Preparation of Proteins and Polypeptides*

Working with amyloidogenic proteins is challenging and there are contradictory reports in the literature on the cytotoxic and conformational properties of these polypeptides. The confusion is likely due, in large part, to differences in the methods used to solubilize and study these polypeptides. A range of methods have been developed to prepare amyloidogenic polypeptides and proteins in (hopefully) initially monomeric states. However low-order oligomers are often detected as soon as the polypeptide is dissolved, meaning that it is very hard to be certain that one is starting a measurement from a monomeric state. The specific protocol used to prepare samples depends on the protein being investigated. The example described here is applicable to IAPP. The reader is referred to the literature for examples of protocols used with other proteins.

1. Dry IAPP powder is dissolved in 100 % hexafluoroisopropanol (HFIP). The peptide stock solution is incubated for several hours until the solution is clear and then filtered through a 0.22 μm filter (*see* **Note 5**).
2. Trace amounts of cosolvents can influence the properties of IAPP and other polypeptides and affect their rate of amyloid formation. Thus aliquots of the IAPP/HFIP stock solution are freeze-dried to remove organic solvents before the samples are used in experiments. The freeze-drying step should be carried out at the coldest temperature under the strongest possible vacuum for a duration of at least 12–24 h to insure the removal of residual organic solvents.

3. To begin the experiment, amyloid formation is initiated by redissolving the dry peptide in a CD-compatible buffer (*see* Subheading 2, **item 3**) containing appropriate choice of salts, at the desired pH (*see* **Note 6**). The rate of amyloid formation by human IAPP (h-IAPP) is strongly pH dependent and is faster when the side chain of the single His residue and N-terminus are neutral. The rate of h-IAPP amyloid formation also depends upon ionic strength and, as noted above, on the choice of the anion. CD spectra can be recorded as a function of time over the course of amyloid formation to monitor the development of α-helical structure preceding formation of β-sheet structure.

### 3.2 Preparation of LUVs for Peptide/Membrane Studies

The membrane used for this example contains 25 % anionic lipids by mole percent and is comprised of a mixture of DOPC and DOPG (*see* **Note 7**).

1. Chloroform stock solutions of DOPC and DOPG are added to a round-bottom glass flask at a 3:1 molar ratio (*see* **Note 8**). Organic solvent is removed by evaporation using a stream of nitrogen gas to generate a film. The lipid film is further dried overnight under a vacuum in order to remove any residual organic solvent.
2. The lipid film is redissolved in the desired buffer and agitated for 1 h using stirring or mild shaking (*see* **Note 9**).
3. After rehydration, the lipid suspension is subjected to ten freeze-thaw cycles and then extruded through filters of the desired pore size at least 15 times (Whatman, GE) (*see* **Notes 10** and **11**).
4. The phospholipid concentration of the resulting LUVs can be determined using the method of Stewart [19].

### 3.3 CD Measurements

The sensitivity of a CD measurement depends on the absorbance of the CD-active molecule of interest (and thus its concentration), the spectral bandwidth used, the time constant, the step size used to scan the spectrum, and the path length. The range of absorbance (from the compound of interest) for CD is typically on the order of 0.5–2.0 A.U. with an OD near 0.87 being optimal (information provided by the AVIV Instruments CD Manual). Of course, the concentration of the sample is also dictated by the biophysics of the system. Too high a concentration may lead to rapid aggregation and hinder the detection of intermediates or alter the mechanism of assembly. Far-UV CD spectra of h-IAPP are typically recorded with a protein concentration of 5–40 μM. The relevant concentration for far-UV CD spectra of proteins is the concentration of peptide bonds; thus the optimal protein concentration for far-UV CD will be different for different proteins. Measurement of far-UV CD spectra of proteins and polypeptides is typically recorded in a 0.1 cm cuvette. A longer path length cell is usually used for near-UV CD, but can result in too strong a background absorbance when used for far-UV CD. One should always check for sample absorption to the walls of the cuvette when studying a new protein or when studying a protein under new conditions. The reader is referenced to the manual for his/her specific type of CD instrument for suggestions on the optimal instrumental parameters to use.

1. Our laboratories typically record CD spectra of 5–40 μM samples of IAPP peptides in low salt using Tris–HCl buffer (pH 7.4) with a 1-s time constant and a 1 nm bandwidth.
2. Aliquots (300 μL) are removed from peptide solutions at different time points over the course of amyloid formation, and transferred to a 0.1 cm quartz cuvette a few minutes prior to data collection.
3. Spectra are typically recorded over a range of 190–260 nm, at 1 nm intervals with an averaging time of 3 s.
4. CD spectra should represent the average of a minimum of three to five repeats. Background spectra should be subtracted from collected data.

### 3.4 Analysis of CD Spectra and Detection of α-Helical Intermediates

CD spectra are acquired as a function of time and the time dependence of the spectrum or of the intensity at a key wavelength is analyzed to detect α-helical intermediates. IAPP, like many amyloidogenic proteins, is classified as "intrinsically disordered" in its monomeric state [14]. The polypeptide is not completely unstructured, but is believed to transiently sample α-helical $\phi-\psi$ angles, as deduced by NMR chemical shift measurements of the non-amyloidogenic rat IAPP sequence; however this dynamic ensemble does not correspond to significant helical structure and does not give rise to an α-helical CD spectrum [20]. The formation of an α-helical intermediate is traditionally detected by changes in the shape of the CD spectrum or by changes in the signal intensity at 222 nm; the classic α-helical CD spectrum has a strong double-signal minima at 208 and 222 nm, while a "random coil" spectrum has no intensity at 222 nm. Hence, the development of α-helical structure can be monitored by following the intensity at 222 nm. β-sheets can however also contribute to some signal intensity in this region of the far UV. It is therefore important to record the full spectrum of CD ranging from 260 to 195 nm or lower, if possible. The presence of partial helical structure can be deduced using spectral deconvolution methods as described below or by singular value decomposition of the entire collection of time-dependent spectra. The amyloid fibril state is rich in β-sheet structure and lacks α-helical structure, giving rise to a CD spectrum that is very different from that of a partially helical ensemble. The physical differences in the CD spectrum produced by random coil conformations and α-helical and β-sheet structures distinguish them from each other and facilitate detection of helical intermediates.

A critical issue is the time resolution of the experiment. This is dictated by the time scale of amyloid formation and the time required to obtain a high S/N CD spectrum. For IAPP, under the conditions used on our laboratories, amyloid formation takes on the order of 15–40 h. Recording a high-quality CD spectrum of IAPP requires less than 15–20 min. Thus, the time resolution is more than adequate to detect a transient α-helical intermediate.

α-Helices are relatively regular elements of secondary structure and have fewer variations in their basic geometry than do β-sheets. Thus, it is reasonable to speak of the CD spectrum of a typical α-helix. In addition, the structural features that can lead to variations in α-helical CD spectra have been well studied [21]. There are two common methods for estimating the helical content of a protein from the CD spectrum. One can deconvolve (fit) the spectrum using a number of freely available programs. Most are accessible on the Web [22]. There are three important considerations when doing so. First, one should strive to record the far-UV CD spectrum to as low a wavelength as possible. Truncation of the spectra at too high a wavelength can lead to ambiguity in deconvolutions. Second, an accurate measurement of the polypeptide concentration is helpful. Third, it is important to avoid the temptation to smooth the spectra. Smoothing does not improve the information content and smoothing noisy spectra can introduce artifacts into the spectrum. The second method for deducing the amount of helical structure relies on the signal intensity, the mean residue ellipticity, at 222 nm, $[\theta]_{222}$. Of course, a reliable estimate requires, as noted above, that other features do not contribute significantly to the CD signal at this wavelength. This is the major disadvantage of the single wavelength approach. The advantage is that because only a single wavelength is being monitored, much better time resolution can be obtained. Signal averaging on the order of 10–30 s ensures excellent signal to noise in our studies of IAPP. The signals for a fully helical peptide, $[\theta]_{\mathrm{H}}$, and for a fully unfolded peptide, $[\theta]_{\mathrm{C}}$, are needed. Both the fully helical intensity and the coil intensity can be temperature dependent and this can be taken into account using empirical relationships [23]. One popular set is

$$[\theta]_{\mathrm{H}} = -40{,}000 \times (1 - 2.5/n) + 100 \times T \tag{1}$$

$$[\theta]_{\mathrm{C}} = 640 - 45 \times T \tag{2}$$

where $n$ is the number of the residues in the peptide, and $T$ is the temperature (°C). A key aspect of the analysis is that one must know the peptide concentration as accurately as possible. Errors in concentration determination translate directly into errors in mean residue ellipticity and directly to errors in the estimated α-helical content. It is important to stress that determining concentration by weight is very imprecise and prone to error, and is not adequate for a quantitative analysis of spectral intensities.

It is important to realize that the estimated amount of helical structure present in a partially structured ensemble deduced by CD can differ from that deduced using NMR secondary chemical shifts. There is no inherent contradiction with the two methods; rather they offer complementary information. NMR secondary shifts are local in origin and provide information on the fraction of the time a particular residue populates the helical region of the Ramachandran plot. The CD spectra of helical peptides are length dependent and a very short helix, or an isolated residue populating $\phi,\psi$ helical angles, will not give rise to significant intensity at 222 nm.

Intermediates can be either on-pathway or off-pathway. An on-pathway intermediate is one that leads productively to amyloid formation, whereas an off-pathway intermediate is one that needs to unfold prior to the system progressing onto amyloid. The simplest example of the two cases is illustrated below. Determining if an intermediate is on- or off-pathway is extraordinary difficult because of the symmetry of the kinetic rate equations; the methods outlined here are not capable of doing so [24]:

$$\text{On}^{-}\text{pathway : Monomer} \leftrightarrow \text{Intermediate} \leftrightarrow \text{Amyloid}$$

$$\text{Off}^{-}\text{pathway : Intermediate} \leftrightarrow \text{Monomer} \leftrightarrow \text{Amyloid}$$

As a final note we stress that the observation of an isodichroic point (isosbestic point) in a set of CD spectra does *not* prove that the transition being monitored is two-state. An isosbestic point is a necessary but not a sufficient condition for a two-state transition. The misinterpretation of isosbestic points in CD spectra is probably one of the most common errors in the literature.

## 4 Notes

1. The IAPP samples used in this example were prepared using Fmoc chemistry. Fmoc-protected pseudoproline dipeptide derivatives were incorporated to facilitate the synthesis. The Cys-2 and Cys-7 disulfide bond in IAPP was generated via oxidation by DMSO in the present example [25].
2. HCl is the preferred ion-pairing agent instead of trifluoroacetic acid (TFA) for HPLC purification of IAPP, even though TFA usually yields better resolution and peak separation. The reason is that excess TFA can cause problems with toxicity in cellular assays and TFA influences the aggregation kinetics of some peptides derived from IAPP [26]. Residual TFA also interferes with IR studies because its IR absorbance spectrum overlaps with the peaks of interest in protein IR spectra. Residual scavengers can be retained during normal HPLC purification, but their levels can be reduced by extraction with HFIP. To do this, purified dry IAPP is dissolved in a minimal

amount of HFIP and incubated at room temperature for several hours. The sample is then directly injected into a C18 column for re-purification. The purified peptide should elude from the HPLC column at the same retention time.

3. If long-term storage is required, lipids should be stored dry at −80 °C to prevent decomposition from exposure to light and air.
4. Glassware rather than plasticware should be used when handling organic solvents such as chloroform.
5. Filtration is recommended to remove large aggregates.
6. The concentration of the peptide solution should be measured to check for any loss during filtration. The concentration can be estimated using the absorbance of the peptide at 280 nm. h-IAPP lacks Trp, but contains two Phe and one Tyr residues. Absorbance measurements can be comprised by modest levels of strong absorbing impurities. Peptide concentrations can also be determined by quantitative amino acid analysis or by using the BCA or Bradford assays [27, 28]. It is important to stress that concentration should not be determined by weight. Dried synthetic peptides contain salts and residual water; this can lead to significant errors in concentration measurements by weight, which in turn impacts the interpretation and analysis of CD data.
7. Different lipid compositions can be used, but the basic protocol is applicable to other symmetric lipid vesicles. We use a 25 % anionic model membrane system as an example here, but more complicated lipid mixtures can be used and there is a large body of literature on the preparation of different types of vesicles.
8. The lipids must be mixed thoroughly to obtain a homogeneous solution.
9. During the hydration step, the lipid suspension needs to be maintained at a temperature above the highest gel-liquid crystal transition temperature ($T_c$) of any of the mixed lipids.
10. Extrusion is used to form large unilamellar vesicles and should be performed at a temperature above the $T_c$ of the mixed lipids. The pore size of the filter used depends on the required size of the lipid vesicles (usually LUVs are in the range of 200–1000 nm). Small unilamellar vesicles (diameter 15–50 nm) can be prepared by sonication.
11. After preparation, LUVs can be stable for up to several days. However, it is recommended that vesicles be prepared fresh on the day of experiments. The uniformity of the lipid vesicles can be checked by cryo-electron microscopy and by turbidity using light scattering.

## Acknowledgements

This work was supported by NIH grants 1F32DK089734-02 (A.A.) and GM078114 (D.P.R.).

### References


1. Selkoe DJ (2004) Cell biology of protein misfolding: the examples of Alzheimer's and Parkinson's diseases. Nat Cell Biol 6: 1054–1061
2. Chiti F, Dobson CM (2006) Protein misfolding, functional amyloid, and human disease. Annu Rev Biochem 75:333–366
3. Sipe JD (1994) Amyloidosis. Crit Rev Clin Lab Sci 31:325–354
4. Westermark P, Andersson A, Westermark GT (2011) Islet amyloid polypeptide, islet amyloid, and diabetes mellitus. Physiol Rev 91:795–826
5. Abedini A, Raleigh DP (2009) A critical assessment of the role of helical intermediates in amyloid formation by natively unfolded proteins and polypeptides. Protein Eng Design Select 22:453–459
6. Taskent-Sezgin H, Marek P, Thomas R, Goldberg D, Chung J, Carrico I, Raleigh DP (2010) Modulation of p-cyanophenylalanine fluorescence by amino acid sidechains and rational design of fluorescence probes of a-helix formation. Biochemistry 49:6290–6295
7. Shim S-H, Gupta R, Ling YL, Strasfeld DB, Raleigh DP, Zanni MT (2009) 2DIR spectroscopy defines the pathway of amyloid formation with residue specific resolution. Proc Natl Acad Sci U S A 106:6614–6619
8. Lukinius A, Wilander E, Westermark GT, Engstrom U, Westermark P (1989) Co-localization of islet amyloid polypeptide and insulin in the beta-cell secretory granules of the human pancreatic-islets. Diabetologia 32:240–244
9. Kahn SE, Dalessio DA, Schwartz MW, Fujimoto WY, Ensinck JW, Taborsky GJ, Porte D (1990) Evidence of cosecretion of islet amyloid polypeptide and insulin by beta-cells. Diabetes 39:634–638
10. Stridsberg M, Sandler S, Wilander E (1993) Cosecretion of islet amyloid polypeptide (IAPP) and insulin from isolated rat pancreatic-islets following stimulation or inhibition of beta-cell Function. Regul Pept 45:363–370
11. Clark A, Wells CA, Buley ID, Cruickshank JK, Vanhegan RI, Matthews DR, Cooper GJS, Holman RR, Turner RC (1988) Islet amyloid, increased alpha-cells, reduced beta-cells and exocrine fibrosis – quantitative changes in the pancreas in type-2 diabetes. Diabetes Res 9:151–159
12. Lorenzo A, Razzaboni B, Weir GC, Yankner BA (1994) Pancreatic-islet cell toxicity of amylin associated with type-2 diabetes-mellitus. Nature 368:756–760
13. Konarkowska B, Aitken JF, Kistler J, Zhang S, Cooper GJ (2006) The aggregation potential of human amylin determines its cytotoxicity towards islet beta-cells. FEBS J 273: 3614–3624
14. Cao P, Marek P, Noor H, Patsalo V, Tu LH, Wang H, Abedini A, Raleigh DP (2013) Islet amyloid: from fundamental biophysics to mechanisms of cytotoxicity. FEBS Lett 587:1106–1118
15. Hebda JA, Miranker AD (2009) The interplay of catalysis and toxicity by amyloid intermediates on lipid bilayers: insights from type II diabetes. Annu Rev Biophys 38:125–152
16. Cao P, Abedini A, Wang H, Tu LH, Zhang XX, Schmidt AM, Raleigh DP (2013) Islet amyloid polypeptide toxicity and membrane interactions. Proc Natl Acad Sci U S A 110:19279–19284
17. Dunkelberger EB, Buchanan LE, Marek P, Cao P, Raleigh DP, Zanni MT (2012) Deamidation accelerates amyloid formation and alters amylin fiber structure. J Am Chem Soc 134:12658–12667
18. Marek P, Pastalo V, Green D, Raleigh DP (2012) Ionic strength effects on amyloid formation by amylin are a complicated interplay among Debye screening, ion selectivity, and Hofmeister effects. Biochemistry 51:8478–8490
19. Stewart JCM (1980) Colorimetric determination of phospholipids with ammonium ferrothiocyanate. Anal Biochem 104:10–14
20. Williamson JA, Miranker AD (2007) Direct detection of transient alpha-helical states in islet amyloid polypeptide. Protein Sci 16:110–117
21. Manning MC, Woody RW (1991) Theoretical CD studies of polypeptide helices-examination of important electronic and geometric factors. Biopolymers 31:569–586

22. Whitmore L, Wallace BA (2008) Protein secondary structure analyses from circular dichroism spectroscopy: methods and reference databases. Biopolymers 89:392–400
23. Scholtz JM, Baldwin RL (1992) Ann Rev Biophys Biomol Struct 21:95–118
24. Ling YL, Strasfeld DB, Shim SH, Raleigh DP, Zanni MT (2009) Two-dimensional infrared spectroscopy provides evidence of an intermediate in the membrane-catalyzed assembly of diabetic amyloid. J Phys Chem B 113: 2498–2505
25. Abedini A, Singh G, Raleigh DP (2006) Recovery and purification of highly aggregation-prone disulfide-containing peptides: application to islet amyloid polypeptide. Anal Biochem 351:181–186
26. Nilsson MR, Raleigh DP (1999) Analysis of amylin cleavage products provides new insights into the amyloidogenic region of human amylin. J Mol Biol 294:1375–1385
27. Bradford MM (1976) A rapid and sensitive method for the quantitation of microgram quantities of protein utilizing the principle of protein-dye binding. Anal Biochem 72: 248–254
28. Zor T, Selinger Z (1996) Linearization of the Bradford protein assay increases its sensitivity: theoretical and experimental studies. Anal Biochem 236:302–308

# Chapter 5

# Fluorescence Correlation Spectroscopy: A Tool to Study Protein Oligomerization and Aggregation In Vitro and In Vivo

**Bankanidhi Sahoo, Kenneth W. Drombosky, and Ronald Wetzel**

## Abstract

Fluorescence correlation spectroscopy (FCS) is a highly sensitive analytical technique used to measure dynamic molecular parameters, such as diffusion time (from which particle size can be calculated), conformation, and concentration of fluorescent molecules. It has been particularly powerful in characterizing size distributions in molecular associations (e.g., dimer/multimer formation) both in well-behaved thermodynamically equilibrated systems in vitro as well as in more complex environments in vivo. Protein aggregation reactions like amyloid formation, in contrast, are complex, often involving a series of uniquely structured aggregation intermediates appearing at different time scales. Nonetheless, FCS can be used in appropriate cases to characterize the early stages of some aggregation reactions. Here are described step-by-step protocols and experimental procedures for the study of molecular complex formation in aggregation systems as observed in simple buffer systems, cell extracts, and living cells. The methods described are illustrated with examples from studies of the self-assembly of huntingtin fragments, but in principle can be adapted for any aggregating system.

**Key words:** Fluorescence correlation spectroscopy, Diffusion, Brightness, Particle size, Protein aggregation, Amyloid, Huntington's disease, Oligomer

## 1 Introduction

Fluorescence Correlation Spectroscopy (FCS), introduced four decades ago [1, 2], has become more sensitive and more accessible with the advancement of optical microscopy, highly sensitive detectors, data analysis algorithms, as well as many innovative protocols [3–5]. FCS uses signal fluctuations from a laser-confined fluorescent sample to extract information about underlying properties and processes, including translational diffusion, concentration, rotational diffusion, and brightness. Self-association can be monitored very accurately and at concentrations as low as sub-nanomolar. Although it is possible to characterize multimer size by other techniques, such as size exclusion chromatography (SEC) [6], analytical

David Eliezer (ed.), *Protein Amyloid Aggregation: Methods and Protocols*, Methods in Molecular Biology, vol. 1345,
DOI 10.1007/978-1-4939-2978-8_5, © Springer Science+Business Media New York 2016

ultracentrifugation (AUC) [7], and dynamic light scattering [8], these techniques have their own limitations. One advantage of FCS is the ability to measure samples in native conditions without exposing the sample to physical perturbations such as the potential for column effects in SEC. Molecules must be made detectible by fluorescence, which often requires chemical modification of the molecule of interest and can potentially lead to artifacts if the modification alters molecular properties in unanticipated ways. However, fluorescence also can bestow important additional advantages, such as sensitivity and the ability to study the behavior of molecules in complex mixtures. For example, this allows FCS, with proper modification, to be used inside living cells [9, 10] when combined with the increasingly popular use of fluorescent fusion proteins.

In spite of its potential for studying protein aggregation, to date there are only limited examples in the literature of this application of FCS. There are a number of possible reasons for this. Aggregation reactions that quickly proceed to highly complex mixtures of aggregates may not be amenable to FCS, unless conditions can be found to slow the process. This is because effective FCS analysis requires that the analyte be in equilibrium or "quasi-equilibrium," since otherwise fluorescence fluctuations may in part reflect on-going chemical or physical changes in the sample being measured. In addition, FCS instruments historically were homemade and were considerably expensive to build, and the technique was not user friendly, requiring special training in complex tasks of data acquisition and analysis. However, in the last two decades, FCS instruments have become commercially available, are more easily constructed, and the costs of specialized detectors and optics have decreased, leading to FCS becoming more popular. The above caveats notwithstanding, FCS can be particularly sensitive for studying solutions of heterogeneous mixtures of particles formed in complex pathways, such as amyloid formation. FCS has been utilized to study the formation and stability of amyloid beta oligomers in vitro [11–13] and in cells [14], and polyglutamine oligomers in cells [15]. Given the importance of protein misfolding and aggregation in human disease, there is great potential for using FCS to identify aggregates and to characterize underlying mechanisms. Here, we detail example protocols for studying intermediates (monomeric, oligomeric, aggregated, and heterogeneous) in the aggregation of biologically important huntingtin fragments both in vitro and in living cells.

### 1.1 Theory of FCS

The fluorescence signal, emitted from a small, well-defined optical volume of a solution in equilibrium with its surroundings, is monitored as a function of time with high sensitivity and at high temporal resolution. A general schematic of the FCS instrument components is shown in Fig. 1a. The mean recorded fluorescence emission signal is directly proportional to the number of fluorescent

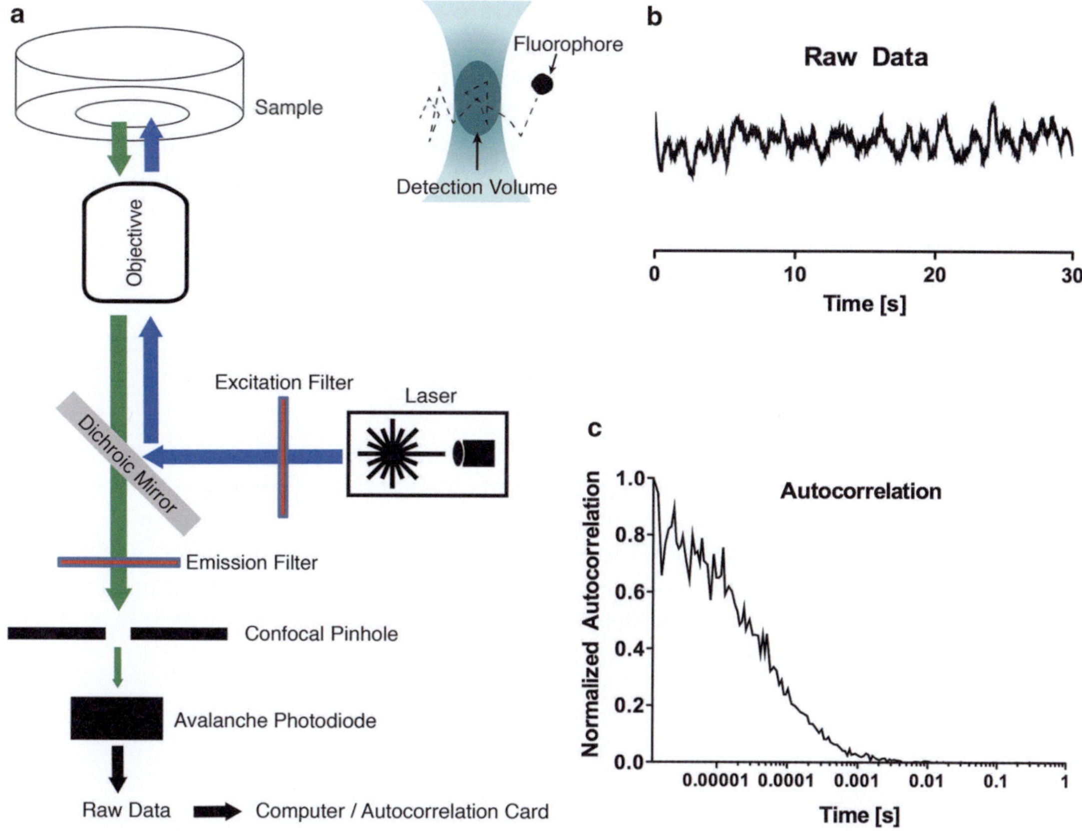


**Fig. 1** FCS setup. (**a**) Typical fluorescence microscope setup with major change of the avalance photodiode detector (APD) and a correlation card to process correlation. (**b**) Typical raw data (*black*), which fluctuates around the mean over time as molecules traverse into and out of the focal volume. (**c**) A typical live (real time) autocorrelation

molecules in the probe volume. The fluorescence as a function of time (Fig. 1b), however, fluctuates about the mean value as molecules diffuse into and out of the optical volume. Although this fluctuation appears random, it contains valuable information about the fluorescent particles [2], such as particle size, molecular shape, and concentration. The fluctuations are quantified by the autocorrelation function Eq. 1 which can be calculated from the raw data (Fig. 1c). The autocorrelation function contains information about different parameters that contribute to fluorescence fluctuations associated with chemical reaction kinetics, coefficients of diffusion, and other phenomena. For a homogeneous solution in a pure diffusional environment, Eq. 1 can be solved to generate Eq. 2, which describes a homogeneous solution containing one kind of particle (i.e., a homogeneously dispersed monomer). Equation 2 is often used to fit the autocorrelation to derive key parameters like the average number of fluorescent particles in the focal volume and their diffusion constant. Equation 3 can be used to translate the measured diffusion time into molecular size information, based on the parallel measurement of a standard and the assumption that all particles are spherical. Equation 1 can also be solved to generate Eq. 4 applicable to systems featuring two components of different size. Equation 5 is used to calculate the size of the observation volume using a fluorescent dye of known diffusion constant.

### 1.2 Equations

Equation 1. The Autocorrelation Function, in which $G(\tau)$ = correlation function, $\delta F(t)$ = fluctuation of signal from the mean, $t$ = time, $<F(t)>$ = average signal, and $\tau$ = delay time.

$$G(\tau) = \frac{<\delta F(t) \times \delta F(t+\tau)>}{<F(t)>^2} \tag{1}$$

Equation 2. Normal diffusion and one-component fitting, in which $G(\tau)$ = correlation function, $N$ = number of particles detected in the focal point, $\tau$ = delay time (the $x$-axis of the autocorrelation graph), $\tau_D$ = diffusion time, $\omega = l/r$, the structure parameter (which is the ratio of long axis ($l$) to short axis ($r$) of the Gaussian focal volume and must be determined using a standard dye for each instrument use (*see* **Note 1**)), $c$ = constant.

$$G(\tau) = \frac{1}{N}\left(\frac{1}{1+\frac{\tau}{\tau_D}}\right)\left(\frac{1}{1+\frac{\tau}{\omega^2 \tau_D}}\right)^{1/2} + c \tag{2}$$

Equation 3. Correlation between molecular weight and diffusion time of two samples assuming a spherical model, in which $\mathrm{MW_A}$ = molecular weight of molecule A (typically a standard free dye (GFP, alexa, etc.)), $\mathrm{MW_B}$ = molecular weight of fluorescent molecule of interest, $\tau_{Da}$ = diffusion time of the standard free dye, $\tau_{Db}$ = diffusion time of fluorescent molecule of interest.

$$\frac{\mathrm{MW_A}}{\mathrm{MW_B}} = \left(\frac{\tau_{Da}}{\tau_{Db}}\right)^3 \tag{3}$$

Equation 4. Normal diffusion and two-component fitting, in which symbols are as in Eq. 2, but with additional terms $\tau_{D1}$ and $\tau_{D2}$ to account for two fluorescent populations each with distinct diffusion times, and $g1$ and $g2$ which are proportional to the concentrations and the square of the brightness of each population.

$$G(\tau) = g1\left(\frac{1}{1+\frac{\tau}{\tau_{D1}}}\right)\left(\frac{1}{1+\frac{\tau}{\omega^2 \tau_{D1}}}\right)^{1/2} + g2\left(\frac{1}{1+\frac{\tau}{\tau_{D2}}}\right)\left(\frac{1}{1+\frac{\tau}{\omega^2 \tau_{D2}}}\right)^{1/2} + c \tag{4}$$

Equation 5. Short axis of the Gaussian volume ($r$) is related to the diffusion time $\tau_D$, where $D$ is diffusion constant of the fluorophore.

$$r^2 = 4D\tau_D \tag{5}$$

## 2 Materials

All the reagents should be of high quality to reduce fluorescent impurities.

### 2.1 Measurements in Simple, Well-Defined Solutions

1. A fluorescent dye with high quantum yield and low sensitivity to photobleaching. Rhodamine derivatives, Alexa Fluors family, and cyanine family (cy3 and cy5) are excellent dyes for FCS. An extensive list of fluorophores with their fluorescent properties can be found in reference [16].
2. A protein of interest labeled with a suitable fluorophore. In this example, a huntingtin analog peptide MATLEKLMKAFE SLKSF$Q_{23}P_{10}C^*K_2$ ($HTT^{NT}Q_{23}P_{10}C^*K_2$) is used, where C* refers to a cysteine chemically modified to incorporate a fluorophore (*see* **Note 2**).
3. 1:1 1,1,1-Trifluoroacetic acid (TFA):1,1,1,3,3,3-hexafluoroisopropanol (HFIP) solution.
4. Water–TFA mixture at pH 3 (*see* **Note 3**).
5. Dilution buffer: A nondenaturing buffer should be used, such as 1× PBS (1.05 mM potassium phosphate monobasic, 155 mM sodium chloride, 2.96 mM sodium dihydrogen phosphate, pH 7.4).
6. Coverglass slides. 0.13–0.18 mm thickness (typical for objective lens used for FCS).
7. Bovine serum albumin (BSA) solution: 1 mg/ml BSA in PBS for coating coverglass slides.

### 2.2 Cell Culture (Cell Lysates)

1. Enhanced Green Fluorescent Protein (EGFP), as a fusion partner to the protein of interest, is an acceptable fluorescent protein for imaging live cells and cell extracts. There are a number of other such proteins that have similar or better quantum yields and photostabilities.
2. An inducible expression system for the protein of interest fused to a fluorescent protein. Stably transfected cells are preferable, but transiently transfected cells are also suitable.
3. Nondenaturing lysis buffer. Use a nondenaturing lysis buffer with the following composition: 0.05 M Tris–HCl, 0.15 M NaCl, 0.5 % v/v Triton X100 + protease inhibitors, pH 7.4 (*see* **Note 4**).

4. Diluted and clarified cell lysate. Pellet two to ten million cells expressing the fluorescently labeled protein and freeze the pellet until needed. Lyse the cell pellet by adding 100–400 μl nondenaturing lysis buffer plus a protease inhibitor cocktail and vortex for 30 s. Centrifuge the lysed cells at 425 × *g* in a bench-top centrifuge and collect the 2/3 of the top supernatant (cell lysates) as samples for FCS measurements. Keep the cell lysates on ice. They are now ready for immediate use on FCS. Stability of lysates will vary with the system. In our experiments on HTT exon1 we found that cell lysate responses remained unchanged up to 3–4 h of storage on ice, but decayed significantly if analyzed after 24 h.

### 2.3 Cell Culture (Live Cells)

1. Adherent cells expressing fluorescently tagged protein of interest can be grown in media directly onto coverglass slides (Subheading 2.1, **item** 7). After inducing the expression of the fluorescent protein, data can be acquired from live cells directly on the glass slide in media. Make sure to use appropriate media and coverglass coatings (*see* **Note 5**).

### 2.4 Microscope Components

A standard confocal microscope can be modified for FCS use by securing the following components:

1. Select lasers appropriate for fluorophores to be used.
2. A highly sensitive and fast detector capable of detecting single photons and producing digital signals, such as an avalanche photodiode (APD), is required. The photomultiplier tubes (PMTs) more typically found on confocal microscopes are sufficiently sensitive to view the relatively bright objects visualized in cell microscopy, but are not sensitive enough for single photon detection by FCS and are also typically analog devices that cannot count single photons.
3. An adjustable pinhole set to 40–70 μm (~0.7–1 Airy units). Optical fibers of different diameter can also be used.
4. A software program for FCS data acquisition and analysis (Zen software, ALV-MultiCorr, etc.). Generally, autocorrelator cards come supplied with data acquisition software. Software for data analysis may or may not be included.
5. A dedicated autocorrelator card for acquisition and real-time analysis of FCS data. Specialized computer components such as the autocorrelator can be readily added via PCI slots to most desktop computers.
6. A C-Apochromat 40×/1.2 W or higher water-immersion objective. A water-immersion objective is necessary for live cell or cell lysate imaging, which are in aqueous environments.
7. For the methods and results described here, we used an automated system Zeiss LSM-510 ConfoCor3.

## 3 Methods

### *3.1 Data Acquisition: Measurements in Simple, Well-Defined Solutions*

1. Label the peptide at an appropriate position following suitable labeling chemistry (*see* **Notes 2** and **6**).
2. Disaggregate (*see* **Note** 7) the fluorescently labeled huntingtin fragment by dissolving 0.2–0.5 mg of labeled peptide in 4 ml of 1:1 TFA:HFIP solution overnight. Dry the sample the following morning under a stream of nitrogen gas and subsequently under vacuum for at least one additional hour. Dissolve the sample (typically a film) in a water–TFA mixture (pH 3) and centrifuge for 2–3 h at 386,000 × *g*. Aliquot the supernatant into 50 μl samples. Snap-freeze each aliquot in liquid nitrogen or dry ice/acetone and store at −80 °C (*see* **Note 8**). The frozen sample can be used, after thawing (*see* **Note 8**), as a soluble monomer stock solution of the labeled peptide. Measure the concentration of the peptide either by fluorescence or by absorbance.
3. Prepare a stock concentration (such as 100 μM) of the standard dye. The concentration can be measured in an absorption spectrophotometer. Dilute the dye to around 50–100 nM or lower for FCS standardization (*see* **Note 9**).
4. Coat the coverglass slides with a 1 mg/ml BSA solution. Pour the BSA solution onto the coverglass and leave for 30 min. Wash the slides gently with deionized water and let them dry at room temperature. The slides can be prepared in advance, but should be used within a couple of days (*see* **Note 5**).
5. Calibrate the instrument using an appropriate dye. Each experiment should be started by calibrating the instrument and comparing parameters from the last experiment (*see* **Note 1**).
6. Focus the instrument on the sample. First, adjust the sample stage X and Y to position the sample above the objective lens. Adjust the height of the objective lens to focus the laser spot well inside the sample volume, i.e., well away from the very top and bottom surfaces of the glass slide which will reflect light and thus give a pseudo-signal. Finding an acceptable interior focus can be done as follows: First, turn on the laser of choice and begin collecting data. Second, explore the *z*-axis and ensure that the major reflections observed are coming from glass boundary reflections by comparing the thickness measured between the two major positions of high signal with the manufacturer's stated thickness between the top and bottom of the slide (typically, 0.12–0.18 mm). Third, return the focus to the site of top reflection. Fourth, from this position move the objective 200 μm towards the slide bottom and into the sample (*see* **Note 10**).

7. Adjust the pinhole of the instrument to 40–70 μm (~0.7–1 Airy units) and maximize fluorescence by adjusting the pinhole position in the X and Y axis. The X and Y position of the pinhole can be changed manually or, in some software, automatically. Scan the sample, changing the X and Y positions, until the count rate (measured as kHz) is maximized.
8. Acquire data. Three 20-s acquisitions is a good starting point (*see* **Note 11**). Most software packages will display a live feed of the count rate (in counts per second, displayed as Hz or kHz) and the per particle brightness (in counts per molecule per second (CPMS)). In the FCS field, the term CPMS is used synonymously with "per particle brightness" and we use CPMS throughout this article to refer to this factor. The CPMS of the standard dye should change very little from day to day, but this value can change over longer period of time between measurements (*see* **Note 12**).
9. Continue to Subheading 3.4 for Data Analysis.

### 3.2 Data Acquisition: Cell Lysates

The steps for cell lysate FCS are nearly identical to that of a simple dye in Subheading 3.1. The exception is to use a starting material of 50–80 μl of cell lysates prepared in Subheading 2.2 instead of a simple dye in buffer.

1. Grow cells and prepare lysates as described in Subheading 2.2, **item 4**. *See* **Note 13** for discussion of the choice of centrifugation settings and lysate stability.
2. Calibrate the instrument using an appropriate dye. Each experiment should be started by calibrating the instrument and comparing parameters from the last experiment (*see* **Note 1**).
3. Acquire data for the standard (e.g., supernatant of lysed cells expressing EGFP alone) and unknown (e.g., experimental cells with an EGFP tagged protein) in the same way as described in Subheading 3.1.

### 3.3 Data Acquisition: Live Cells

1. Control and experimental cells should be grown, and the expression of the fusion of the protein of interest with a fluorescent protein (*see* **Note 6**) induced, directly on glass coverslides coated with collagen, poly D-lysine, or BSA (*see* **Note 5**). Calibrate the instrument using an appropriate dye. Each experiment should be started by calibrating the instrument and comparing parameters from the last experiment (*see* **Note 1**).
2. Place the glass coverslide containing the cells and cell media onto the FCS instrument. When doing experiments on live cells, especially for long periods of time, control the temperature and $CO_2$ levels. Several stage-top incubation systems (e.g. LiveCell™ Pathology Devices) are designed to control temperature and $CO_2$ around the sample while still allowing the FCS instrument to collect data.

3. Find the layer of cells in the *Z* plane by eye or via a live view feed via the software. The live feed image of cells should look comparable to a confocal image. In fact, taking an image of the plane of cells will be beneficial for an experimental record.
4. Most microscope software will allow cropping and zooming to fit the image of a single cell as shown (Fig. 2a). If there is more than one cell morphology present, decide whether to survey all cell types or focus on cells with particular characteristics, then choose representative cell(s) for data acquisition based on the selection strategy.
5. Position the laser for FCS. Use the FCS software to position the laser for data collection. In the ZEN software (Zeiss), this is as easy as placing cross-hairs on a point in the cell (Fig. 2a).
6. Start a live feed of data to examine count rate. A minimally acceptable count rate to distinguish signal from noise is ~10 kHz (assuming a background noise of ~500 Hz or less). For FCS measurements from live cell producing an EGFP-tagged protein, aim for at least 10–20 kHz. Finally, check the correlation and CPMS. A good correlation should be above the baseline as shown in Fig. 2b, which has been normalized. If the correlation is a flat line essentially superimposed on the base line, this means the sample is not decaying properly and is not yielding a useful correlation (*see* **Note 14**).
7. Record data. Three independent 10 s acquisitions are usually adequate. Depending on the tendency of the dye to photobleach (*see* **Note 15**), acquisition time may be increased or decreased. Monitor the count rate (*see* **Note 16**)—if it is

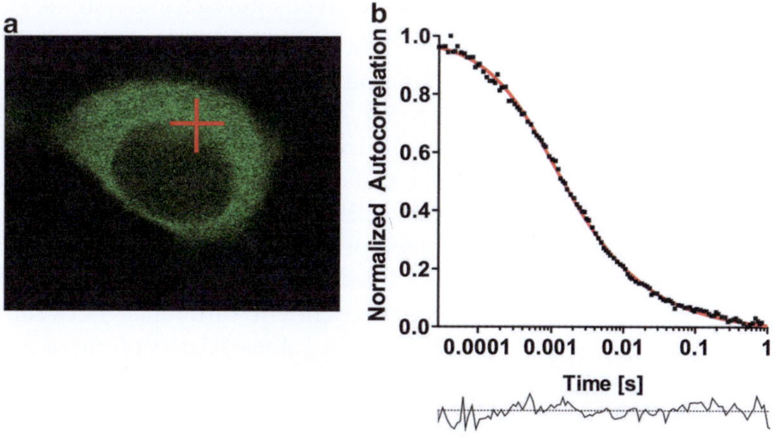


**Fig. 2** FCS analysis of a living cell. (**a**) Confocal image of PC12 cells producing EGFP-tagged full-length HTT exon1-Q25, with a crosshair showing the spot where FCS data were recorded. (**b**) Autocorrelation fit, with residuals, recorded from the spot shown in panel (**a**). The diffusion time, compared with the diffusion time of cellular EGFP alone (not shown), indicates a monomer

noticeably decreasing over time (a decrease of >10 % of the count rate over the entire 30 s of acquisition), photobleaching and/or the movement of the cells (*see* **Note 5**) is likely occurring (*see* **Note 14**).

8. Continue collecting data from different points in the cell and/or from other cells. Some software can be programmed for automatic sequential collection of such data.
9. As a reference standard, make FCS measurements from a cell expressing the isolated fluorescent protein (*see* **Note 17**).

### 3.4 Data Analysis (One- and Two-Component Modeling)

1. Import the autocorrelation data (*x*-variable delay time, in seconds; *y*-variable autocorrelation) into graphing software of choice, such as Origin, SigmaPlot, or Graphpad Prism.
2. Although most graphing software comes with pre-built fitting equations for common mathematical models (e.g., linear regression, quadratic, and power), they are not equipped with equations necessary to fit an autocorrelation function for FCS. Manually input Eq. 2 (Subheading 1.3) into the software for one-component fitting. One-component fitting Eq. 2 is best used to describe a uniform distribution of a fluorescent protein (Fig. 3a, b). A homogeneous starting solution containing only a single form of fluorophore fits a single-component model. Use two-component modeling Eq. 4 to fit heterogeneous solutions (Fig. 3a, c). Two-component modeling will attempt to categorize a heterogeneous fluorescent population into two distinct fluorescent particle sizes (e.g., a solution containing both a low molecular weight monomer and a high molecular weight multimer of the fluorescent particle). Two-component fitting will be appropriate for some well-behaved aggregating systems (Fig. 3c) (*see* **Note 18**).
3. Use a nonlinear least-squares algorithm to fit the one- or two-component model to the autocorrelation data. We employ several hundred iterations of a Levenberg-Marquardt algorithm to minimize $\chi^2$ using Origin Pro 7.5. When complete, the fitted line (Figs. 2b and 4a, red line) should visually overlay well with the experimental data points (Fig. 2b and 4a, black pixels). The goodness of the fit can be judged by the randomness of the distribution of the residuals (a plot of the time-dependent difference between the experimental values and the fit values). An example of a good fit with good residuals is shown in Fig. 4b, where the experimental data points overlap very well with the estimated values from the fit. The reduced $\chi^2$ should be as close as to 1 as possible. A bad fit will yield a nonrandom distribution of residuals, such as shown in Fig. 4c. If the residual plot looks nonrandom, it may indicate poor modeling, or poor quality data (due, for example, to photobleaching, or to an ongoing aggregation reaction, dur-

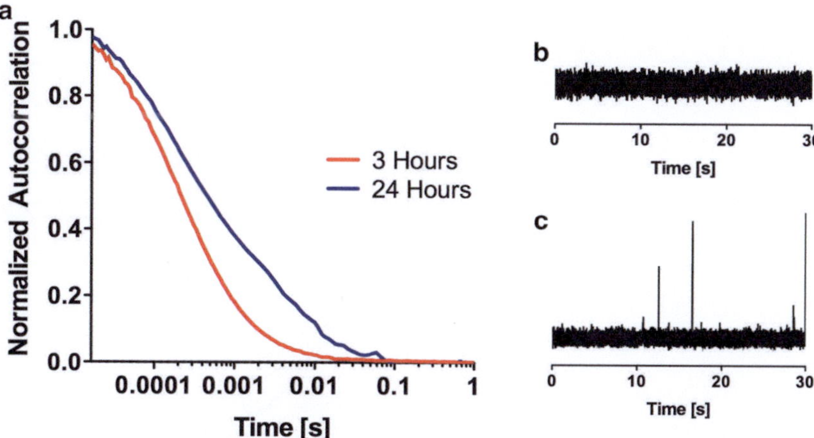


**Fig. 3** FCS analysis of cell lysates. (**a**) Autocorrelations of clarified lysates of cells producing EGFP-labeled, full length HTT exon1-Q97 harvested at 3 h (*red*) and 24 h (*blue*). (**b**) Raw data from 3 h indicating a homogeneously populated, low molecular weight species. (**c**) Raw data at 24 h indicating, in addition to the low molecular weight species, spikes of fluorescent intensity corresponding to larger, oligomeric species. The inclusion of these oligomers shifts the correlation for the 24 h data to the right (Panel (**a**), *blue* curve)

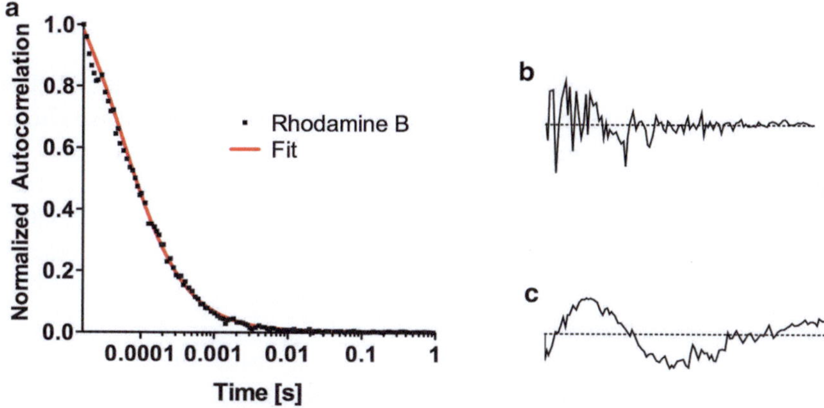


**Fig. 4** Standardization of the FCS instrument using rhodamine B. (**a**) Single-component autocorrelation fit of rhodamine B data, yielding a value for diffusion time $\tau_D$ of 50 μs, a value for the structure parameter $\omega$ of 5.9, and a quality of fit (reduced chi square) of 0.9999. (**b**) The residuals (difference between the data and the fitted values) for the fit shown in panel (**a**) indicate a good fit to the single-component analysis. (**c**) Typical poor, nonrandom residuals for a fit to an inappropriate equation. This case shows residuals from a single-component fit to a suspension of a mixture of a small MW form and aggregated forms of a molecule (correlation not shown)

ing the measurement). Explore different model equations to optimize the quality of the fit and ensure the best interpretation of the data (*see* **Notes 18** and **19**).

4. The fit line yields estimates of the critical variables $N$ (number of particles in the focal volume) and $\tau_D$ (diffusion time). For homogeneous solutions such as a standard free dye in a simple

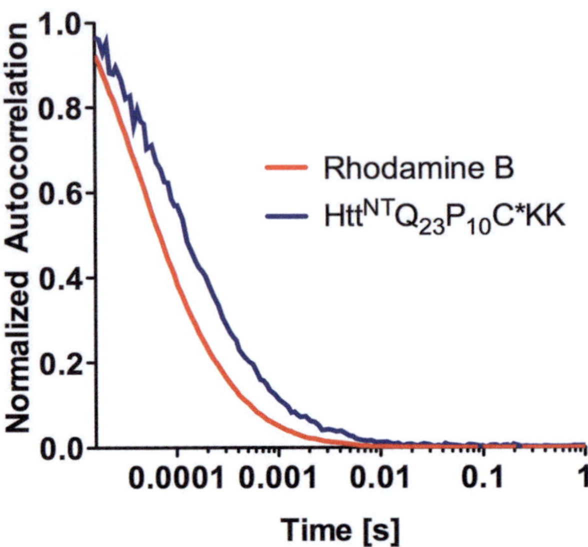


**Fig. 5** FCS autocorrelation function of a well-behaved molecule. Shown are autocorrelations for free dye rhodamine B (*red*), and monomeric, Alexa555-labeled $HTT^{NT}Q_{23}P_{10}C^{*}KK$ (*blue*). An autocorrelation function shifted to the right indicates a larger particle size

buffer, $\tau_D$ can be obtained by fitting with Eq. 2. Using the diffusion time and known diffusion constant for the free dye, the observation focal volume can be determined using Eq. 5. Figure 5 shows autocorrelation analysis of data for rhodamine B (red) and Alexa555-labeled $HTT^{NT}Q_{23}P_{10}C^{*}K_2$ (blue). When fitted as a single component, rhodamine B and Alexa555-$HTT^{N}TQ_23P_10C^{*}K_2$ give diffusion times ($\tau_D$) of 50 and 118 μs, respectively.

5. Determine the particle size using Eq. 3. Input $MW_A$ (known molecular weight of the free standard dye), $\tau_{Da}$ (diffusion time of the free dye), and $\tau_{Db}$ (diffusion time of sample of interest, which can be estimated as described in Subheading 3.4, **step 4**). Use these three values to determine $MW_B$ (the molecular weight of a single molecule of the protein of interest, including the associated fluorophore). Compare the $MW_B$ that was determined empirically from Eq. 3 to the theoretically expected $MW_B$. In a homogeneous solution with no other high MW components, or in a complex solution where it is known that hetero-oligomerization is not taking place, the ratio of $MW_B$ observed to $MW_B$ theoretical is the complex size. For example, from Fig. 5, the theoretical $MW_B$ of $HTT^{NT}Q_{23}P_{10}C^{*}K_2$ is 7.3 kDa and the observed $MW_b$ is also ~7 kDa: the protein of interest is behaving as a monomer (*see* **Note 20**).

6. If the single-component model gives poor fitting parameters, or if visual inspection of the raw data (*see* Fig. 3c) indicates more than one component, fit the autocorrelation function to a two-component equation Eq. 4. Two-component fitting of the autocorrelation data will naturally generate two $\tau_D$ values—for example, one for the aggregates (the "spikes" in the raw data in Fig. 3c) and one for the low MW portion. If the quality of fit of the two-component analysis is also poor, it may indicate additional heterogeneity. This might happen, for example, if the aggregates are too diverse in size or too large. To test for this, it is possible to manually recalculate autocorrelations from the raw data for a better fit by trimming contiguous portions of the data to focus on particular species of interest. Of course, if this is done it must be kept in mind that some components of the solution were ignored in the analysis. Finally, even with a good fit, it is possible that over-simplification has occurred. For example, in an aggregation time point featuring high MW diffusible aggregates as well as a mixture of tetramers and monomers, a two-component fit may yield an average MW for the diffusible aggregates and an average value for the mixture of monomers and tetramers.
7. Diffusion times higher than what is expected for the monomer may be indicative of a major shape change within the monomer or of complex formation that could be consistent with either homo-multimerization (as in the aggregation analysis described above) or hetero-complex formation (as in the binding of the fluorescently tagged protein to another cellular protein such as a molecular chaperone). There are several ways to confirm multimerization. First, if technically possible, one can analyze the protein of interest in homogeneous solution without any additional high MW species present and see if nonmonomer sizes continue to be obtained. Secondly, one can carry out molecular brightness analysis (*see* **Note 21**), which can be used to place a value on the number of fluorophores in each particle identified by FCS.
8. There are several ways to carry out a molecular brightness analysis (*see* **Note 21**). The simplest method, only valid if there is only one fluorescent species present, is to use the FCS autocorrelation function to obtain molecular brightness (*see* **Note 22**). Alternatively, if the analyte is clearly heterogeneous in particle size, then additional analysis is required to determine the brightness of each type of particle. This can be done either by enabling the instrument to construct a photon counting histogram (PCH) directly during data collection, or by analyzing saved raw data using appropriate software. Here, we have shown some standard data for brightness analysis using the later method. Figure 6 shows the PCH curves for rhodamine B in PBS and in 10 % Ficoll solution. The brightness analysis shows that rhodamine B is

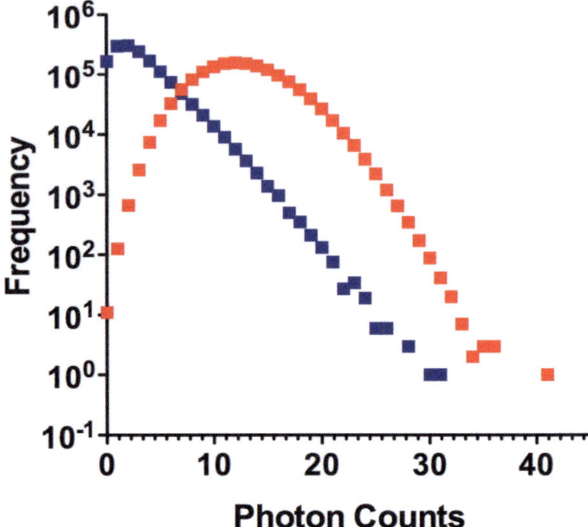


**Fig. 6** Standard photon counting histogram (PCH) for brightness analysis. In the mathematical analysis of PCH data, the width of the PCH curve at half-height is proportional to the number of fluors per particle (brightness). The *red* curve, from a PBS solution of monomeric rhodamine B, gives a brightness of 43,000 cpms (counts per molecule per second). The *blue* curve, for a solution of rhodamine B in 10 % Ficoll, which induces an oligomeric form of rhodamine B, gives a brightness of 318,000 cpms, indicating that there are approximately 318,000/43,000, or 7.4 molecules of rhodamine B per particle in 10 % Ficoll

about 7.4 times brighter in Ficoll than in PBS, making the average particle size a 7-mer to 8-mer (*see* **Note 21**).

## 4 Notes

1. Calibration of the FCS instrument is very important. The most important parameters for calibration are the diffusion time of the standard dye, the per particle brightness (CPMS) and the structure parameter $\omega$ (Eq. 2). The structure parameter $\omega$ for a typical FCS setup should be in the range of 4–8. An unusual value of $\omega$ outside this range may indicate that the laser confined focal volume is not a perfect Gaussian and therefore the analysis will not give accurate diffusion parameters [17]. The collection pinhole size and the thickness of the coverglass might also affect these parameters. Users of the Zeiss LSM-510 Confocor3 can use the following tutorial to adjust the pinhole size and measure the thickness of the coverglass [18]. CPMS and diffusion time should be approximately the same from experiment to experiment for a particular laser power or at saturation. Rhodamine B is a standard dye with diffusion constant $D = 2.8 \times 10^{-6}$ $cm^2/s$ and was used for cali-

bration for the experiments with Alexa555 labeled peptides. In contrast, fluorescein was used to calibrate the instrument when the fluorescent probe fused to the protein of interest was EGFP. In general, any stable dye of known diffusion constant and similar fluorescence properties to the fluorescent probe in the experiments can be used to calibrate the instrument. Figure 4a shows an example of a good correlation of rhodamine B, with fitted parameters described in the legend.

2. The two common types of chemistry used for labeling synthetic proteins with fluorophores in vitro exploit either the reactive sulfhydryl of cysteine or the reactive amino groups of lysine and the protein N-terminus. We have used cysteine labeling for all of our work. Detailed procedures can be found in the Molecular Probes Handbook [19]. Fluorescently labeling aggregation-prone proteins presents particular challenges. This can be managed by selecting solvents for the labeling reaction, such as Gdn.HCl solutions, that have been determined to suppress aggregation of the molecule of interest. Reaction mixtures should be purified as much as possible. Care must be taken to remove any unconjugated dye, which otherwise will compromise FCS. Incompletely labeled protein, on the other hand, may or may not introduce problems depending on the situation. If the molecule is a monomer, only labeled molecules will be detected, so there will be no error in $\tau_D$ but $N$, the number of particles in the focal volume, will be underestimated. If the molecule is a multimer, diffusion times may not be significantly affected, but brightness analysis could yield a low value for multimer size.
3. The water–TFA solution should be made by titrating TFA into water until the pH drops to approximately 3.
4. Nondenaturing lysis buffer is required, since harsh reagents, such as SDS, can solubilize aggregates and/or oligomers and thus compromise the sample and affect the results.
5. Coverglass slides coated with BSA, collagen or poly-D-Lysine should be made fresh. Cells do not adhere as well to coated coverglass slides older than 1–2 weeks. Poorly adhered cells will move while imaging and thereby disrupt data acquisition and give distorted data. In order to minimize disturbing the cells, grow and incubate cells in physical proximity to the FCS instrument. Some media contain pH dyes and other dyes that may create noise with FCS readings. Make sure the fluorescent dye used for FCS does not spectrally overlap with any dyes present in the media. Coatings to increase adhesion of cells onto the coverglass slide, such as collagen, poly-D-lysine, or BSA, do not disrupt live-cell imaging.
6. A suitable dye should be chosen for FCS experiments, mostly based on photostability and quantum yield. Rhodamine and

its derivatives, as well as the class of Alexa dyes, are very good for FCS. When fluorescent fusion proteins are used, choose a fluorophore with a high quantum yield and low sensitivity to photobleaching. EGFP and comparable fluorescent proteins also work well. Other dyes can be chosen based on suitable fluorescence properties [16].

7. For in vitro studies, if the protein of interest is aggregation prone, the peptide should be disaggregated properly in order to remove any preexisting oligomers or aggregates that can act as an aggregation seed. A detailed method for disaggregation, as well as for measuring monomeric stock protein concentration, is described by O'Nuallain et al. [20]. Different aggregating proteins may require different disaggregation protocols and storage conditions.

8. Snap-freezing should involve a single step of freezing using liquid nitrogen or dry ice/acetone and samples should be immediately stored at −80 °C. Likewise, thawing such frozen solutions for analysis should be done relatively quickly in a warm (~37 °C) water bath followed immediately by storage on ice until analysis. Aggregation of some protein sequences, such as polyglutamine, can be stimulated in the frozen state of aqueous solutions when incubated at temperatures (such as −20 °C) above the solute eutectic points [21, 22]. Therefore this situation should be avoided both in the freezing direction and in the thawing direction. This is particularly important with a highly sensitive analysis method like FCS, which is capable of sensing low levels of aggregates.

9. Working concentrations of fluorophores should be low enough to get fluorescence below half of the detector's saturation limit. FCS is a very sensitive method and all APDs have an upper limit before the signal becomes saturated (typically around 1 MHz; check the individual APD). If saturation becomes a problem under cell lysis or in vitro conditions, simply dilute the sample with an appropriate nondenaturing buffer. Occasionally, saturation occurs in live-cell imaging, especially with cells expressing high levels of a fluorophore. In this scenario, one must choose cells or growth conditions that express the fluorophore-tagged protein at an appropriate level. The fluorescence should be less than 400–500 kHz. Another way to decrease signal is to reduce laser power, however if this is done the instrument must be recalibrated with the standard dye under the new conditions.

10. Most modern objective lenses come with an adjustable collar to accommodate the thickness of the coverglass slide. The collar must be adjusted once the thickness of the coverglass slide is established.

11. Acquisition time should be kept as short as possible in order to minimize photobleaching and triplet state formation. If a stable dye is used, the data acquisition time can be increased to get better quality data. However, overly long acquisition times also can introduce problems, such as depletion of sample, sticking of the sample to the coverglass, or even concentration changes due to drying of the sample. The problem of proteins sticking to the coverglass surface is especially evident at lower analyte concentrations in simple buffers that promote adsorption. Glass slides pretreated with BSA or other coatings can suppress adsorption (*see* **Note 5**).
12. The CPMS value is dependent on an individual microscope's optics, the quantum yield of the dye, laser power, filter settings, and other factors. In our experiments, rhodamine B typically yields a CPMS of 12–15,000 Hz and EGFP of 7–8000 Hz. CPMS values of sensitive fluorophores might be as high as 100,000–250,000 Hz at saturating laser power. However, except for determining the sensitivity of the instrument, saturating laser power should be avoided during measurements since it has the potential to introduce photobleaching and triplet state formation. Although it is possible to collect data at laser power up to 50 % that of saturation, it is normally advisable to conduct experiments at much lower power, perhaps five- to tenfold below saturating laser power. Background noise typically has a count rate of ~100 Hz. In order to collect robust data, the signal to noise ratio should be at least one or two orders of magnitude.
13. Cell lysates should not be left on ice for prolonged periods before measurements and should contain a protease inhibitor cocktail. Under these conditions, fresh cell lysates remain stable over 3–4 h. Centrifugation speeds can be adjusted depending on the types of particles being studied. If one is only focusing on monomers or small oligomers it may be advantageous to centrifuge at a higher speed to clear more cell debris. The 425 × *g* centrifugation was designed to clear large inclusions (which in any case are not observable by FCS) while allowing diffusible aggregates to remain in the supernatant.
14. A strong fluorescence signal can sometimes yield no correlation. This could be because of the cell membrane creating random light scattering. Adjust the *z*-axis of the sample to better focus the laser inside the cytoplasm. Users of the Zeiss LSM-510 Confocor3 can refer to the tutorial [18].
15. Photobleaching can be a major concern for FCS studies in cells since, because of viscosity and molecule size, the residence time of molecules in the laser focal volume can be high. One way to avoid photobleaching is to reduce the laser power

to the lowest possible setting that still gives a good correlation after about 10 s of data acquisition. Another way to suppress photobleaching is to reduce the focal volume of the laser spot, which can be done by using different objective lenses and/or by changing excitation optics [23].

16. During measurement, the count rate should fluctuate randomly around a mean. If the count rate is decreasing systematically over time, the fluorophore is likely photobleaching and/or depleting from the solution by preferential surface binding to the glass. Consider collecting for a shorter period of time, reducing the laser power, or switching to a more photostable fluorophore and coating the coverglass with BSA.
17. Each experiment must be internally standardized with an appropriate fluorophore. If the unknown is a well-characterized peptide chemically tagged with a fluorophore, then either a relatively small fluorophore like rhodamine B, or a previously characterized related fluorescently labeled peptide, could be used as the standard. On the other hand, if the unknown is a fusion of EGFP with a protein of interest, for example, then the standard should be similarly expressed and processed EGFP itself. In the former case, the rhodamine B could be used both to calibrate the instrument and as the standard for comparison. In the latter case, EGFP cannot be used to calibrate the instrument unless its diffusion constant under identical experimental conditions were known to precision (which is unlikely). Instead, a small MW fluorophore with similar fluorescence properties to EGFP, like fluorescein, should be used for calibration, and EGFP used as the standard.
18. One- and two-component fitting will be useful and appropriate in most of the situations. However, in the case of highly heterogeneous samples where these analysis do not give good fits, continuous distribution fitting methods are available, such as MEMFCS described elsewhere [13].
19. Potential sources of poor fits include improper calibration of the instrument, employing the wrong model, and significant loss of signal due to either photobleaching or adsorption to the glass surface. The equations described here are best used in situations where fluctuations of the fluorescent signal are due only to diffusion. If this is not the case, either because of artifacts such as those cited above or because of an ongoing physical or chemical change taking place during the measurement (such as, for example, continued aggregation during the measurement), then these equations will not perform accurately.
20. Equation 3 is precise only for spherical particles. However, many molecules that deviate from spherical shape can still be well-described by this equation. In the example, neither rhoda-

mine B (a largely planar heteroaromatic ring system) nor HTT exon1 (a linear composite of three distinct intrinsically disordered segments [24] favoring disorder ($HTT^{NT}$), compact coil (polyglutamine), and polyproline type II helix (proline-rich domain)) are expected to be spherical, yet the FCS data with Eq. 3 strongly indicates that the Q25 version of HTT exon1 behaves as a monomer under the experimental conditions. In spite of the good agreement found with nonspherical particles in this example, it is advisable to generally exercise caution when using Eq. 3 to evaluate macromolecular behavior, and to use it in combination with other data. For example, if Eq. 3 suggests that a protein is behaving as a multimer, it should be possible to carry out brightness analysis (Subheading 3.3, **step 8**) to independently confirm the multimer. Additionally, if the multimer is the product of a reversible oligomerization, it may be possible to carry out a concentration dependent analysis showing a transition from monomer to oligomer.

21. Brightness analysis can be performed in a number of different ways, including the FCS autocorrelation curve method (*see* **Note 22**), or by Number and Brightness analysis of rasterized images of fluorescent arrays, or by photon counting histogram (PCH) analysis [18]. For the data shown here we used ImageJ (NIH) with the PCH analysis plugin provided by Jay Unruh of the Stowers Institute for Medical Research in Kansas City, MO. Brightness analysis can be used in combination with FCS diffusion time estimates of MW to determine a shape factor of large molecules. While the analysis of autocorrelation assumes the particles to be spherical, this may not be true for some proteins and protein aggregates like individual amyloid fibrils. If MW estimates by brightness analysis and by FCS diffusion time analysis do not correlate, this means the particles are not spherical. In principle, in this situation the FCS diffusion time analysis can be adjusted, by introducing a nonspherical shape factor parameter [25], to both better understand the morphology of the aggregates and to obtain agreement between diffusion time and brightness analysis.

22. For homogeneous solutions only, molecular brightness can be determined from the FCS autocorrelation function itself. Dividing the count rate in a particular experiment on the molecule of interest by the particle concentration (i.e., $N$) from the autocorrelation function gives the fluorescence intensity per particle in the sample. Carrying out an equivalent calculation on an FCS experiment on the isolated probe (i.e., EGFP or the Alexa dye used to modify the protein) gives the fluorescence intensity of the probe molecule. The ratio of the two molecular brightness intensities gives directly the average number of probe molecules per particle.

## References


1. Elson EL, Magde D (1974) Fluorescence correlation spectroscopy. I. Conceptual basis and theory. Biopolymers 13:1–27
2. Magde D, Elson EL, Webb WW (1974) Fluorescence correlation spectroscopy. II. An experimental realization. Biopolymers 13: 29–61
3. Eigen M, Rigler R (1994) Sorting single molecules: application to diagnostics and evolutionary biotechnology. Proc Natl Acad Sci U S A 91(13):5740–5747
4. Kohl T, Schwille P (2005) Fluorescence correlation spectroscopy with autofluorescent proteins. Adv Biochem Eng Biotechnol 95:107–142
5. Orden AV, Jung J (2008) Review fluorescence correlation spectroscopy for probing the kinetics and mechanisms of DNA hairpin formation. Biopolymers 89(1):1–16. doi:10.1002/bip.20826
6. Folta-Stogniew E (2006) Oligomeric states of proteins determined by size-exclusion chromatography coupled with light scattering, absorbance, and refractive index detectors. Methods Mol Biol 328:97–112. doi:10.1385/1-59745-026-X:97
7. Mok YF, Howlett GJ (2006) Sedimentation velocity analysis of amyloid oligomers and fibrils. Methods Enzymol 413:199–217. doi:10.1016/S0076-6879(06)13011-6
8. Gast K, Fiedler C (2012) Dynamic and static light scattering of intrinsically disordered proteins. In: Uversky VN, Dunker AK (eds) Intrinsically disordered protein analysis, vol 896, Methods in molecular biology. Springer, New York, pp 137–161. doi:10.1007/978-1-4614-3704-8_9
9. Politz JC, Browne ES, Wolf DE, Pederson T (1998) Intranuclear diffusion and hybridization state of oligonucleotides measured by fluorescence correlation spectroscopy in living cells. Proc Natl Acad Sci U S A 95(11): 6043–6048
10. Brock R, Vamosi G, Vereb G, Jovin TM (1999) Rapid characterization of green fluorescent protein fusion proteins on the molecular and cellular level by fluorescence correlation microscopy. Proc Natl Acad Sci U S A 96(18):10123–10128
11. Garai K, Sahoo B, Sengupta P, Maiti S (2008) Quasihomogeneous nucleation of amyloid beta yields numerical bounds for the critical radius, the surface tension, and the free energy barrier for nucleus formation. J Chem Phys 128(4):045102. doi:10.1063/1.2822322
12. Garai K, Sengupta P, Sahoo B, Maiti S (2006) Selective destabilization of soluble amyloid beta oligomers by divalent metal ions. Biochem Biophys Res Commun 345(1):210–215. doi:10.1016/j.bbrc.2006.04.056
13. Sengupta P, Garaj K, Balaji J, Periasamy N, Maiti S (2003) Measuring size distribution in highly heterogeneous systems with fluorescence correlation spectroscopy. Biophys J 84:1977–1984
14. Kitamura A, Kubota H (2010) Amyloid oligomers: dynamics and toxicity in the cytosol and nucleus. FEBS J 277(6):1369–1379. doi:10.1111/j.1742-4658.2010.07570.x
15. Takahashi Y, Okamoto Y, Popiel HA, Fujikake N, Toda T, Kinjo M, Nagai Y (2007) Detection of polyglutamine protein oligomers in cells by fluorescence correlation spectroscopy. J Biol Chem 282(33):24039–24048. doi:10.1074/jbc.M704789200
16. Lavis LD, Raines RT (2008) Brights ideas for chemical biology. ACS Chem Biol 3: 142–155
17. Masuda A, Ushida K, Okamoto T (2005) New fluorescence correlation spectroscopy enabling direct observation of spatiotemporal dependence of diffusion constants as an evidence of anomalous transport in extracellular matrices. Biophys J 88(5):3584–3591. doi:10.1529/biophysj.104.048009
18. Youker RT, Teng H (2014) Measuring protein dynamics in live cells: protocols and practical considerations for fluorescence fluctuation microscopy. J Biomed Opt 19(9):90801. doi:10.1117/1.JBO.19.9.090801
19. The Molecular Probes® Handbook (2010) A guide to fluorescent probes and labeling technologies. Doi: http://www.lifetechnologies.com/us/en/home/references/molecular-probes-the-handbook.html
20. O'Nuallain B, Thakur AK, Williams AD, Bhattacharyya AM, Chen S, Thiagarajan G, Wetzel R (2006) Kinetics and thermodynamics of amyloid assembly using a high-performance liquid chromatography-based sedimentation assay. Methods Enzymol 413:34–74. doi:10.1016/S0076-6879(06)13003-7
21. Chen S, Berthelier V, Hamilton JB, O'Nuallain B, Wetzel R (2002) Amyloid-like features of polyglutamine aggregates and their assembly kinetics. Biochemistry 41(23): 7391–7399
22. Kar K, Arduini I, Drombosky KW, van der Wel PC, Wetzel R (2014) D-polyglutamine amyloid recruits L-polyglutamine monomers and kills cells. J Mol Biol 426(4):816–829. doi:10.1016/j.jmb.2013.11.019

23. Hess ST, Webb WW (2002) Focal volume optics and experimental artifacts in confocal fluorescence correlation spectroscopy. Biophys J 83(4):2300–2317. doi:10.1016/S0006-3495(02)73990-8
24. Wetzel R, Mishra R (2014) Structural biology: order, disorder, and conformational flux. In: Bates G, Tabrizi SJ, Jones L (eds) Huntington's disease, 4th edn. Oxford University Press, Oxford, UK, pp 274–322
25. Cantor CR, Schimmel PR (1980) Biophysical chemistry, part II. Techniques for study of biological structure and function. W.H. Freeman and Company, San Francisco, pp 539–590

# Chapter 6

# Deep UV Resonance Raman Spectroscopy for Characterizing Amyloid Aggregation

**Joseph D. Handen and Igor K. Lednev**

## Abstract

Deep UV resonance Raman spectroscopy is a powerful technique for probing the structure and formation mechanism of protein fibrils, which are traditionally difficult to study with other techniques owing to their low solubility and noncrystalline arrangement. Utilizing a tunable deep UV Raman system allows for selective enhancement of different chromophores in protein fibrils, which provides detailed information on different aspects of the fibrils' structure and formation. Additional information can be extracted with the use of advanced data treatment such as chemometrics and 2D correlation spectroscopy. In this chapter we give an overview of several techniques for utilizing deep UV resonance Raman spectroscopy to study the structure and mechanism of formation of protein fibrils. Clever use of hydrogen-deuterium exchange can elucidate the structure of the fibril core. Selective enhancement of aromatic amino acid side chains provides information about the local environment and protein tertiary structure. The mechanism of protein fibril formation can be investigated with kinetic experiments and advanced chemometrics.

**Key words** Protein folding, UV resonance Raman spectroscopy, Chemometrics, 2DCoS, Protein (un)folding kinetics, Fibrils, Hydrogen-deuterium exchange

## 1 Introduction

Since the fibrillation of numerous peptides and proteins has been linked to various human diseases, investigation of the structure and mechanism of formation of these fibers has become increasingly important [1]. Full-length fibrils are typically noncrystalline and insoluble, making the use of the two standard techniques used in structural biology, X-ray crystallography and conventional NMR, impractical [2]. Deep UV resonance Raman (DUVRR) is uniquely suited to probe many aspects of protein structures and fibrillation, including dihedral angle distributions, carbonyl and amino hydrogen-bonding, and kinetic parameters [1–11]. The described techniques provide label-free, sensitive, and selective methods with simple sample preparation. Additional data treatment with chemometrics [12–17] or two-dimensional correlation

David Eliezer (ed.), *Protein Amyloid Aggregation: Methods and Protocols*, Methods in Molecular Biology, vol. 1345,
DOI 10.1007/978-1-4939-2978-8_6, © Springer Science+Business Media New York 2016

spectroscopy (2DCoS) can dramatically increase the information gained from experiments.

Raman spectroscopy is a scattering phenomenon wherein incident light interacts with a molecule and there is a transfer of vibrational energy between the light and the molecule. The energy difference between the incident and inelastically scattered light provides information about the vibrational characteristics of the molecule the light interacted with. Typically, vibrational spectra are made of many overlapping bands. In resonance Raman spectroscopy, the frequency of the incident light is within an electronic absorption band of a selected chromophore which enhances the signal as much as $10^6$-fold for selected vibrational modes of the chromophore. The latter simplifies significantly the interpretation of Raman spectra and makes Raman spectroscopy a selective technique. A particular chromophore in a complex sample could be chosen and selectively characterized due to the resonance enhancement by adjusting the excitation laser wavelength. For example, by using 195–205 nm excitation, the bands resulting from the protein amide chromophore are selectively enhanced. The amide chromophore is a building block of the polypeptide backbone. Raman spectra of the amide chromophore, including DUVRR spectra, depend strongly on the polypeptide backbone dihedral angles and report on the protein three-dimensional structure [18]. Using a 195 nm excitation results in resonance enhancement of Raman scattering from the phenylalanine side chain [1, 19]. Tyrosine and tryptophan amino acid residues dominate the Raman spectrum of proteins when the excitation wavelength is around 230 nm [20]. The resonance Raman spectra of all three amino acid residues depend on the local environment and report on the tertiary structural changes of proteins [18]. Visible light excitation results in resonance enhancement of Raman scattering from the protein heme group and can be used for monitoring redox-coupled processes in heme proteins [21, 22].

### 1.1 Elucidating the Structure of the Fibril Core

By combining hydrogen-deuterium exchange (HX) with DUVRR spectroscopy, it is possible to probe the secondary structure of a fibril cross-$\beta$ core. It has been shown that in an amino acid residue, only the main chain NH group and O-, N-, and S-bound protons are easily exchanged [23]. Additionally, strong hydrogen bonding will greatly reduce the rate of HX [24], meaning a fibril core will remain protonated during exchange (Fig. 1). Upon exchange, there will be a downshift of the amide II DUVRR band from ~1555 to ~1450 cm$^{-1}$ (amide II′) and the virtual disappearance of the amide III band in an unordered protein [11, 25]. For example, Fig. 2 illustrates these spectral changes in DUVRR spectra of lysozyme in both fibrillar and unordered forms. The resulting spectra will have overlapping bands from the deuterated unordered fraction and the protonated fibril core, which need to be resolved for detailed analysis.

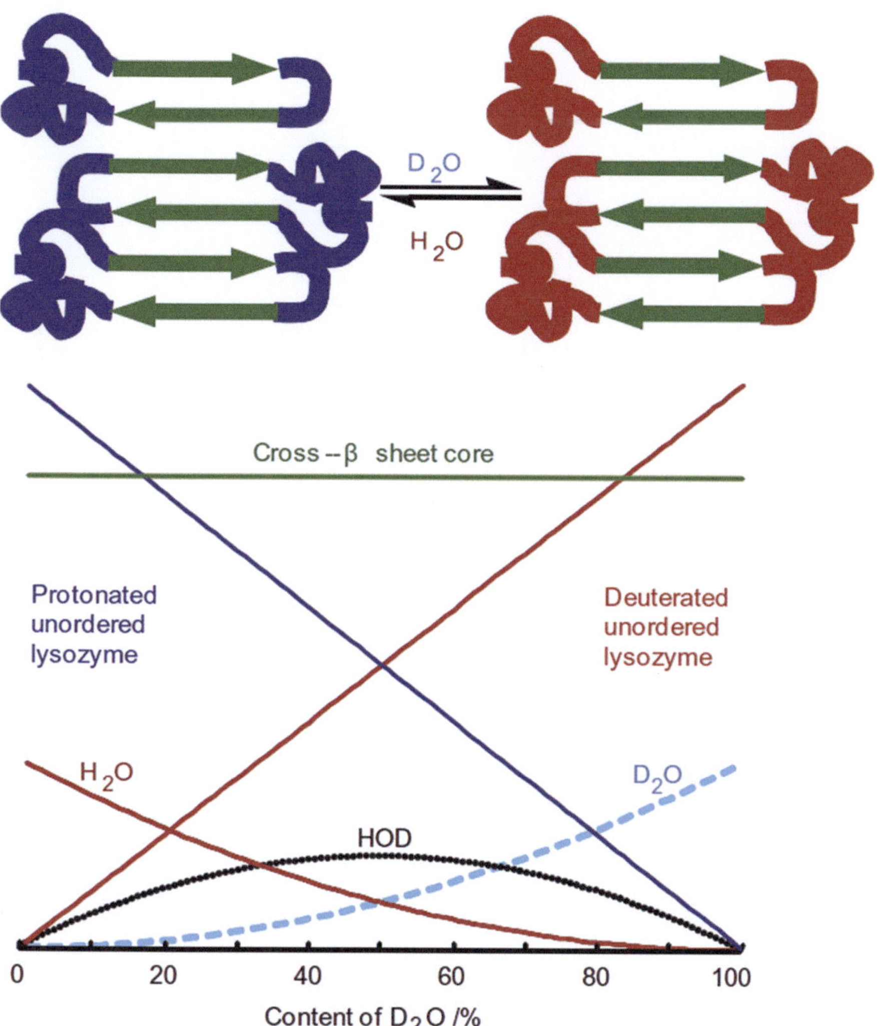


**Fig. 1** Hydrogen-deuterium exchange (HX) was used to allow for the extraction of the Raman signature of the fibril core (*top*). Schematic representation of HX: the protonated fibril core is protected from HX, while the unordered fraction is not (*bottom*). Expected concentration profile of major components versus the fraction of $D_2O$: protonated unordered lysozyme (*blue*), cross-$\beta$-sheet core (*green*), deuterated unordered lysozyme (*red*), $H_2O$ (*brown* solid), HOD (*black* solid), and $D_2O$ (*light blue dashed*). Reprinted with permission from [9] with permission from Elsevier. Copyright 2010 Elsevier

With the DUVRR spectrum of the fibril core isolated, it is possible to calculate the distribution of the $\Psi$ dihedral angle of the cross-$\beta$ core [26]. The Bayesian approach [27] is able to outperform blind source separation algorithms such as independent component analysis (ICA) or pure variable methods for resolving contributions from the fibril core and unordered fractions. Specifically, a Bayesian signal dictionary approach can be used to incorporate a priori information about characteristic spectral bands by using a reference band library built by fitting DUVRR spectra of fibrils [9].

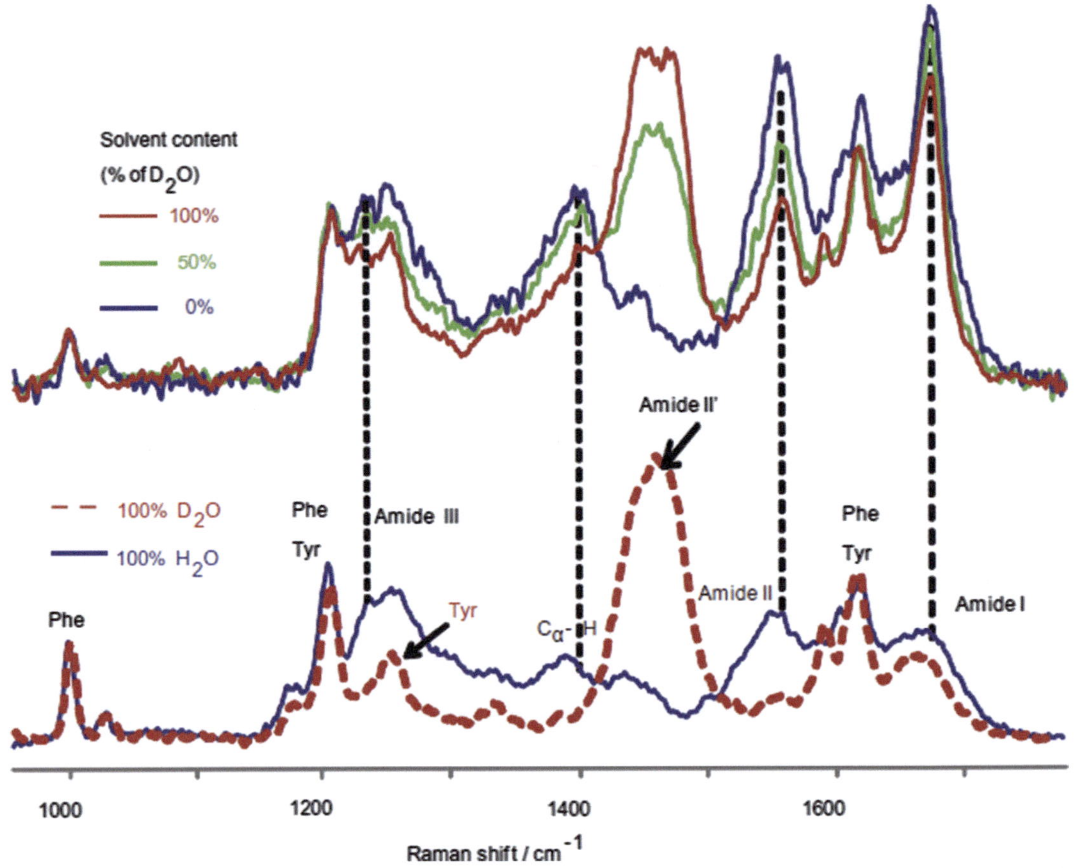


**Fig. 2** Deep UVRR spectral changes in fibrillar and unordered hen egg white lysozyme after hydrogen-deuterium exchange provide evidence of a still-protonated core. (**a**) DUVRR spectra of HEWL fibrils in $H_2O$ (*blue*), $D_2O$ (*red*), and 50/50 % $H_2O/D_2O$ mixture (*green*). (**b**) DUVRR spectrum of unordered lysozyme in $H_2O$ (*blue*), and $D_2O$ (*red*). Reprinted with permission from [9] with permission from Elsevier. Copyright 2010 Elsevier

### *1.2 Aromatic Amino Acids as Intrinsic Probes of Local Environment*

The aromatic amino acids tyrosine, tryptophan, and phenylalanine are useful as intrinsic probes of the local environment and protein tertiary structure when using resonance Raman spectroscopy. Raman spectroscopy of phenylalanine is especially valuable biochemical tool because, in contrast to tyrosine and tryptophan, phenylalanine does not work as a fluorescent probe of the local environment. The phenyl ring stretching responsible for the 1000 $cm^{-1}$ phenylalanine band has a Raman cross section that is strongly dependent on exposure to water [1]. Figure 3 illustrates this dependence with spectra recorded of solution of *N*-acetyl-L-phenylalanine ethyl ester with varying concentrations of acetonitrile in water. This change has been proven to be useful in the study of amyloid fibrils. For example, Fig. 4 shows the change in the intensity of the 1000 $cm^{-1}$ phenylalanine band during the fibrillation of hen egg white lysozyme as a monoexponential decrease, indicating that phenylalanine is being increasingly exposed to water as fibrillation occurs [1].

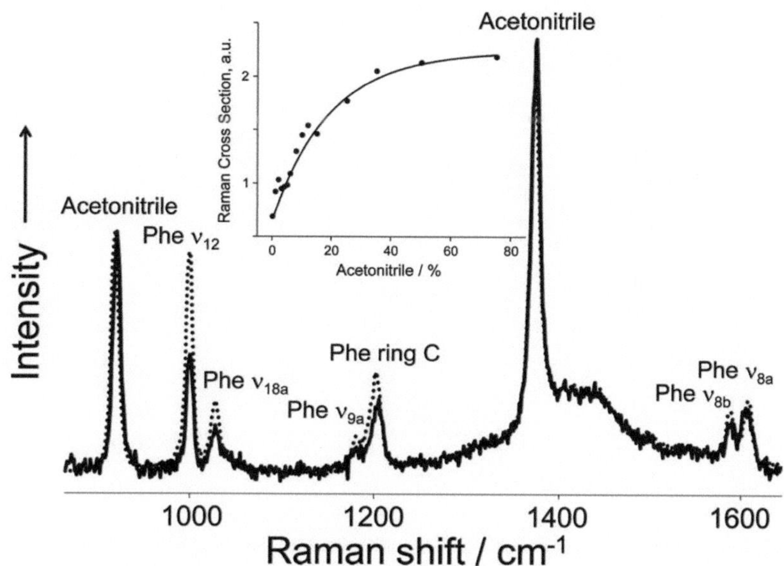


**Fig. 3** The Raman cross section of the 1000 cm$^{-1}$ band phenylalanine band was shown to vary as a function of water exposure, making it a useful probe of local environment. Deep UV resonance Raman spectra of *N*-acetyl-L-phenylalanine ethyl ester in water and 50 % acetonitrile (*dotted line*) and 5 % acetonitrile (*solid line*). Inset: Raman cross section of the 1000 cm$^{-1}$ band as a function of acetonitrile concentration. Reproduced with permission from [1]. Copyright Wiley-VCH Verlag GmbH & Co. KGaA

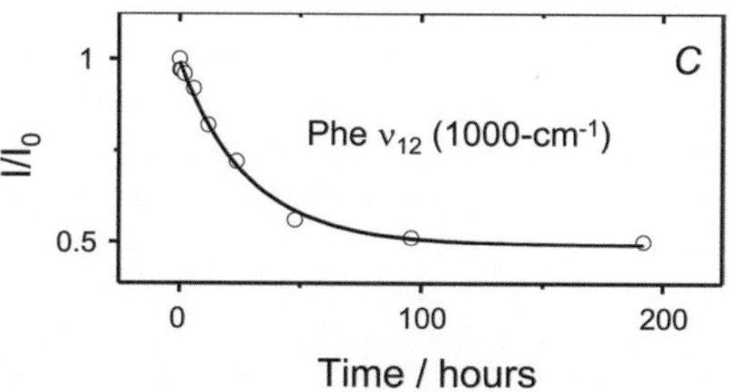


**Fig. 4** The local environment of hen egg white lysozyme was probed using phenylalanine as an intrinsic reporter. The relative intensities of the 1000 cm$^{-1}$ bands of phenylalanine from the Raman spectra of HEWL were plotted as a function of incubation time, indicating an increasing exposure of phenylalanine to water as fibrillation progresses. Reproduced with permission from [1]. Copyright Wiley-VCH Verlag GmbH & Co. KGaA

### 1.3 Elucidating the Mechanism of Fibrillation

By combining 2DCoS with DUVRR spectroscopy, it is possible to resolve the often heavily overlapped bands found in vibrational spectroscopy arising from secondary structure motifs, polypeptide backbone and amino acid side chains. Notably, 2DCoS is able to extract information about structural transitions and the kinetic

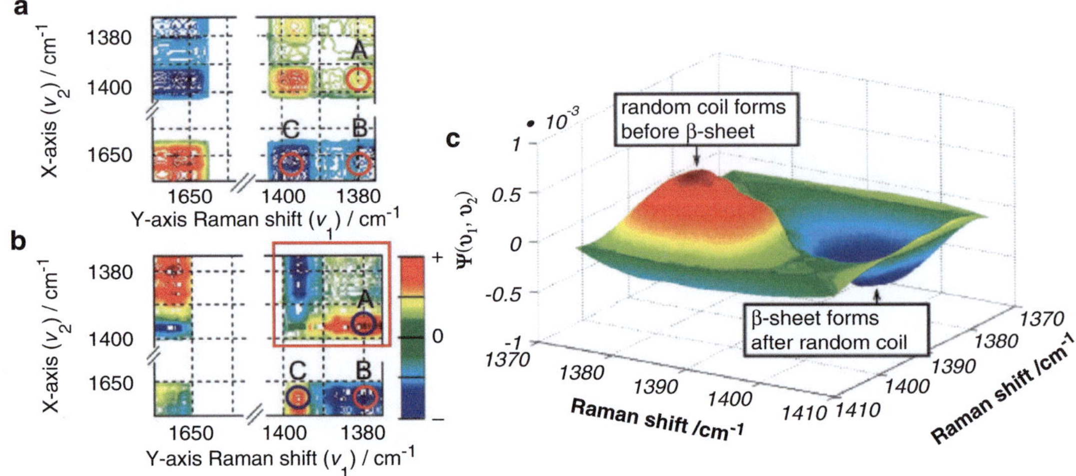


**Fig. 5** 2D correlation spectroscopy was used to establish sequential order of events during the fibrillation of hen egg white lysozyme. The synchronous (**a**) and asynchronous (**b**) 2D-DUVRR correlation maps of the $C_\alpha$–H bending region and α–helix sub-band of the Amide I region suggest the order of structural change to be: random coil → $\alpha$-helix → $\beta$-sheet (adapted with permission from [6]. Copyright 2008 American Chemical Society). The 3D perspective plot (**c**) of the asynchronous $C_\alpha$–H bending region (*red* rectangle in (**b**)) illustrates the implication that random coil forms before $\beta$-sheet (adapted with permission from [5]. Copyright 2007 American Chemical Society). Reprinted with permission from [18]. Copyright 2012 American Chemical Society

reaction mechanisms [6]. By collecting DUVRR spectra over the course of fibrillation, Noda's approach [28] can be used to calculate the 2D correlation DUVRR spectra [5, 7]. Correlation between the intensities of the DUVRR bands gives clues as to the temporal order of transitions. For example, Fig. 5 shows two-dimensional correlation maps for the fibrillation of hen egg white lysozyme (HEWL). Cross-peak A represents the correlation of $\beta$-sheet and disordered structure; cross-peak B represents the correlation of $\alpha$-helix and $\beta$-sheet; and cross-peak C represents the correlation of $\alpha$-helix and disordered structure. Positive peaks on the synchronous map indicate that the spectral changes are positively correlated, while negative peaks indicate a negative correlation. The matching peaks of the asynchronous map give information regarding the average rates of changes; and when considered with the synchronous map, the sequential order of events [6]. The characteristic times for structural changes can be extracted from the two-dimensional data by *kv* correlation analysis, which involves calculating the asynchronous correlation of the experimental decaying intensity data with a set of reference exponentially decaying intensities [29].

## 2 Materials

### 2.1 Instrument

A homebuilt instrument [30] is used for the procedures below. *See* Fig. 6 for a block diagram of the Raman spectroscopic apparatus. The instrument comprises:

1. *Indigo S* laser system (Coherent): A Ti:Sa oscillator is pumped by the intra-cavity frequency-doubled output of a diode pumped and Q-switched Nd:YLF *Evolution 15* laser (527-nm wavelength, 5 kHz repetition rate, 140 ns pulse duration, up to 10 W average power). The radiation output of the system is tunable in the range of 772–820 nm. The *Indigo S* system harmonics package can convert the fundamental of the Ti:Sa oscillator to a harmonic with the desired wavelength output. The second harmonic generation (SHG) provides a range of 386–410 nm, the third harmonic generation (THG) provides a range of 257–273 nm, and the fourth harmonic provides a range of 193–205 nm [9] (*see* **Note 1**). Laser power on the sample is typically kept between 0.5 and 1.0 mW.

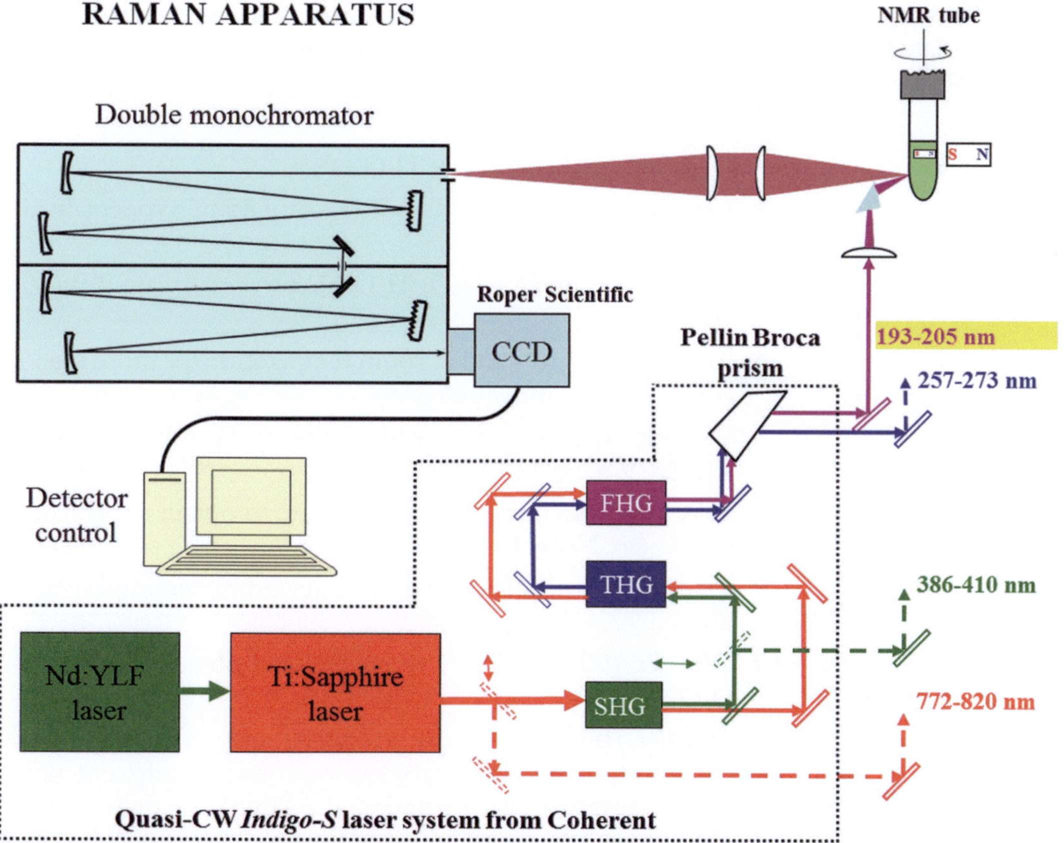


**Fig. 6** Block diagram of the tunable deep UV resonance Raman spectrographic apparatus. Reproduced from [30] with kind permission from Springer Science and Business Media

2. Sample cell: Suprasil NMR tube, magnetic stirbar, rotating cuvette holder (*see* **Note 2**), stationary external magnet.
3. Monochromator: A double spectrograph with a Czerny-Turner configuration.
4. Camera: A Princeton Instruments Spec-10:400B digital CCD camera (Roper Scientific), cooled to −120 °C with liquid nitrogen.
5. Software: GRAMS/AI (7.01) (Thermo Galactic) for processing, WinSpec 32 for collection, MATLAB for calculations

### 2.2 Sample Components

1. Hen egg white lysozyme: 14 mg/mL of lysozyme in aqueous solution, pH 2.0. Weigh about 56 mg of lysozyme powder into a 5-mL Eppendorf tube. Add 4 mL of distilled water, mix, and adjust pH with HCl. Store at −20 °C (*see* **Note 3**).
2. $D_2O/H_2O$ mixtures: Prepare 51 mixtures of $D_2O/H_2O$ from 0 % $D_2O$ to 100 % $D_2O$ at 2 % intervals.

## 3 Methods

### 3.1 Determining the Raman Spectrum of the Fibril Core

1. Incubate Eppendorf tubes of the hen egg white lysozyme solution at 65 °C for 6 days (*see* **Note 3**).
2. Centrifuge tubes at 16,100 × *g* for 30 min to separate out the gelatinous phase (fibrils).
3. Wash fibrils with water or $D_2O$ twice.
4. Centrifuge tubes again at 16,100 × *g* for 45 min to separate out the fibrils.
5. Resuspend fibrils in $D_2O/H_2O$ mixtures to create 2 mg/mL samples.
6. Allow samples to stand at room temperature for more than 4 h before making measurements.
7. Calibrate Raman system with Teflon using the peaks at 732 and 1379.5 $cm^{-1}$.
8. Take a measurement of empty cuvette to obtain a spectrum of quartz (*see* **Note 4**).
9. Transfer 100 μL of water to cuvette and obtain a spectrum of water.
10. Transfer 100 μL of $D_2O$ to cuvette and obtain a spectrum of $D_2O$.
11. Transfer 100 μL of sample to cuvette and add a magnetic stir bar. Position the stationary external magnet adjacent to the sample cuvette. Begin rotating the sample to reduce sample photodegradation and localized heating.

12. Collect spectra as six 30-s accumulations and average them to obtain a single spectrum of the sample. Software was used to automatically detect and remove interference from cosmic rays.
13. Repeat **steps 11** and **12** to obtain a spectrum of each sample.
14. Assemble the 51 spectra of sample in the various $D_2O$ concentrations into a matrix (which will be referred to as Data). These spectra are the results of contributions from the cross-$\beta$ fibril core, the protonated and deuterated unordered portions of the fibrils, $H_2O$, HOD, $D_2O$, quartz, and oxygen. Bayesian source separation is used to extract the spectrum of the core.
15. The problem is presented as Data $= C \times S + E$; where $C$ is the concentration matrix, $S$ is the matrix of pure component spectra, and $E$ is any systematic or random error. To solve the problem, determining either $C$ or $S$ is sufficient, but it is easier to obtain the matrix $C$ (also, *see* **Note 5**). The probability of the concentration matrix is $P(C \mid \text{Data}, I) - \int ds \times \prod_i \delta(\text{Data}_i - C_{ik} \times S_k) \times \prod_l P_l(S_l)$ (or, *see* **Note 6**).
16. Solve using a genetic algorithm [9].

### *3.2 Determine the Kinetic Mechanism of Fibril Formation*

1. Incubate Eppendorf tubes of the hen egg white lysozyme solution at 65 °C for 48–60 h at 1 h intervals (13 samples total).
2. Centrifuge tubes at 16,000 × *g* for 30 min to separate out the gelatinous phase from the soluble phase. Keep the soluble phase for measurements.
3. Calibrate Raman system with Teflon using the peaks at 732 and 1379.5 wavenumbers.
4. Take a measurement of empty cuvette and obtain a spectrum of quartz (*see* **Note 4**).
5. Transfer 100 μL of water to cuvette and obtain a spectrum of water.
6. Transfer 100 μL of sample to a clean cuvette and add a magnetic stir bar. Position the stationary external magnet adjacent to the sample cuvette. Begin rotating the sample to reduce sample photodegradation and localized heating.
7. Collect spectra as six 30-s accumulations and average them to obtain a single spectrum the sample. Software was used to automatically detect and remove interference from cosmic rays.
8. Repeat **steps 6** and **7** to obtain a spectrum of each of the 13 samples.
9. Subtract the quartz spectrum from the solvent spectrum, and then subtract the resulting solvent spectrum from the sample spectrum. Subtract the quartz spectrum from the sample spectrum (*see* **Note 7**).

10. Calculate the synchronous $\Phi(\nu 1,\nu 2)$ and asynchronous $\Psi(\nu 1,\nu 2)$ 2D-Raman spectra using Noda's approach:

$$\Phi(v1,v2)+i\Psi(v1,v2)=\frac{1}{\pi(T_{\max}-T_{\min})}\times\int_0^{\infty}\tilde{Y}_1(\omega)\tilde{Y}_2^*(\omega)d\omega$$

where $\tilde{Y}_1(\omega)$ and $\tilde{Y}_2^*(\omega)$ are the forward and conjugate Fourier transforms, respectively, of the experimental spectral intensities $\tilde{Y}(\nu,t)$ for all frequencies $\nu$ from $\nu 1$ to $\nu 2$, and incubation times $t$, during the observation period from $T_{\min}$ to $T_{\max}$ [5] (*see* **Note 8**).

11. Compare cross-peaks from signals of interest in both the synchronous and asynchronous 2D-Raman spectra. *See* Subheading 1.3 for guidance on interpretation.

## 4 Notes

1. While the fourth harmonic includes wavelengths lower than 195 nm, UV absorption of molecular oxygen interferes with data collection closer to 193 nm (Fig. 7).
2. A modified hand drill was used to hold and rotate the sample cell. A tube with an inner diameter approximately equal to the outer diameter of the sample cell was held in the drill chuck. A notch was taken out of the tube and a rubber band wrapped around the tube over the notch so as to hold the sample cell by friction when it is inserted into the tube.
3. Typically, in vitro fibrillation is performed at low pH and elevated temperature.

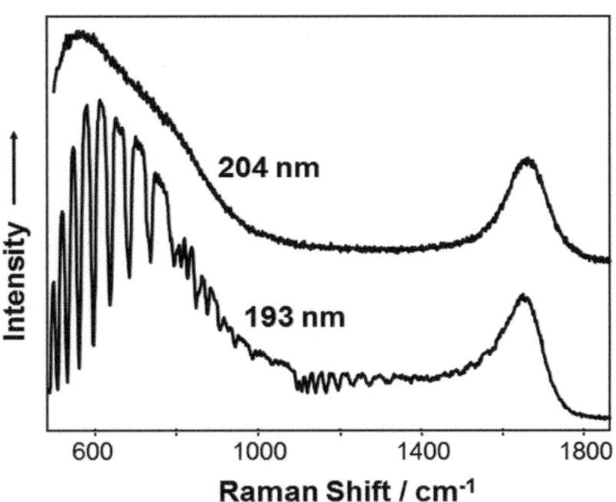


**Fig. 7** Raman spectra of water collected using 193 and 204 nm excitation wavelengths. The interference due to molecular oxygen is apparent in the lower wavenumber range. Reproduced from [30] with kind permission from Springer Science and Business Media

4. Keep laser focus where the sample will be, not focused on the surface of the cuvette.
5. Additional prior information about the concentration matrix can be included by estimating hyper-parameters $\alpha i$ for each of the eight contributing components so that $Cj = \alpha j \times Tij$; where $j = 1{:}8$, and $T$ is the template matrix represented by Fig. 1. The contributions of quartz and oxygen can be assumed to be constant but unknown in each sample, and refined with matrix least squares after relevant parameters are found.
6. This probability can be reduced to $P(C|\text{Data}, I) = \log(\det(W)) + \sum l \log(\text{P}l(Sl))$, if the data is noise free, where $W$ is the separation matrix such that $S = W \times \text{Data}$.
7. Background subtraction of quartz and solvent must be performed separately. The quartz and solvent will have different contributions to the sample spectrum than the combined solvent and quartz spectrum.
8. Software developed by Shigeaki Morita is available for free on the web which can produce the synchronous and asynchronous correlation maps. The download can be found at http://sci-tech.ksc.kwansei.ac.jp/~ozaki/NIR2DCorl_e.html.

## Acknowledgments

This work was supported by the National Science Foundation under Grant No. CHE-1152752 (IKL).

### References

1. Xu M, Ermolenkov V, Uversky V et al (2008) Hen egg white lysozyme fibrillation: a deep-UV resonance Raman spectroscopic study. J Biophotonics 1(3):215–229
2. Lednev I, Shashilov V, Xu M (2009) Ultraviolet Raman spectroscopy is uniquely suitable for studying amyloid diseases. Curr Sci 97(2): 180–185
3. Kurouski D, Lauro W, Lednev I (2010) Amyloid fibrils are "alive": spontaneous refolding from one polymorph to another. Chem Commun 46(24):4249–4251. doi:10.1039/b926758a
4. Popova L, Kodali R, Wetzel R et al (2010) Structural variations in the cross-beta core of amyloid beta fibrils revealed by deep UV resonance Raman spectroscopy. J Am Chem Soc 132(18):6324–6328
5. Shashilov V, Xu M, Ermolenkov V et al (2007) Probing a fibrillation nucleus directly by deep ultraviolet Raman spectroscopy. J Am Chem Soc 129(22):6972–6973
6. Shashilov V, Lednev I (2008) 2D correlation deep UV resonance Raman spectroscopy of early events of lysozyme fibrillation: kinetic mechanism and potential interpretation pitfalls. J Am Chem Soc 130(1):309–317
7. Shashilov V, Lednev I (2009) Two-dimensional correlation Raman spectroscopy for characterizing protein structure and dynamics. J Raman Spectrosc 40(12):1749–1758
8. Shashilov V, Lednev I (2010) Advanced statistical and numerical methods for spectroscopic characterization of protein structural evolution. Chem Rev 110(10):5692–5713
9. Shashilov V, Sikirzhytski V, Popova L et al (2010) Quantitative methods for structural characterization of proteins based on deep UV resonance Raman spectroscopy. Methods 52(1):23–37
10. Sikirzhytski V, Topilina N, Higashiya S et al (2008) Genetic engineering combined with deep UV resonance Raman spectroscopy for

structural characterization of amyloid-like fibrils. J Am Chem Soc 130(18):5852–5853
11. Xu M, Shashilov V, Lednev I (2007) Probing the cross-beta core structure of amyloid fibrils by hydrogen-deuterium exchange deep ultraviolet resonance Raman spectroscopy. J Am Chem Soc 129(36):11002–11003. doi:10.1021/ja073798w
12. Brereton R (2003) Chemometrics: data analysis for the laboratory and chemical plant. Wiley, Chinchester (England)
13. Hyvärinen A, Oja E (2000) Independent component analysis: algorithms and applications. Neural Netw 13(4-5):411–430. doi:10.1016/s0893-6080(00)00026-5
14. Hyvärinen A, Karhunen J, Oja E (2002) What is independent component analysis? Wiley, New York, pp 145–164
15. Shao X, Wang G, Wang S et al (2004) Extraction of mass spectra and chromatographic profiles from overlapping GC/MS signal with background. Anal Chem 76(17): 5143–5148. doi:10.1021/ac035521u
16. Cichocki A, Amari S (2005) Adaptive blind signal and image processing: learning algorithms and applications. Reprinted with corrections. Wiley, New York
17. Bell A, Sejnowski T (1995) An information-maximization approach to blind separation and blind deconvolution. Neural Comput 7(6): 1129–1159. doi:10.1162/neco.1995.7.6.1129
18. Oladepo S, Xiong K, Hong Z et al (2012) UV resonance Raman investigations of peptide and protein structure and dynamics. Chem Rev 112(5):2604–2628. doi:10.1021/cr200198a
19. Huang C, Balakrishnan G, Spiro T (2006) Protein secondary structure from deep-UV resonance Raman spectroscopy. J Raman Spectrosc 37(1–3):277–282. doi:10.1002/jrs.1440
20. Chi Z, Asher S (1998) UV Raman determination of the environment and solvent exposure of Tyr and Trp residues. J Phys Chem B 102(47):9595–9602. doi:10.1021/jp9828336
21. Balakrishnan G, Hu Y, Oyerinde O et al (2007) A conformational switch to beta-sheet structure in cytochrome c leads to heme exposure. Implications for cardiolipin peroxidation and apoptosis. J Am Chem Soc 129(3):504–505. doi:10.1021/ja0678727
22. Ray G, Copeland R, Lee C et al (1990) Resonance Raman evidence for low-spin iron(2+) heme a3 in energized cytochrome c oxidase: implications for the inhibition of oxygen reduction. Biochemistry 29:3208–3213
23. Englander S, Sosnick T, Englander J et al (1996) Mechanisms and uses of hydrogen exchange. Curr Opin Struct Biol 6(1):18–23. doi:10.1016/s0959-440x(96)80090-x
24. DeFlores L, Tokmakoff A (2006) Water penetration into protein secondary structure revealed by hydrogen-deuterium exchange two-dimensional infrared spectroscopy. J Am Chem Soc 128(51):16520–16521. doi:10.1021/ja067723o
25. Mikhonin A, Asher S (2005) Uncoupled peptide bond vibrations in alpha-helical and polyproline II conformations of polyalanine peptides. J Phys Chem B 109(7):3047–3052. doi:10.1021/jp0460442
26. Asher S, Ianoul A, Mix G et al (2001) Dihedral psi angle dependence of the amide III vibration: a uniquely sensitive UV resonance Raman secondary structural probe. J Am Chem Soc 123(47):11775–11781
27. Knuth, K. H. (1998) Bayesian source separation and localization. SPIE, 3459:147–158. doi:10.1117/12.323794
28. Noda I, Ozaki Y (2004) Two-dimensional correlation spectroscopy: applications in vibrational and optical spectroscopy. Wiley, Chichester
29. Shanmukh S, Dluhy R (2004) k$\nu$ Correlation analysis. A quantitative two-dimensional IR correlation method for analysis of rate processes with exponential functions. J Phys Chem A 108(26):5625–5634. doi:10.1021/jp049689a
30. Lednev I, Ermolenkov V, He W et al (2005) Deep-UV Raman spectrometer tunable between 193 and 205 nm for structural characterization of proteins. Anal Bioanal Chem 381(2):431–437

# Chapter 7

# Analyzing Tau Aggregation with Electron Microscopy

**Carol J. Huseby and Jeff Kuret**

## Abstract

Conversion of monomeric tau protein into filamentous aggregates is a defining event in the pathogenesis of Alzheimer's disease. To gain insight into disease pathogenesis, the mechanisms that trigger and mediate tau aggregation are under intense investigation. Characterization efforts have relied primarily on recombinant tau protein preparations and high-throughput solution-based detection methods such as thioflavin-dye fluorescence and laser-light-scattering spectroscopies. Transmission electron microscopy (TEM) is a static imaging tool that complements these approaches by detecting individual tau filaments at nanometer resolution. In doing so, it can provide unique insight into the quality, quantity, and composition of synthetic tau filament populations. Here we describe protocols for analysis of tau filament populations by TEM for purposes of dissecting aggregation mechanism.

**Key words** Aggregation, Electron microscopy, Kinetic analysis, Length distribution, Immunogold labeling

## 1 Introduction

Transmission electron microscopy (TEM) is a powerful tool for characterizing tau aggregate quality, quantity, and composition. First, its ability to capture morphology at nanometer resolution allows one to distinguish mature filaments from amorphous aggregation products, and to determine whether they most closely resemble the paired-helical, straight, or hemifilament forms isolated from disease tissue [1–3]. Second, TEM captures length information that can be leveraged to assay filament formation and disaggregation. Although low throughput and only semiquantitative in nature, length measurements become a powerful tool for assessing aggregation mechanism when collected as a function of time or protein concentration and subjected to regression analysis. Fundamental aggregation parameters can be estimated by this approach, including the minimal concentration needed to support aggregation, lag times of nucleation-dependent reactions, and dissociation rates of mature filaments [4]. Moreover, filament length

David Eliezer (ed.), *Protein Amyloid Aggregation: Methods and Protocols*, Methods in Molecular Biology, vol. 1345, DOI 10.1007/978-1-4939-2978-8_7, 

distributions can be leveraged to detect the presence of secondary aggregation pathways and also to provide an independent check on rate constants deduced from time-dependent evolution of filament mass [5]. Finally, when combined with immunogold labeling methods, TEM provides information on aggregate composition. This approach can be used to confirm that filamentous aggregates contain tau protein [6, 7], to determine whether specific tau epitopes are exposed or sequestered in the aggregated state [7], and to clarify whether tau aggregates associate with heterologous proteins [8].

Here we summarize two protocols for TEM analysis of tau aggregates. The basic protocol (*see* Subheading 3.2) details adsorption of tau fibrils onto TEM grids and measurement of filament length. It then summarizes methods for analyzing length data to obtain aggregation parameters such as minimal concentration, aggregation rates, and filament dissociation rates (*see* Subheadings 3.3.1–3.3.4). It also describes measurement of length distributions. The second protocol (*see* Subheading 3.4) describes immunogold labeling of tau filaments for assessment of composition. Previous reviews of electron microscopy methods applied to amyloid aggregates, including assessment of structure by cryo-electron and scanning transmission electron microscopies, can be consulted for additional approaches [9, 10].

## 2 Materials

### 2.1 Tau Filaments

All buffers and reagents are made with ultrapure water (18.2 MΩ cm at 25 °C) and filtered prior to use (pore size ≤0.22 μm).

1. Recombinant tau proteins: These are expressed in *E. coli* and purified by liquid chromatography as described previously [11, 12] (*see* **Note 1**).
2. Aggregation inducer: Thiazine Red (TR; also known as Geranine G, Chemical Abstract Service registry number 2150-33-6) (*see* **Note 2**).
3. Assembly buffer: 10 mM HEPES, pH 7.4, 100 mM NaCl, 1 mM DTT.

### 2.2 Transmission Electron Microscopy

1. Transmission electron microscope: We use a Tecnai G2 Spirit BioTWIN transmission electron microscope (FEI Company, USA) operated at 80 kV acceleration voltage and equipped with digital image capture.
2. 2 % (w/v) Uranyl acetate (UA) solution in water (*see* **Note 3**).
3. Copper grids 300-mesh formvar/carbon-coated (Electron Microscopy Sciences Cat. No. FCF300-CU): These commercial grids are supplied with film laid on the shiny side. They can be used directly without glow discharging.

4. Hydrophobic laboratory film (e.g., Parafilm): Cut into 4 × 4 in. squares for easiest handling.
5. 25 % w/v Glutaraldehyde in water.
6. Fine-tipped forceps for handling grids (110 mm, Structure Probe Inc., West Chester, PA).
7. Cellulose filter paper (e.g., Whatman Qualitative No. 2 filter paper) for blotting off excess liquids from grids: Cut into small squares for easiest manipulation.
8. Grid box (Electron Microscopy Sciences) for storage and transport of grids (*see* **Note 4**).

### 2.3 Image Analysis

1. ImageJ or equivalent image analysis software: ImageJ can be downloaded for free from the website http://imagej.nih.gov/ij/.
2. Microsoft Excel or equivalent spread sheet software for manipulating filament length data.
3. SigmaPlot (Systat Software, Inc., San Jose, CA, USA) or equivalent graphics software for curve fitting and regression analysis.

### 2.4 Immunogold Labeling

This method requires a primary antibody capable of binding recombinant tau proteins with high affinity. Here we illustrate the method using a commercially available rabbit polyclonal antibody raised against the V5 epitope (GKPIPNPLLGLDST). Tau proteins tagged at the N-terminus with the V5 epitope bind strongly to anti-V5 antibodies in both monomeric and polymeric states.

1. Primary V5 antibody, Rabbit Polyclonal (Bethyl Laboratories, Inc. Cat. No. A190-120A, Montgomery, TX).
2. Secondary goat anti-rabbit IgG (H + L), 12 nm gold-conjugated, EM grade (Jackson Immunoresearch Laboratories, Inc. Cat. No. 111-205-144, West Grove, PA).
3. Phosphate-buffered saline (PBS): 137 mM NaCl, 2.7 mM KCl, 10 mM $Na_2HPO_4$, 2 mM $KH_2PO_4$.
4. Blocking buffer: 1 % bovine serum albumin (w/v) in PBS.
5. 96-Well, flat-bottom, low-protein-binding assay plate (e.g., Corning Inc, polystyrene plate #3641).

## 3 Methods

### 3.1 Tau Aggregation

1. Initiate aggregation of tau protein at 37 °C without agitation in assembly buffer containing 100 μM TR.
2. Stop reactions by gently adding glutaraldehyde to 1 % (w/v) final concentration (*see* **Note 5**).

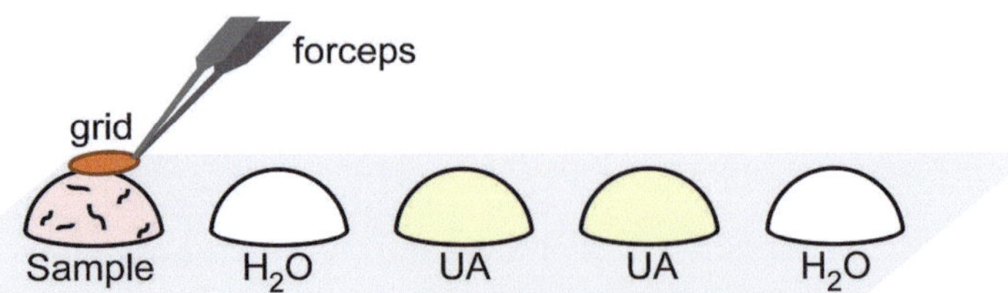


**Fig. 1** Technique for applying tau samples to grids. A sheet of parafilm (shown as *grey* surface) is tacked down to the benchtop by scraping a blunt surface across each corner, creating an immobile hydrophobic surface onto which 50 μL aliquots of aggregated sample, water, and UA are placed. Using microforceps, the grid is sequentially contacted with each solution before drying and storage

### 3.2 Basic Protocol

All steps can be carried out at room temperature (RT).

1. Dispense 50 μL each of the aggregate sample, two drops water, and two drops UA onto a sheet of parafilm (Fig. 1).
2. With tweezers, carefully pick the grid up by the edge, being careful not to damage the grid, and place the grid, shiny-side down, onto the sample drop for 1 min (*see* **Note 6**). Do not completely submerge: surface tension will support the grid while fibrils diffuse and absorb to the grid surface.
3. Again using the tweezers, carefully remove the grid from the sample drop and blot off excess sample by gently touching filter paper with the grid edge.
4. Wash the grid briefly by dipping it, shiny-side down, on top of the first water drop and again carefully blot off excess liquid with filter paper.
5. Rinse briefly in the first UA stain drop, blot off excess stain with filter paper, and place it on the second drop of UA, shiny-side down for 1 min (*see* **Note 7**).
6. Remove grid from the UA, blot with filter paper, wash again on the second water drop, blot with filter paper, and finally leave grid shiny-side up on filter paper to completely dry. Stained grids can be stored at room temperature for weeks.
7. Acquire images on transmission electron microscope, and save images for analysis. Be sure to record the magnification scale of all images (*see* **Note 8**).

### 3.3 Filament Length Measurement

1. Load TEM images into ImageJ (NIH). Tutorials are available at the ImageJ website.
2. Calibrate length scale: In ImageJ, draw a straight line over the TEM calibration scale bar. Under the analyze tab, use "set scale" to define "units" and "known distance" so that they match the scale bar marked on each image (Fig. 2a).

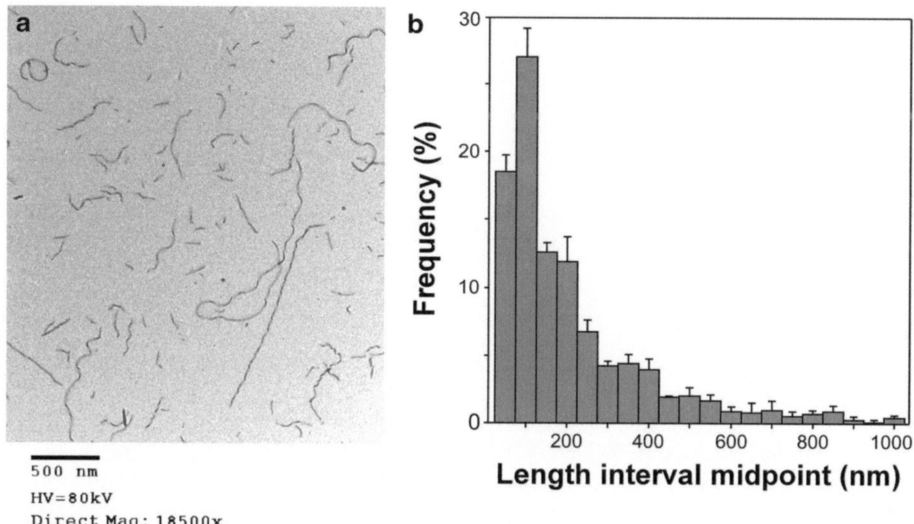


**Fig. 2** Electron micrographs of tau filaments. Recombinant human 2N4R tau was prepared, aggregated (1 μM) in the presence of TR inducer, and subjected to the basic TEM protocol (*see* Subheading 3.2). (**a**) This image, captured at 18,500-fold magnification, shows that filaments are well resolved under these conditions. The lengths of filaments having both ends visible in the field are quantified using ImageJ. (**b**) Length distribution calculated with bin size set to 50 nm, where each bar represents the average of Panel (**a**) and two technical replicates ± S.D. Filaments shorter than 25 nm were not resolved at this magnification and were not included in the distribution

3. Measure only the lengths of filaments where both ends are fully resolved in the field (Fig. 2a). Transfer the length data into Microsoft Excel or similar software for analysis of replicates and statistics.

#### 3.3.1 Filament Length Distribution

1. Measure the lengths of a filament population (*see* Subheading 3.3). Typically, three fields or technical replicates per assay are sufficient for analysis of each assay condition.
2. Choose a bin size for filament lengths, and count the number of filaments per bin. Typically, bin size is chosen so that each contains multiple observations. Often this corresponds to 10–20 bins per distribution (*see* **Note 9**).
3. Plot the average frequency (i.e., the number of filaments per bin divided by total number of filaments in all bins) as a function of bin number (Fig. 2b).

#### 3.3.2 Time Series

1. Measure total filament length/field (*see* Subheading 3.3) as a function of time and constant tau monomer concentration (*see* **Note 10**).
2. Plot time series in SigmaPlot or other graphics program capable of regression analysis.

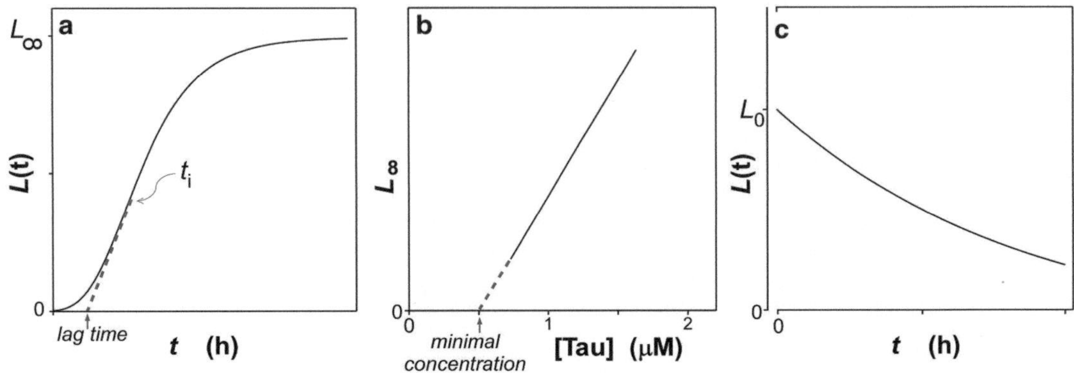


**Fig. 3** Graphical depiction of parameter estimation. (**a**) Hypothetical time series subjected to Gompertz regression, where *L*(*t*) corresponds to total filament length/field measured as a function of time *t* after addition of aggregation inducer. Calculated parameters useful for assessing mechanism include $t_i$, the time of maximal growth rate; $k_{app}$, an estimate of the underlying first-order growth rate constant; and $L_\infty$, an estimate of the maximum total length/field at reaction plateau. Extrapolation from $t_i$ to the abscissa intercept (*dashed line*) yields an estimate of lag time, which for a nucleation-dependent aggregation is inversely proportional to nucleation rate. (**b**) Tau concentration dependence of total filament length/field measured at aggregation plateau ($L_\infty$). Linear regression and extrapolation to the abscissa intercept (*dashed line*) allow quantification of the minimal tau concentration needed to support filament formation. (**c**) Hypothetical dissociation time series modeled as simple exponential decay, where *L*(*t*) corresponds to total filament length/field measured as a function of time *t* after beginning disaggregation by dilution. Exponential regression analysis yields $k_{app}$, the apparent first-order decay rate constant for filament length as a function of incubation time. Conversion of $k_{app}$ to the filament dissociation rate constant, a measure of filament stability, requires knowledge of mass per unit length, the filament length/field at the start of disaggregation ($L_0$), and the number of filament ends at the start of disaggregation

3. Fit to Gompertz [13], logistic [14], or other growth models appropriate for analyzing sigmoidal time series. These models yield estimates of $t_i$, the inflection point corresponding to maximal growth rate, $k_{app}$, an estimate of the underlying first-order growth rate constant, and an estimate of the maximum total length/field at reaction plateau (Fig. 3a).
4. Lag time is when the tangent to the point of maximum growth rate (i.e., $t_i$) intersects the abscissa of the sigmoidal curve [14]. In Gompertz regression, this time corresponds to $t_i - 1/k_{app}$ [13].

#### *3.3.3 Tau Concentration Dependence*

1. Measure total filament length/field (*see* Subheading 3.3) as a function of tau monomer concentration. For this experiment, incubation time is held constant and is greater or equal to the time of reaction plateau (*see* Subheading 3.3.2).
2. Plot total filament length/field versus tau monomer concentration in SigmaPlot or other graphics program capable of regression analysis (Fig. 3b).
3. Perform linear regression. Minimal concentration corresponds to abscissa intercept (*see* **Note 11**).

### 3.3.4 Filament Dissociation Rate

1. Aggregate tau protein (*see* Subheading 3.1) until at least plateau is attained (as determined in Subheading 3.3.2).
2. Dilute the aggregates tenfold with assembly buffer containing TR and continue incubation at 37 °C.
3. Withdraw aliquots as a function of time (up to 5 h post-dilution), treat with glutaraldehyde, and subject to TEM as in Subheading 3.2.
4. Measure total filament length/field (L) and filament number for each sample.
5. Fit to exponential decay function $L = L_0 e^{-k_{app} t}$ to obtain $k_{app}$, the pseudo-first-order rate constant describing the time-dependent decrease in filament length, and $L_0$, the total filament length at time zero (Fig. 3c). Dissociation rate constant $k_{e-}$ is estimated as the initial velocity of the decrease in total filament length ($L_0$ $k_{app}$) multiplied by the number of tau protomers per unit length (*see* **Note 12**) and normalized for the number of filaments measured at time zero [4, 15, 16].

## 3.4 Immunogold Labeling Protocol

1. Dispense 50 μL each of sample and blocking buffer onto a sheet of parafilm.
2. Adsorb the sample to the grid as in **steps 2** and **3** of the basic protocol (*see* Subheading 3.2).
3. With tweezers, carefully pick the grid up by the edge, blot off excess sample with filter paper, and place it shiny-side down onto the drop of blocking buffer. Incubate for 5 min (*see* **Note 13**).
4. Place 50–200 μL of primary antibody diluted in blocking buffer in the wells of a 96-well plate (*see* **Note 14**).
5. Carefully remove grid from the sample, blot off excess blocking buffer with filter paper, and place the grid sample-side down in primary antibody dispensed in the wells of the 96-well plate. Incubate for 4 h at 4 °C with agitation.
6. Carefully remove the grid from the primary antibody, and wash twice by floating on 50 μL drops of blocking buffer interspersed with blotting off of excess liquid.
7. Dispense secondary antibody diluted in blocking buffer into 96-well plate.
8. Carefully place the grid in a well containing gold-labeled secondary antibody. Incubate with agitation for 2 h at 4 °C.
9. After removing the grid from the well of secondary antibody, wash four times by floating on successive drops for 5 min each as follows: 2× PBS and 2× $H_2O$. Excess liquid is blotted off with filter paper between transfers.

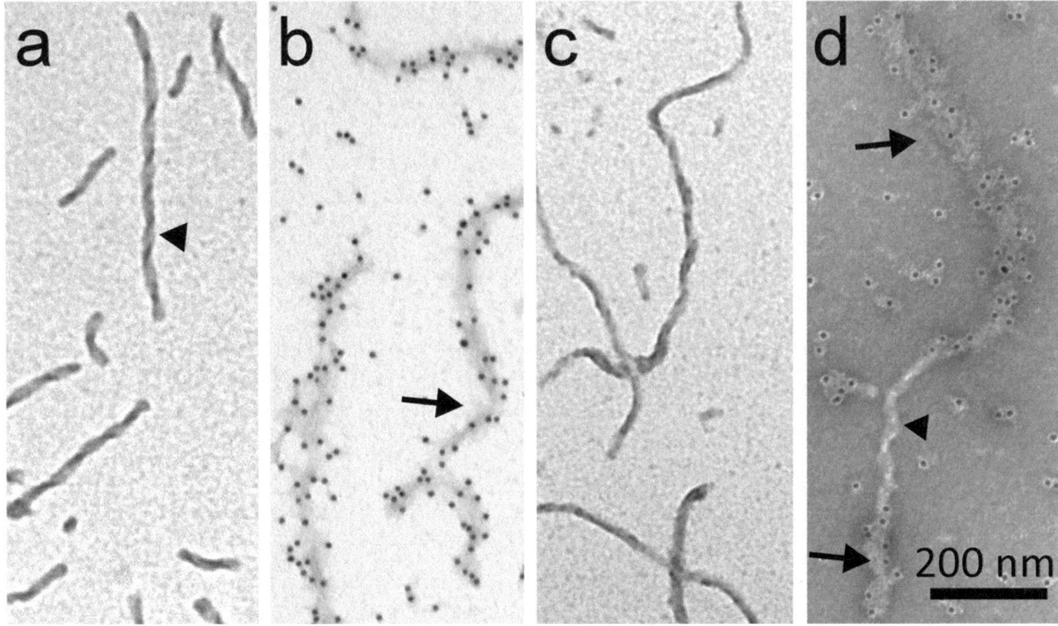


**Fig. 4** Immunogold labeling of epitope-tagged tau filaments. Recombinant human 2N4R tau and V5-tagged 2N4R tau were prepared and aggregated in the presence of TR inducer. (**a**) V5-2N4R filaments were analyzed by basic protocol (*see* Subheading 3.2). The filaments retained the length and morphology characteristics of non-tagged 2N4R tau (*arrowhead* points to one example). (**b**) V5-2N4R filaments stained with polyclonal anti-V5 primary and 12 nm-labeled secondary antibodies using the immunogold labeling protocol (*see* Subheading 3.4). Extensive decoration of filaments with these antibodies obscures filament morphology, but preserves length and provides clear evidence for the presence of V5-labeled tau (*arrow* points to one example). (**c**) 2N4R tau filaments subjected to the same protocol as in Panel (**b**). These filaments lack the V5 epitope, and do not label with the anti-V5/secondary antibody pair. Morphological information is mostly retained. (**d**) 2N4R tau filaments were prepared, mixed with V5-2N4R tau, and then incubated for an additional 16 h. Products were then subjected to the immunogold protocol as in Panels (**b**) and (**c**). This image was captured in negative stain. It shows a filament containing a well-resolved, unlabeled central segment corresponding to 2N4R tau (*arrowhead*) flanked by extensively gold-labeled ends (*arrows*) composed at least in part of V5-2N4R tau. These data indicate that filament extension in the presence of TR inducer proceeds from both filament ends, and illustrate the utility of gold labeling for detecting tau composition in a filament

10. Samples are stained with UA and imaged as described in **steps 5** and **6** of the basic protocol (*see* Subheading 3.2).
11. Subject to TEM, and capture images. An example of immunogold labeling of tau filaments is shown in Fig. 4.

## 4 Notes

1. This purification method has been applied to different tau isoforms [4], missense mutants [17], and posttranslational modifications [16, 18]. Additional purification steps, such as boiling, can be included without changing aggregation or TEM assay performance [19]. Purified tau is aliquoted and stored at −80 °C.

2. Spontaneous fibrillization of tau protein at physiological concentrations in vitro does not occur over experimentally tractable time periods [20]. However, tau aggregation can be accelerated by addition of anionic inducers, such as heparin [7], anionic surfactants including arachidonic acid or alkyl sulfate detergents [21, 22], and small-molecule dyes, such as thiazine red (TR) [23]. This protocol leverages TR as tau aggregation inducer.
3. UA is light sensitive and toxic. The solid material should be stored under the hood in a dark metal container protected from moisture. Care should be taken to avoid contact with and exposure to the material and subsequent solutions containing UA. Used UA and lab disposables that come in contact with UA are disposed of per institutional guidelines and appropriate Materials Safety Data Sheet.
4. Secure the lid of the grid box with a small piece of labeling tape while transporting grids to prevent the inadvertent opening of the box and losing and/or mixing of the grids.
5. Glutaraldehyde is toxic and care should be taken not to breathe the fumes or allow contact with skin. Under the hood with gloves, add a drop of glutaraldehyde to the inside wall of the tube just above the sample. Carefully allow the drop to mix with the sample by gentle flicking of the tube. Aggressive mixing or vortexing can cause clumping of tau filaments.
6. Filament adsorption is time dependent [24]. Therefore, it is important to accurately maintain constant adsorption time for all samples.
7. UA most frequently interacts with tau filaments to generate a positive staining effect (Fig. 2a), where filaments appear dark against a light background. This staining is adequate for quantification of filament length. Less frequently, UA fosters negative staining in certain areas of the grid (Fig. 4d). Negative stain is especially valuable for interrogating filament morphology.
8. Three or more random images from each grid are captured typically at 8000- to 100,000-fold magnification. High magnification better captures morphological features, whereas lower magnification is necessary for quantification of filament length. A typical magnification for quantification of tau filament lengths in the presence of TR ranges from 18,000- to 35,000-fold.
9. Measurement of lengths <20 nm is difficult at lower magnifications (e.g., Fig. 2a). As a result, frequency measurements can be biased toward higher relative occupancy of longer lengths. In addition, all length measurements will underestimate total filament length/field. However, the error is predicted to be modest when tau aggregation proceeds under nucleation-dependent conditions owing to low occupancy of short-length bins [5].

Indeed, aggregation characteristics are similar when quantified by TEM methods or laser light scattering, a solution-based technique [13].

10. Because inducer-mediated tau aggregation follows a nucleation-elongation mechanism [25], the dependence of total filament length/field on incubation time at constant tau monomer concentration is especially informative. In particular, measurement of lag time is valuable because this parameter is proportional to nucleation rate [26]. Nonetheless, care must be taken when interpreting lag times because they can reflect factors other than nucleation under certain conditions [27, 28].
11. The dependence of total filament length/field on tau concentration is linear and frequently intercepts the abscissa at positive values of tau concentration. Therefore, inverse prediction methods are used to solve for the intercept [4]. This parameter represents the minimal concentration of tau monomer needed to support aggregation. Because tau filament adsorption to grids is sensitive to tau concentration [24], it is important to employ tau concentrations that vary no more than two- to threefold above minimal concentration for this measurement. Extrapolating over large tau concentrations can yield nonlinear plots.
12. Calculation of dissociation rates requires knowledge of filament mass per unit length, so that changes in length can be related to changes in tau protomer number. Mass per unit length measurements have been reported for synthetic human TR- [5], arachidonic acid- [29], and heparin-induced [30] tau filaments (though for only a limited number of isoforms and filament morphologies).
13. When blocking and washing samples on the benchtop, it is prudent to prevent dust particles in the air from contacting the drops. Covering the drops with a lid from a 96-well plate works well for this purpose.
14. Volumes of at least 100 μL of antibody solution per well are preferred, because smaller volumes complicate the placement and removal of grids from 96-well plates.

## Acknowledgment

This work was supported by NIH grant AG14452. Images presented in this report were generated using an instrument at the Campus Microscopy and Imaging Facility, The Ohio State University, Columbus, OH.

## References


1. Crowther RA, Olesen OF, Jakes R, Goedert M (1992) The microtubule binding repeats of tau protein assemble into filaments like those found in Alzheimer's disease. FEBS Lett 309: 199–202
2. Barghorn S, Davies P, Mandelkow E (2004) Tau paired helical filaments from Alzheimer's disease brain and assembled in vitro are based on beta-structure in the core domain. Biochemistry 43:1694–1703
3. King ME, Ahuja V, Binder LI, Kuret J (1999) Ligand-dependent tau filament formation: implications for Alzheimer's disease progression. Biochemistry 38:14851–14859
4. Zhong Q, Congdon EE, Nagaraja HN, Kuret J (2012) Tau isoform composition influences rate and extent of filament formation. J Biol Chem 287:20711–20719
5. Congdon EE, Kim S, Bonchak J, Songrug T, Matzavinos A, Kuret J (2008) Nucleation-dependent tau filament formation: the importance of dimerization and an estimation of elementary rate constants. J Biol Chem 283: 13806–13816
6. Montejo de Garcini E, Avila J (1987) In vitro conditions for the self-polymerization of the microtubule-associated protein, tau factor. J Biochem 102:1415–1421
7. Goedert M, Jakes R, Spillantini MG, Hasegawa M, Smith MJ, Crowther RA (1996) Assembly of microtubule-associated protein tau into Alzheimer-like filaments induced by sulphated glycosaminoglycans. Nature 383:550–553
8. Giasson BI, Forman MS, Higuchi M, Golbe LI, Graves CL, Kotzbauer PT, Trojanowski JQ, Lee VM (2003) Initiation and synergistic fibrillization of tau and α-synuclein. Science 300: 636–640
9. Gras SL, Waddington LJ, Goldie KN (2011) Transmission electron microscopy of amyloid fibrils. Methods Mol Biol 752:197–214
10. Sousa AA, Leapman RD (2013) Mass mapping of amyloid fibrils in the electron microscope using STEM imaging. Methods Mol Biol 950: 195–207
11. Carmel G, Mager EM, Binder LI, Kuret J (1996) The structural basis of monoclonal antibody Alz50's selectivity for Alzheimer's disease pathology. J Biol Chem 271:32789–32795
12. Carmel G, Leichus B, Cheng X, Patterson SD, Mirza U, Chait BT, Kuret J (1994) Expression, purification, crystallization, and preliminary x-ray analysis of casein kinase-1 from Schizosaccharomyces pombe. J Biol Chem 269: 7304–7309
13. Necula M, Kuret J (2004) A static laser light scattering assay for surfactant-induced tau fibrillization. Anal Biochem 333:205–215
14. Nielsen L, Khurana R, Coats A, Frokjaer S, Brange J, Vyas S, Uversky VN, Fink AL (2001) Effect of environmental factors on the kinetics of insulin fibril formation: elucidation of the molecular mechanism. Biochemistry 40:6036–6046
15. Kristofferson D, Karr TL, Purich DL (1980) Dynamics of linear protein polymer disassembly. J Biol Chem 255:8567–8572
16. Necula M, Kuret J (2005) Site-specific pseudophosphorylation modulates the rate of tau filament dissociation. FEBS Lett 579:1453–1457
17. Chang E, Kim S, Yin H, Nagaraja HN, Kuret J (2008) Pathogenic missense *MAPT* mutations differentially modulate tau aggregation propensity at nucleation and extension steps. J Neurochem 107:1113–1123
18. Funk KE, Thomas SN, Schafer KN, Cooper GL, Liao Z, Clark DJ, Yang AJ, Kuret J (2014) Lysine methylation is an endogenous post-translational modification of tau protein in human brain and a modulator of aggregation propensity. Biochem J 462:77–88
19. Schafer KN, Cisek K, Huseby CJ, Chang E, Kuret J (2013) Structural determinants of Tau aggregation inhibitor potency. J Biol Chem 288:32599–32611
20. Jarrett JT, Lansbury PT Jr (1993) Seeding "one-dimensional crystallization" of amyloid: a pathogenic mechanism in Alzheimer's disease and scrapie? Cell 73:1055–1058
21. Chang E, Congdon EE, Honson NS, Duff KE, Kuret J (2009) Structure-activity relationship of cyanine tau aggregation inhibitors. J Med Chem 52:3539–3547
22. Chirita CN, Necula M, Kuret J (2003) Anionic micelles and vesicles induce tau fibrillization in vitro. J Biol Chem 278:25644–25650
23. Chirita CN, Congdon EE, Yin H, Kuret J (2005) Triggers of full-length tau aggregation: a role for partially folded intermediates. Biochemistry 44:5862–5872
24. Necula M, Kuret J (2004) Electron microscopy as a quantitative method for investigating tau fibrillization. Anal Biochem 329: 238–246

25. Friedhoff P, von Bergen M, Mandelkow EM, Davies P, Mandelkow E (1998) A nucleated assembly mechanism of Alzheimer paired helical filaments. Proc Natl Acad Sci U S A 95:15712–15717
26. Zhao D, Moore JS (2003) Nucleation-elongation: a mechanism for cooperative supramolecular polymerization. Org Biomol Chem 1:3471–3491
27. Pappu RV, Wang X, Vitalis A, Crick SL (2008) A polymer physics perspective on driving forces and mechanisms for protein aggregation. Arch Biochem Biophys 469:132–141
28. Ferrone F (1999) Analysis of protein aggregation kinetics. Methods Enzymol 309:256–274
29. King ME, Ghoshal N, Wall JS, Binder LI, Ksiezak-Reding H (2001) Structural analysis of Pick's disease-derived and in vitro-assembled tau filaments. Am J Pathol 158:1481–1490
30. von Bergen M, Barghorn S, Muller SA, Pickhardt M, Biernat J, Mandelkow EM, Davies P, Aebi U, Mandelkow E (2006) The core of tau-paired helical filaments studied by scanning transmission electron microscopy and limited proteolysis. Biochemistry 45: 6446–6457

# Part III

## Oligomers

# Chapter 8

# Characterization of Amyloid Oligomers by Electrospray Ionization-Ion Mobility Spectrometry-Mass Spectrometry (ESI-IMS-MS)

**Charlotte A. Scarff, Alison E. Ashcroft, and Sheena E. Radford**

## Abstract

Soluble oligomers formed during the self-assembly of amyloidogenic peptide and protein species are generally thought to be highly toxic. Consequently, thorough characterization of these species is of much interest in the quest for effective therapeutics and for an enhanced understanding of amyloid fibrillation pathways. The structural characterization of oligomeric species, however, is challenging as they are often transiently and lowly populated, and highly heterogeneous. Electrospray ionization-ion mobility spectrometry-mass spectrometry (ESI-IMS-MS) is a powerful technique which is able to detect individual ion species populated within a complex heterogeneous mixture and characterize them in terms of shape, stoichiometry, ligand binding capability, and relative stability. Herein, we describe the use of ESI-IMS-MS to characterize the size and shape of oligomers of beta-2-microglobulin through use of data calibration and the derivation of models. This enables information about the range of oligomeric species populated *en route* to amyloid formation and the mode of oligomer growth to be obtained.

**Key words** Protein aggregation, Amyloid, Oligomerization, Native mass spectrometry, Ion mobility spectrometry-mass spectrometry

## 1 Introduction

The identification and characterization of oligomers populated *en route* to amyloid fibril formation is a major challenge. In the early stages of protein aggregation multiple, rapidly converting, transient and lowly populated species are co-populated in solution, so the detection and characterization of individual species is extremely difficult [1]. Mass spectrometry (MS) is one technique which lends itself to the study of such heterogeneous mixtures as it enables the detection of multiple ions within the same sample, at femtomolar concentrations, and their identification based on their mass-to-charge ratios ($m/z$). Nano-electrospray ionization (nESI)-MS allows for the analysis of noncovalently bound species and there is good evidence to support the view that complexes observed in the

David Eliezer (ed.), *Protein Amyloid Aggregation: Methods and Protocols*, Methods in Molecular Biology, vol. 1345, DOI 10.1007/978-1-4939-2978-8_8, 

gas phase are reflective of species populated in solution [2–4]. Ion mobility spectrometry (IMS)-MS, which separates ions based on their mobility through an inert gas under the influence of a weak electric field, allows ion species of the same *m/z* but different shapes to be separated such that the presence of different protein conformational states and oligomeric species can be distinguished confidently [5, 6]. Oligomer formation can be followed over time and any changes in oligomer distribution or protein conformation can be monitored [7]. Potential small molecule inhibitors can also be added to the protein and binding to specific protein conformers or oligomers detected [8, 9]. Ion mobility measurements obtained on the Synapt HDMS traveling-wave ESI-IMS-MS instrument [10], as used herein, can be converted by use of a suitable calibration to estimate rotationally averaged collision cross sections (CCSs) for individual ion species [11–13]. These can then be compared with modeled structures of monomers and/or oligomers and insights into the mechanism of oligomeric growth obtained [14]. Traditional drift tube IMS-MS instruments can also be used for analyses of this type and measurements obtained on these instruments can be directly converted to CCSs [15]. ESI-IMS-MS can also be used to study the relative stability of individual ion species by accelerating these species through the instrument under different voltages and recording their unfolding and dissociation profiles [16]. Additionally, subunit dynamics can be studied by mixing isotopically labeled species with nonlabeled species (e.g., $^{15}$N and $^{14}$N labeled proteins) and analyzing the rate of subunit exchange in real time [5, 17].

Here ESI-IMS-MS is applied to the study of oligomers of a variant of beta-2-microglobulin (β2m) named ΔN6, a truncated variant without the first six N-terminal residues, which undergoes aggregation at neutral pH into amyloid fibrils [18, 19]. β2m is the causative agent of dialysis-related amyloidosis and both wild-type b2m and ΔN6 have been found in amyloid plaques [20]. Ion mobility measurements obtained are calibrated by use of protein calibrants (with known CCSs) to produce estimated CCSs for each ion species. Estimated CCSs are compared with those obtained from model structures to allow for an understanding of the mechanism of oligomer growth.

## 2 Materials

### 2.1 Samples for ESI-IMS-MS Analysis

1. Caesium iodide (CsI).
2. Beta-2-microglobulin variant ΔN6.
3. Cytochrome c.

4. Bovine serum albumin (BSA).
5. Concanavalin A.
6. Alcohol dehydrogenase (ADH).

### 2.2 Solvents, Chemicals, and Buffers for ESI-IMS-MS Analysis

1. Desalting columns or buffer-exchange devices.
2. Gold/palladium-coated nanoflow needles.
3. GELoader tips (Eppendorf).
4. Buffer A: 50 mM ammonium bicarbonate, 120 mM ammonium acetate, pH 6.2.
5. CsI solution: Dissolve 2 mg/mL CsI in a 50 % (v/v) water/isopropanol mixture.

### 2.3 Mass Spectrometry Instrumentation and Data Acquisition, Analysis, and Interpretation

1. Synapt HDMS instrument equipped with a nanoflow ESI source and needle holder (Waters Corporation).
2. Borosilicate glass capillaries.
3. Micropipette puller (Model P-97, Sutter Instrument Co.).
4. Sputter coater with a gold/palladium target (Emitech Sc7620).
5. MassLynx 4.1with Driftscope (Waters Corporation).
6. MOBCAL software (http://www.indiana.edu/~nano/software.html).
7. FORTRAN compiler and editor.
8. TextPad.

## 3 Methods

### 3.1 Sample Preparation

For the study of protein oligomers by ESI-MS, solution conditions must be found not only in which the protein undergoes the aggregation process of interest on a suitable time scale but also that are MS-compatible (*see* **Note 1**). Most buffers used for in vitro biochemical experiments, such as Tris–HCl and MOPS, are incompatible with ESI-MS analysis as they are largely nonvolatile, resulting in suppression of ionization and/or extensive adduct formation [21]. Proteins purified or stored in these types of buffers can be buffer-exchanged to allow for ESI-MS analysis by use of buffer-exchange devices or desalting columns (*see* **Note 2**). Adequate removal of nonvolatile buffer components is often the most critical parameter governing spectral quality and so buffer exchange must be stringent. Typically, proteins are buffer-exchanged into aqueous volatile buffer solutions such as ammonium acetate, ammonium formate, or ammonium bicarbonate solution. The choice of buffer will be dependent on the protein and aggregation properties under study and the desired ionic strength and pH (*see* **Notes 3** and **4**). If buffer additives such as metal ions, cofactors, or reducing agents are required to maintain protein stability or to maintain aggregation properties, these can be added up to approximately 1 mM concentration without significantly influencing spectral quality [21].

For the study of aggregation of ΔN6 at pH 6.2, the following sample preparation procedure is undertaken:

1. Lyophilized ΔN6 (1–2 mg) is resuspended in buffer A to a final volume of 1 mL.
2. 100 μL is desalted by use of a 7 kDa MWCO Spin Desalting Column (Zeba™, ThermoFisher Scientific) equilibrated with buffer A.
3. The concentration of the protein solution is determined by measurement of the absorbance at 280 nm using a molar extinction coefficient 20,065 $M^{-1}$ $cm^{-1}$. The sample is diluted with buffer A to a working concentration of 40 μM.
4. Protein calibrants, cytochrome c, BSA, concanavalin A, and ADH are prepared in 200 mM ammonium acetate solution to a working concentration of 10–20 μM following desalting by use of Spin Desalting Columns.
5. CsI solution is prepared freshly for mass calibration (*see* **Note 5**).

### 3.2 Acquisition of ESI-IMS-MS Data of ΔN6 Oligomeric Species

ESI-IMS-MS analysis is performed on a Synapt HDMS instrument, which has a quadrupole/traveling-wave ion mobility (TWIM)/orthogonal time-of-flight geometry. The instrument is equipped with a nESI source and a 32,000 *m/z* range RF generator. nESI allows for the analysis of small sample volumes, enhanced desolvation efficiency of protein molecules from droplets, improved sensitivity, and increased tolerance to buffer salts in comparison to conventional ESI [21].

Samples are introduced into the instrument by the use of in-house prepared capillaries. These are borosilicate glass capillaries with a tapered edge coated with a mixture of gold and palladium. Capillaries were prepared by use of a micropipette puller and coated using a sputter coater. The puller must be programmed to produce capillary tips with acceptable shapes and this is a trial-and-error process [21]. The diameter of the capillaries is critical to be able to obtain a stable spray and requires optimization. Capillary tips often need to be trimmed to obtain the ideal length and orifice diameter for spraying. This is sample dependent and again a trial-and-error process. Capillaries can also be purchased from various sources, such as Waters Corporation and Proxeon Biosystems, but may not yield the same spray properties as those prepared in-house.

1. CsI solution is used for tuning and mass calibration of the instrument.

**Table 1**
**Instrument parameters of importance for protein oligomer analysis**

| | |
|---|---|
| Backing pressure | 3–7 mbar |
| Capillary voltage | 0.8–1.7 kV |
| Sample cone | 20–170 V, above approximately 60 V monomeric protein may be activated and start to unfold yet higher-order oligomers may only be observed at higher cone voltage |
| Extraction cone | 0–10 V, again if this is too high monomer may be activated but at low values oligomeric species may not be transferred into the instrument effectively |
| Trap collision energy | 6–40 V, higher values will improve mass accuracy but may induce gas-phase unfolding/dissociation |
| Transfer collision energy | 4–40 V, at higher voltages dissociation/unfolding may occur but transmission of higher-order species may be improved |
| Trap/transfer pressure | 0.01–0.05 mbar (Ar) |
| Ion mobility pressure | 0.5 mbar ($N_2$) |
| Ion mobility mode | |
| Trap DC entrance | 3.0 V |
| Trap DC bias | 16–24 V, protein dependent |
| Tri-wave IMS | |
| Wave height | 5–25 V |
| Wave velocity | 200–400 m/s |
| Quadrupole MS profile | This should be optimized for the desired mass range |

2. ΔN6 is analyzed under various instrument parameters and optimal instrument conditions for analysis obtained (discussed in further detail below).
3. Data are acquired under optimized conditions at three different wave heights to allow for CCS calibration (discussed in further detail below).

### 3.3 Optimization of Instrument Parameters for the Acquisition of Spectra of ΔN6 Oligomeric Species

The optimal parameters for the study of oligomeric species may not be the same as for the study of monomeric species and so data often need to be obtained under various instrument parameters and compared and contrasted. Instrument parameters of particular importance to data acquisition are listed in Table 1 and discussed in more detail below. Further discussion of instrument parameters for the study of noncovalent complexes in general can be found elsewhere [11, 21].

A key parameter to optimize for the efficient transfer of noncovalent oligomers into the mass spectrometer is the backing pressure in the source region of the instrument. Increasing the backing pressure results in collisional cooling of the ions and therefore aids in the transmission of ions with high *m/z* [22]. The backing pressure can be altered by partially closing the Speedivalve (isolation valve) for the roughing pump and therefore changing the conductance of the source vacuum line [21].

Another important parameter that requires optimization is the cone voltage. Figure 1 shows spectra of ΔN6 obtained under two different cone voltages (Fig. 1a, b). At a higher cone voltage (170 V), spectral quality is improved significantly due to increased desolvation of protein species, resulting in an increased signal-to-noise ratio and increased mass accuracy, thus allowing for confident assignments of ion species based on their *m/z*. However, use of a higher cone voltage can lead to gas-phase unfolding (as determined by ion mobility) and charge stripping of some ion species (Fig. 1c–e). Figure 1c, d shows driftscope plots corresponding to the spectra shown in Fig. 1a, b respectively and Fig. 1e shows extracted arrival time distributions (ATDs) for ions with *m/z* 2784. The driftscope plots show drift time on the *x* axis, *m/z* on the *y* axis, and relative ion intensity on the *z* axis. Gas-phase activation and unfolding of the ion species monomer (4+), dimer (8+), and trimer (12+) occurs at the higher cone voltage (170 V) that is not observed at the lower cone voltage (30 V) (Fig. 1c, d, inset dashed box and Fig. 1e) yet the dimeric ion species (5+) at the higher *m/z* 4455 is unaffected by this increased voltage (Fig. 1c, d, inset dotted box). The instrument parameters that yield the highest quality mass spectra in terms of signal-to-noise and mass accuracy are therefore not those that best preserve solution-phase properties. Acquisition of data under both of these contrasting instrument conditions, however, allows peaks to be accurately assigned and ion mobility measurements to be performed on gas-phase structures that are most likely to be reflective of those present in solution.

One of the advantages of the use of ion mobility coupled with MS is that multiple oligomeric species present at the same *m/z* value can be separated, identified, and characterized in terms of their shape (as illustrated in Fig. 1e). Ion mobility separation depends on mass, charge, and shape of an ion species. Ions with a more compact structure will experience fewer collisions with the buffer gas in comparison to ions with a more extended structure and thus will exit the ion mobility cell faster. Ions with higher charges will also traverse the drift cell faster so higher-order oligomeric species will normally have shorter drift times than lower-order oligomeric species present at the same *m/z*. This, however, is not always the case and so carbon isotope distributions and the

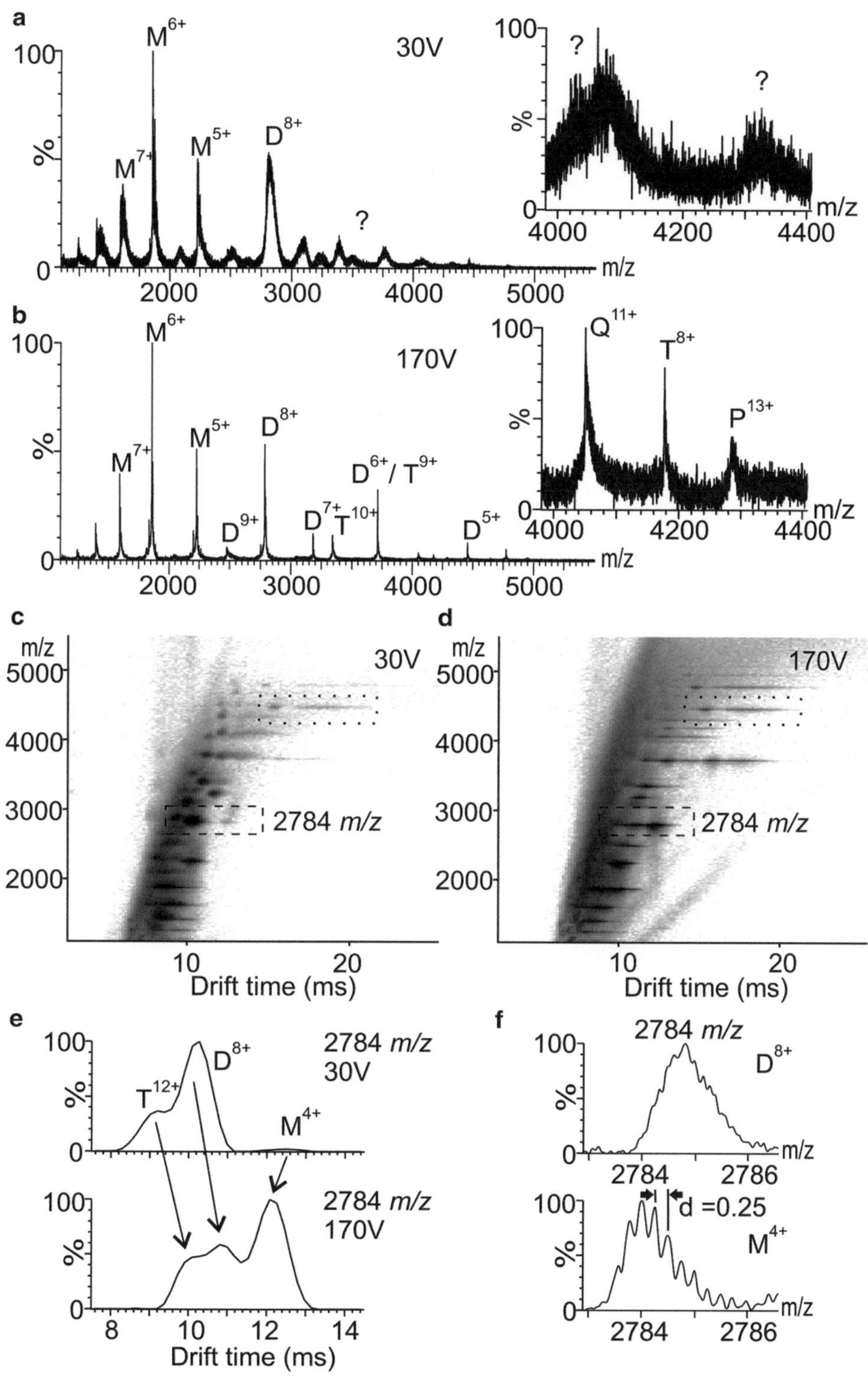


**Fig. 1** Mass spectra of ΔN6 (40 μM) in 50 mM ammonium bicarbonate 120 mM ammonium acetate pH 6.2. Obtained at a cone voltage of (**a**) 30 V and (**b**) 170 V with inset *m/z* region 3950–4400 magnified, M = monomer, D = dimer, T = trimer, Q = tetramer, P = pentamer; (**c**) and (**d**) corresponding ESI-IMS-MS driftscope plots respectively with *m/z* 2784 highlighted in *dashed boxes*; (**e**) extracted arrival time distributions for *m/z* 2784 at 30 and 170 V cone voltage; (**f**) mass spectra corresponding to $M^{4+}$ and $D^{8+}$ extracted from arrival time distribution peaks labeled in (**e**)

*m/z* spacing of sodiated peaks can aid in the determination of oligomer number from arrival time distributions. For *m/z* 2784, three peaks are observed in the ATD (170 V) (shown in Fig. 1e) centered approximately at drift times of 10 ms, 11 ms, and 12 ms respectively. Extraction of the mass spectrum for the last peak (centered at 12 ms) results in an *m/z* 2784 species with a carbon isotope spacing of 0.25, indicating that this ion possesses four charges and hence has a molecular mass of approximately 11.1 kDa and therefore corresponds to a monomeric species (i.e., $M^{4+}$) (Fig. 1f, lower panel). Extraction of the mass spectrum for the peak centered at 11 ms indicates that this ion is more highly charged as the isotope spacing is not discernible ($D^{8+}$) (Fig. 1f, upper panel). It is also interesting to note that within the mass spectrum the centroid mass shifts to the right between the monomeric and dimeric ion (Fig. 1f). This is because for the monomeric ion species the monoisotopic peak is visible but for the dimer the monoisotopic peak is no longer observable as the percentage of species that contain only $C^{12}$ atoms is negligible.

Optimization of ion mobility parameters is of great importance to obtain good separation of ion species. Three traveling-wave ion guides (TWIGs), the trap, ion mobility cell, and transfer region, form the TWIM device [10, 23]. The trap periodically gates a packet of ions into the ion mobility cell. This packet of ions is then separated on account of the different mobilities of the ions, with the time it takes each ion to traverse the mobility cell and reach the TOF pusher recorded. For each gated pulse, 200 orthogonal acceleration pushes of the TOF pusher are recorded to form one ion mobility experiment. The drift time of an ion species is therefore proportional to the pusher frequency, which is dependent on *m/z* acquisition range. Figure 2 shows the total ion chromatograms obtained upon analysis of ΔN6 under three different sets of ion mobility conditions over the 500–8000 *m/z* mass range. At 300 m/s wave velocity (WV) and 7 V wave height (WH) little separation of ion species is obtained. Much better separation of ion species is obtained by use of a ramped WH from 5 to 15 V. At 400 m/s WV, 7 V WH not all ion species have exited the ion mobility cell and reached the TOF pusher before the next set of ions have been pulsed into the ion mobility cell; this is termed rollover. Rollover is evident when the total ion chromatogram does not reach baseline intensity within one ion mobility experiment (0–25.5 ms in this example). Accurate drift times for individual ion species cannot be determined unless all ions are contained within the same mobility pulse.

To perform the optimization the following steps are undertaken:

1. Approximately 10 μL of sample solution is loaded into a capillary using a GELoader tip and the end of the capillary is positioned perpendicular to the MS sample cone.

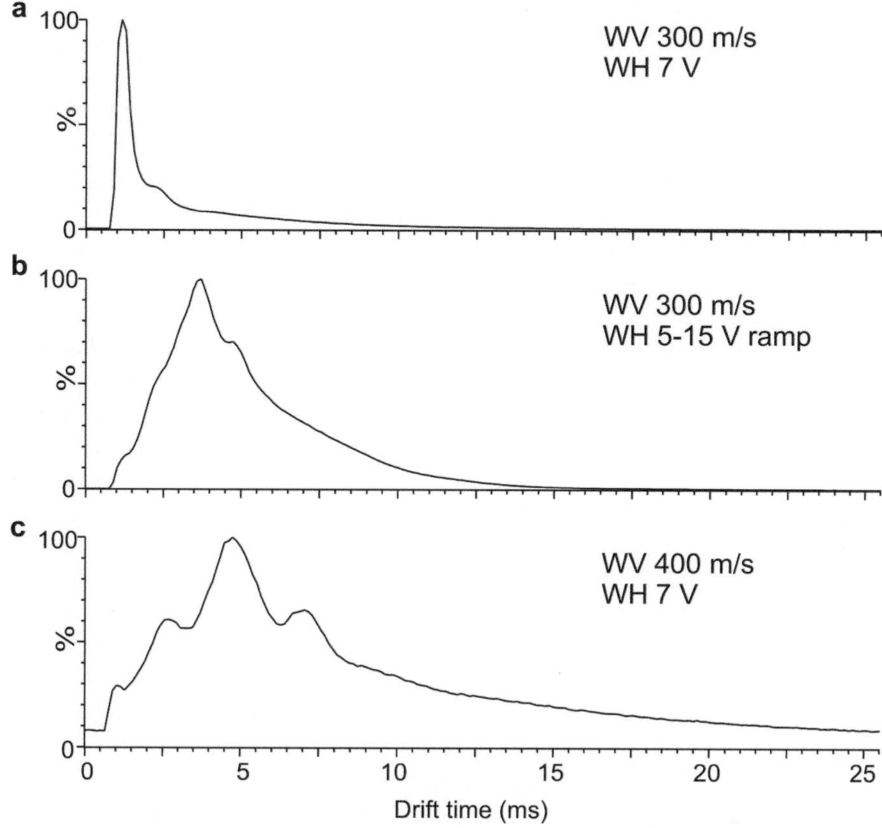


**Fig. 2** Total ion chromatograms obtained upon the ESI-IMS-MS analysis of **Δ**N6 under different ion mobility conditions. (**a**) Wave velocity (WV) 300 m/s wave height (WH) 7 V; (**b**) WV 300 m/s ramped WH 5–15 V; (**c**) WV 400 m/s WH 7 V

2. The position of the capillary relative to the sampling cone and the length of the capillary tip are optimized. A stable spray from the nanoflow capillary is obtained at the lowest capillary voltage possible (*see* **Note 6**).
3. The backing pressure of the mass spectrometer is increased in 0.5 mBar steps until the largest population of higher-order oligomeric species can be seen (this is conducted at both high and low cone voltages so influence of cone voltage on the quality of the resulting spectra can be considered).
4. Extraction cone, trap collision energy, and transfer collision energy are optimized at low cone voltage to achieve the greatest ion transmission without inducing gas-phase unfolding or dissociation.
5. Ion mobility parameters are optimized to provide the greatest separation of ion species without rollover by use of a ramped wave height (Fig. 2, 5–15 V) and spectra are recorded at both low and high cone voltages (30–170 V).

6. The wave height is changed from a ramped to a fixed value and spectra acquired under low and high cone voltage conditions at three fixed wave heights that provided ion mobility separation without rollover.

### 3.4 ESI-IMS-MS Data Calibration

Drift time measurements obtained from ESI-IMS-MS analysis can be used to provide estimates of the CCS of an ion species, which can then be compared with values obtained from atomic resolution structures or estimated values from model structures [12, 14]. On the Synapt ESI-IMS-MS instrument, a direct conversion between drift time and CCS is not possible and so CCS measurements may be estimated by use of a calibration obtained by analysis of protein standards with known CCSs obtained from drift tube ion mobility measurements.

Several choices of calibrants are available and the most appropriate for each analyte of interest must be chosen. Clemmer provides a database of cross-section measurements for denatured proteins (cytochrome c, ubiquitin, lysozyme) and peptides (tryptic digests of ADH and cytochrome c) (http://www.indiana.edu/~clemmer/Research/Cross%20Section%20Database/cs_database.php) whilst Bush et al. provide measurements obtained for proteins analyzed under native conditions, ranging from cytochrome c at 12 kDa with a cross section of 1490 Å$^2$ (in $N_2$) to GroEL at 801 kDa and a CCS of approximately 21,800 Å$^2$ (in $N_2$) [24].

The calibration standards used must be analyzed under identical instrument conditions for all parameters downstream of the trapping ion guide as used for the protein of interest [11]. Ideally, the same class of molecules should be used for calibration as that of the sample of interest, i.e., native protein calibrants used to calibrate data obtained for a protein of interest under native solution conditions and peptide calibrants for peptide data. Due to the nature of the ion mobility separation and the relationship between drift time and CCS, in order to obtain accurate CCSs ion mobility measurements must be made under fixed wave heights and not by use of a wave height ramp. The corrected arrival times of the calibration standards must also bracket those of the analyte for the calibration to be valid [25].

To perform the calibration the following steps are undertaken:

1. The modal arrival time ($t_d$) at which each calibrant ion arrives at the detector is extracted from the ion mobility data (*see* **Note 7**).
2. The arrival time is corrected for *m/z* dependent flight time. The *m/z* dependent flight time is proportional to the square root of the *m/z*. It must be subtracted to obtain the corrected effective drift time ($t'_d$), i.e., the time taken to traverse the mobility cell. The corrected effective drift time ($t'_d$) is given by:

$$t'_d = t_d - \frac{c\sqrt{m/z}}{1000},$$

where $c$ is the enhanced duty cycle (EDC) delay coefficient found in the instrument settings (*see* **Note 8**).

3. Calibration coefficients are obtained from published absolute cross-section data ($\sigma$). Published cross sections are corrected to take into account the effects of reduced mass and charge state. Where $e$ is the charge on the ion, $m_i$ is the mass of the ion, and $m_n$ is the mass of the mobility gas, normalized cross sections ($\sigma'$) are given by:

$$\sigma' = \frac{\sigma}{e \times \sqrt{\left(\frac{1}{m_i} + \frac{1}{m_n}\right)}}.$$

4. $\sigma'$ is plotted against $t'_d$.
A power series fit ($y = Ax^B$) or a linear series fit ($y = Ax + B$) to the data points is applied. A power series fit has been shown to provide a more reliable calibration for large compounds, such as proteins, whereas a linear relationship has been found to be more appropriate for smaller molecules, such as peptides [26]. For denatured calibrants an $r^2$ fit should be >0.98 and for native calibrants >0.95 for the calibration to be acceptable.

5. Experimental T-Wave mobility measurements obtained for an analyte are converted into estimated CCSs by correction for reduced mass and charge and application of the power series fit or the linear series fit as appropriate.

$$\text{CCS} = A \times t_d'^{\,B} \times e \times \sqrt{\left(\frac{1}{m_i} \times \frac{1}{m_n}\right)},$$

or

$$\text{CCS} = \left[\left(A \times t_d'\right) + B\right] \times e \times \sqrt{\left(\frac{1}{m_i} \times \frac{1}{m_n}\right)}.$$

Plotting estimated CCS against charge state for individual oligomeric species can be used to help ensure the correct assignments have been made. CCS should increase with an increase in charge state and if this is not the case it is likely that an incorrect assignment has been made.

### 3.5 Calculation of CCSs for ΔN6 Oligomeric Species

1. Data are acquired for the calibrants cytochrome c, BSA, concanavalin A, and ADH under the optimized instrument conditions (from the trap onwards with low cone voltage) used for analysis of ΔN6 (at fixed wave heights).

2. Modal drift times for each calibrant ion are extracted from ATDs and used to produce a calibration with a power fit following the procedure detailed above for each wave height. A typical calibration is shown in Fig. 3.

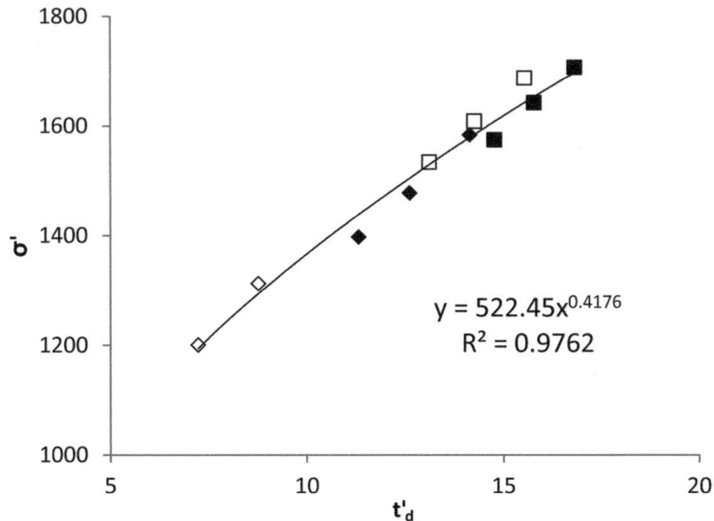


**Fig. 3** Ion mobility data calibration with absolute CCSs corrected for reduced mass and charge ($\sigma'$). Data for cytochrome c (*open diamonds*), BSA (*filled diamonds*), concanavalin A (*open squares*), and ADH (*filled squares*) are plotted against corrected drift times ($t'_d$) for calibrant ions. A power series fit is shown

3. Modal drift times for each ΔN6 species are extracted from their ATDs and their CCSs calculated by use of the appropriate calibration.
4. Average CCS measurements for each oligomeric species and each charge state are calculated by averaging measurements obtained from three replicate experiments. The error in the calibration measurement is usually in the range of 5–8 % (*see* **Note 9**) [11].

### 3.6 Modeling

CCS values estimated from ion mobility data can be compared to calculated CCS values from available high-resolution structures or coarse-grain models. This can allow for comparison of CCS measurements obtained in the gas phase to solution-phase measurements. CCS values calculated for atomic structures are generally in good agreement with those calculated by IMS. The lowest charged ion species are generally thought to be most reflective of solution-phase structure and so measurements are most often compared to these.

Calculation of CCS values from atomic structures or coarse-grain models can be performed by the use of MOBCAL. MOBCAL, an open-source program to calculate mobilities [27, 28], facilitates the use of three approximations to calculate CCSs. The simplest method is the projection approximation (PA). This replaces the CCS of an ion with its projection (shadow) and averages the projections created by every orientation of that ion [29]. The PA is an adequate approximation for small molecules but tends to underestimate the CCS of protein ions with highly convex structures where buffer gas interactions become important [28]. The trajectory method (TM) gives the most reliable estimate, incorporating all interactions but is computationally intense (taking in excess of a week to calculate the CCS from an atomic structure for a 10 kDa protein on a single processor). A compromise is to use a third model, the exact hard sphere scattering (EHSS) method. This ignores electrostatic interactions so requires substantially less computational time, and can calculate CCSs to within a few percent of values obtained by the trajectory method [28]. More recently, the Bowers and Bleiholder groups have developed the projected superposition approximation (PSA) method to calculate CCSs from structures (http://luschka.bic.ucsb.edu:8080/WebPSA/) [30–33]. This is a more accurate version of the PA that takes into account a shape factor and so has been shown to provide more reliable estimates of CCS than the PA or EHSS approach but in significantly less time than required to run the TM.

For coarse-grain modeling of oligomer structures, isotropic and linear growth can be estimated by the use of equations. In isotropic growth, $\sigma_n = \sigma_m \times n^{2/3}$, where $n$ = oligomer number, $\sigma_n$ is the CCS of oligomer number $n$, and $\sigma_m$ is the CCS of the monomer [14]. Similarly, linear growth in one direction (fibril growth) can be estimated by $\sigma_n = a \times n + k$, where $a$ describes the CCS of a monomer within a fibril and $k$ is the size of the fibril cap. For more complex models of oligomer growth, spheres representative of the shape and size of a single subunit (monomer) within an oligomer can be arranged in three-dimensional space to build models. A CCS for these more complex models can be calculated by use of the MOBCAL software and the PA method. The mass and radius of a single monomer subunit is required along with the $x,y,z$ coordinates of each monomer center contained within the model.

Here, isotropic and linear growth pathways of ΔN6 oligomers were modeled by use of the equations given above and MOBCAL was used to produce a ring model of oligomeric growth. For the linear growth model, estimated monomeric and dimeric CCSs (for the lowest charge states of each species observed) from ion mobility data were used to determine $a$ and $k$ by solving the two simultaneous equations $\sigma_1 = a \times 1 + k$ and $\sigma_2 = a \times 2 + k$. To use MOBCAL to calculate CCSs, input files for MOBCAL need to be generated (.mfj) and the MOBCAL code needs to be modified. Force 3.0 (free distribution software) was used to compile and edit the FORTRAN script and input files were generated in TextPad. The input file needs to be in a specific format and contain the Cartesian coordinates of each atom or monomer within the model structure. Further description of the layout of .mfj files is given below and can be found in the "read me" information provided with the MOBCAL script at (http://www.indiana.edu/~nano/software.html).

For calculation of the ring model of oligomeric growth by use of MOBCAL, the following steps were undertaken:

1. The MOBCAL script was edited by use of the FORTRAN editor and compiler Force 3.0 in two sections of the code that define atom mass and radii (lines ~580 and ~2600). Atom mass and atom radius were replaced with ΔN6 monomer mass and ΔN6 monomer radius. The radius of the ΔN6 monomer was calculated based on a spherical particle that would give rise to the monomeric CCS estimated by ESI-IMS-MS ($\pi r^2$). The script was also edited to stop calculation of the TM by placing a "c" in front of line 338 [11].
2. Coordinate entry .mfj files for the MOBCAL script were generated in TextPad with the following structure:

   Line 1: name of model

   Line 2: number of models (always 1 in this case)

   Line 3: number of monomers in model

   Line 4: ang

   Line 5: none

   Line 6: 1.0000

   Line 7: x,y and z coordinate of first atom (monomer), followed by monomer mass separated by indents.

   Line 7 + n: x,y and z coordinate of (first + n) atom (monomer), followed by monomer mass separated by indents.

   Line 7 + n + 1: number of monomers in model

   e.g.

   ```
   ΔN6__DIMER
   1
   2
   ang
   none
   1.0000
       1.000 1.000 1.000 11136
       1.000 1.000 25.00 11136
   2
   ```
3. The script was compiled and executed using Force 3.0. and the mobcal.run file was edited so it used the generated .mfj file as input.

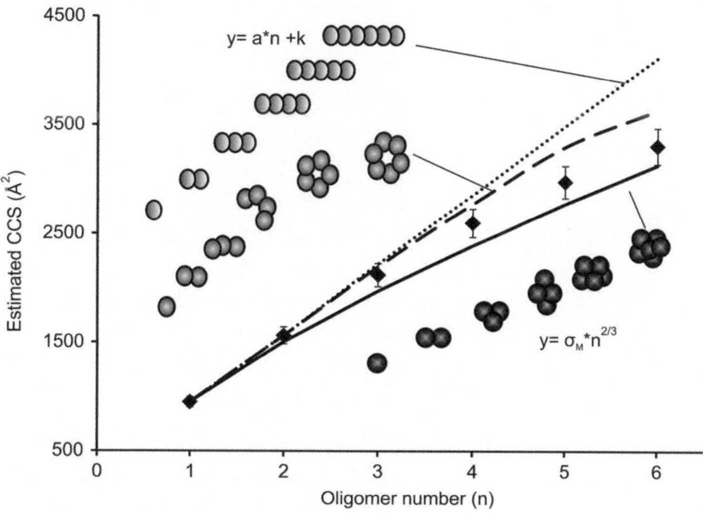


**Fig. 4** Estimated CCS values for ΔN6 oligomers of different oligomer number (*n*) (*filled diamonds*). Various models of oligomer growth are shown: isotropic (*solid line*); ring (*dashed line*); linear (*dotted line*)

4. The appropriate spacing between two monomer centers to use within the models was determined by varying the distance between two monomers in the model for the dimer until the script successfully calculated the estimated dimeric CCS. All models generated in this way use the monomeric and dimeric CCSs estimated from ion mobility data to instruct higher-order oligomer models.
5. Coordinate .mfj files for higher-order oligomers were generated by use of trigonometry to determine the *x*,*y*,*z* coordinates for the center of each monomer within the model of a trimer, tetramer, and pentamer *en route* to a regular hexagon structure.

Figure 4 shows the CCS values estimated from IMS-MS experiments for the lowest charge states of each oligomer of ΔN6 observed alongside CCS values calculated for various models of oligomer growth. Estimated CCSs for oligomers of ΔN6 are larger than expected for an isotropic mode of oligomer growth yet smaller than expected for both the linear and ring models of oligomeric growth. These simple models are thus insufficient to describe the oligomer formation pathway for ΔN6. More complex assembly mechanisms or a switch in mechanism with oligomer size are thus needed to describe the experimental data.

## 4 Notes

1. Simply replacing a nonvolatile buffer with a volatile buffer at the same pH and ionic strength may not yield the same aggregation parameters. The rate and mechanism of aggregation may depend on ion composition as well as ionic strength and pH. The aggregation process must therefore be analyzed by other biophysical techniques such as fluorescence, turbidity, analytical ultracentrifugation, and/or electron microscopy to ensure that the protein aggregation pathway followed is comparable between two different buffer systems (*see* refs. 1, 5, and 7 for further details).
2. Buffer exchange can be performed using desalting columns or buffer-exchange devices, such as Zeba™ Spin Desalting Columns (7K MWCO, ThermoFisher Scientific) or Micro Bio-Spin 6 chromatography columns (6K MWCO, Bio-Rad). Dialysis may also be used but may not be a suitable approach if the protein of interest aggregates on a short time scale at low temperature as aggregation will proceed during the time taken to perform the dialysis. If a protein sample is already at a low concentration, it can be buffer-exchanged and concentrated by use of centrifugal devices, such as Ultrafree-0.5 and Microcon (Millipore), or Vivaspin 500 μL concentrators (Millipore).
3. Buffer salt concentrations from 10 mM to 1 M are routinely used, with higher concentrations reducing the negative effects of any remaining nonvolatile buffer components [21].
4. Due to the pKa values of acetate and bicarbonate, it is difficult to buffer a solution between pH 6 and 8, which is the physiological pH range experienced by most proteins. A mixture of ammonium bicarbonate and ammonium acetate solutions can be used to try and achieve this. Many studies, however, use ammonium acetate solutions at pH 7, which do not buffer well at this pH. In this case, care must be taken to ensure that the pH of the solution is not changing during aggregation as this may influence the nature of aggregates formed and the aggregation pathway followed.
5. The concentration of CsI used can be increased in order to cover a wider $m/z$ calibration range.
6. Experience has shown that the minimum capillary voltage required to generate a stable spray generally produces the optimum MS-spectrum [34].
7. The extracted ATD for each calibrant ion may have multiple features, suggesting multiple conformations of the calibrant ion are present. Multiple $\sigma$'s for a given charge state are published for some calibrant species but not for all. For calibrant

ions, for which only a single $\sigma$ is published the most abundant feature in the ATD, which usually has the shortest drift time of all the features, is usually chosen for calibration purposes. A calibration with a low correlation coefficient may result if an incorrect feature in the ATD for a calibrant ion is used.

8. The EDC value is constant provided that the parameters for the transfer T-Wave guide and the transfer ion optics remain unchanged since EDC calibration [11].
9. The error in the CCS measurements can be estimated by the sum of the reproducibility (standard deviation of three or more replicate measurements), the average error of the calibration curve (<2.5 %), and the error in measurement of the protein standards used to produce the calibration (assumed to be 1 %) [11].

## Acknowledgements

CAS is funded by the Biotechnology and Biological Sciences Research Council (BBSRC; grant number BB/H024875/1). The Synapt HDMS mass spectrometer was purchased with funds from the BBSRC's Research Equipment Initiative (BB/E012558/1). SER's research is supported by an ERC Advanced grant (322408). The authors would like to thank Dr. Anton Calabrese for critical evaluation of the manuscript.

### References

1. Woods LA, Radford SE, Ashcroft AE (2013) Advances in ion mobility spectrometry–mass spectrometry reveal key insights into amyloid assembly. Biochim Biophys Acta 1834: 1257–1268
2. Hilton GR, Benesch JLP (2012) Two decades of studying non-covalent biomolecular assemblies by means of electrospray ionization mass spectrometry. J R Soc Interface 9:801–816
3. Ruotolo BT, Robinson CV (2006) Aspects of native proteins are retained in vacuum. Curr Opin Chem Biol 10:402–408
4. Ruotolo BT, Giles K, Campuzano I et al (2005) Evidence for macromolecular protein rings in the absence of bulk water. Science 310:1658–1661
5. Smith DP, Radford SE, Ashcroft AE (2010) Elongated oligomers in β2-microglobulin amyloid assembly revealed by ion mobility spectrometry-mass spectrometry. Proc Natl Acad Sci U S A 107:6794–6798
6. Smith DP, Giles K, Bateman RH et al (2007) Monitoring copopulated conformational states during protein folding events using electrospray ionization-ion mobility spectrometry-mass spectrometry. J Am Soc Mass Spectrom 18:2180–2190
7. Smith DP, Woods LA, Radford SE et al (2011) Structure and dynamics of oligomeric intermediates in β2-microglobulin self-assembly. Biophys J 101:1238–1247
8. Woods LA, Platt GW, Hellewell AL et al (2011) Ligand binding to distinct states diverts aggregation of an amyloid-forming protein. Nat Chem Biol 7:730–739
9. Young LM, Cao P, Raleigh DP et al (2013) Ion mobility spectrometry–mass spectrometry defines the oligomeric intermediates in amylin amyloid formation and the mode of action of inhibitors. J Am Chem Soc 136:660–670
10. Pringle SD, Giles K, Wildgoose JL et al (2007) An investigation of the mobility separation of some peptide and protein ions using a new hybrid quadrupole/travelling wave ims/oa-tof instrument. Int J Mass Spectrom 261:1–12

11. Ruotolo BT, Benesch JLP, Sandercock AM et al (2008) Ion mobility-mass spectrometry analysis of large protein complexes. Nat Protoc 3:1139–1152
12. Scarff CA, Thalassinos K, Hilton GR et al (2008) Travelling wave ion mobility mass spectrometry studies of protein structure: biological significance and comparison with x-ray crystallography and nuclear magnetic resonance spectroscopy measurements. Rapid Commun Mass Spectrom 22:3297–3304
13. Smith DP, Knapman TW, Campuzano I et al (2009) Deciphering drift time measurements from travelling wave ion mobility spectrometry-mass spectrometry studies. Eur J Mass Spectrom 15:113–130
14. Bleiholder C, Dupuis NF, Wyttenbach T et al (2011) Ion mobility–mass spectrometry reveals a conformational conversion from random assembly to β-sheet in amyloid fibril formation. Nat Chem 3:172–177
15. Lanucara F, Holman SW, Gray CJ et al (2014) The power of ion mobility-mass spectrometry for structural characterization and the study of conformational dynamics. Nat Chem 6:281–294
16. Zhong Y, Han L, Ruotolo BT (2014) Collisional and coulombic unfolding of gas-phase proteins: high correlation to their domain structures in solution. Angew Chem Int Ed Engl 53:9209. doi:10.1002/anie.201403784
17. Leney AC, Pashley CL, Scarff CA et al (2014) Insights into the role of the beta-2 microglobulin d-strand in amyloid propensity revealed by mass spectrometry. Mol Biosyst 10:412–420
18. Eichner T, Kalverda AP, Thompson GS et al (2011) Conformational conversion during amyloid formation at atomic resolution. Mol Cell 41:161–172
19. Sarell CJ, Woods LA, Su Y et al (2013) Expanding the repertoire of amyloid polymorphs by co-polymerization of related protein precursors. J Biol Chem 288:7327–7337
20. Bellotti V, Stoppini M, Mangione P et al (1998) β2-microglobulin can be refolded into a native state from ex vivo amyloid fibrils. Eur J Biochem 258:61–67
21. Hernandez H, Robinson CV (2007) Determining the stoichiometry and interactions of macromolecular assemblies from mass spectrometry. Nat Protoc 2:715–726
22. Chernushevich IV, Thomson BA (2004) Collisional cooling of large ions in electrospray mass spectrometry. Anal Chem 76:1754–1760
23. Giles K, Pringle SD, Worthington KR et al (2004) Applications of a travelling wave-based radio-frequency-only stacked ring ion guide. Rapid Commun Mass Spectrom 18:2401–2414
24. Bush MF, Hall Z, Giles K et al (2010) Collision cross sections of proteins and their complexes: a calibration framework and database for gas-phase structural biology. Anal Chem 82:9557–9565
25. Shvartsburg AA, Smith RD (2008) Fundamentals of traveling wave ion mobility spectrometry. Anal Chem 80:9689–9699
26. Thalassinos K, Grabenauer M, Slade SE et al (2009) Characterization of phosphorylated peptides using traveling wave-based and drift cell ion mobility mass spectrometry. Anal Chem 81:248–254
27. Shvartsburg AA, Jarrold MF (1996) An exact hard-spheres scattering model for the mobilities of polyatomic ions. Chem Phys Lett 261:86–91
28. Jarrold MF (1999) Unfolding, refolding, and hydration of proteins in the gas phase. Acc Chem Res 32:360–367
29. Mack E (1925) Average cross-sectional areas of molecules by gaseous diffusion measurements. J Am Chem Soc 47:2468–2482
30. Bleiholder C, Wyttenbach T, Bowers MT (2011) A novel projection approximation algorithm for the fast and accurate computation of molecular collision cross sections (i). Method. Int J Mass Spectrom 308:1–10
31. Bleiholder C, Contreras S, Do TD et al (2013) A novel projection approximation algorithm for the fast and accurate computation of molecular collision cross sections (ii). Model parameterization and definition of empirical shape factors for proteins. Int J Mass Spectrom 345–347:89–96
32. Bleiholder C, Contreras S, Bowers MT (2013) A novel projection approximation algorithm for the fast and accurate computation of molecular collision cross sections (iv). Application to polypeptides. Int J Mass Spectrom 354–355:275–280
33. Anderson SE, Bleiholder C, Brocker ER et al (2012) A novel projection approximation algorithm for the fast and accurate computation of molecular collision cross sections (iii): Application to supramolecular coordination-driven assemblies with complex shapes. Int J Mass Spectrom 330–332:78–84
34. Campuzano I, Giles K (2011) Nanospray ion mobility mass spectrometry of selected high mass species. In: Toms SA, Weil RJ (eds) Nanoproteomics, vol 790, Methods in molecular biology. Humana Press, Totowa, NJ, pp 57–70. doi:10.1007/978-1-61779-319-6_5

# Chapter 9

# Formation and Characterization of α-Synuclein Oligomers

**Wojciech Paslawski, Nikolai Lorenzen, and Daniel E. Otzen**

## Abstract

The aggregation of α-synuclein (αSN) into oligomeric structures has received increasing interest during the last 10–15 years. The oligomers' potential involvement in Parkinson's disease makes them a promising therapeutic target. Therefore reproducible protocols to prepare and analyze oligomers are very important to allow direct comparison of results obtained by different research groups. In this chapter we present one established method to obtain αSN oligomers from a monomeric ensemble in a relatively easy manner. Also, we briefly discuss a selection of biophysical methods which allow for a quick characterization of oligomer purity and structure.

**Key words** Aggregation, α-synuclein, Biophysics, Chromatography, Oligomer, Protein, Purification

## 1 Introduction

Parkinson's disease (PD) is a neurodegenerative disease characterized amongst others by the loss of dopaminergic neurons in the part of the brain known as the *Substantia nigra*. The 140-residue natively unfolded protein α-synuclein (αSN) is a key component in the development of PD and accumulates as intracellular fibrillar deposits known as Lewy Bodies in the brains of affected PD patients. However, it is widely accepted that the neurotoxic species formed by αSN are not fibrils but rather oligomers [1–4]. Generally, little is known about the formation, structure, and toxicity of oligomeric species probably due to the heterogeneity and instability of many oligomers. Furthermore, several different types of oligomers can be formed, depending on environmental conditions or additives [1, 5–14], making it difficult to compare data from different research groups. In this chapter we will focus on a type of stable αSN oligomers which we have characterized in our laboratory [1, 15–18]. An accompanying chapter by Subramaniam and coworkers deals with oligomers made in a related manner [10, 19–23]. The sizes of the oligomers made by these two different approaches have been determined by independent methods,

David Eliezer (ed.), *Protein Amyloid Aggregation: Methods and Protocols*, Methods in Molecular Biology, vol. 1345, DOI 10.1007/978-1-4939-2978-8_9, 

namely light scattering and small angle X-ray scattering (SAXS) in one case [15] and Poisson distribution of fluorophore-labeled oligomers in another [23]. Nevertheless, all these methods agree on an oligomer size corresponding to around 30 monomers. Reassuringly, the oligomeric species formed in this way has been shown by SAXS to consist of an ellipsoidal core (axes 9.4 and 4.7 nm), presumably stably folded, surrounded by a 5 nm thick outer shell of disordered protein chains [1, 15]. These oligomers do not revert to monomers and inhibit rather than promote fibrillation. Thus, prolonged incubation of oligomers leads to larger non-fibrillar aggregates. Moreover, they are very stable and resist both extreme temperature and extreme pH, and only high urea concentrations induce dissociation into monomers [24]. It should also be noted that the oligomers are complex species and do not represent a single homogeneous state: hydrogen/deuterium exchange coupled with mass spectrometry has revealed the co-existence of structurally and dynamically different oligomers, which however share the same core sequence (Y39-T75) [16].

Here we present a simple and reproducible method to produce and purify αSN oligomers. We base our method on the protocol developed in the laboratory of Peter T. Lansbury who pioneered the research field on αSN oligomers [2, 25]. We have optimized this method to obtain 2–3 % conversion of monomeric αSN into oligomers purified from larger aggregates. Furthermore, we present selected biophysical methods which allow for a simple and fast analysis of purity, structure, and function of the purified oligomers. These methods include SDS-PAGE, circular dichroism, electron microscopy, atomic force microscopy, Fourier-Transform Infrared Spectroscopy, and an assay to detect release of calcein from phospholipid vesicles.

## 2 Materials

Unless stated otherwise, all solutions are prepared using analytical grade reagents and ultrapure water (sensitivity of 18 MΩ cm at 25 °C). All reagents are prepared and stored at room temperature (if not otherwise stated) and waste materials are discarded following disposal regulations.

### 2.1 Monomeric αSN Purification

αSN is expressed recombinantly in *E. coli* using an auto-induction method [26] and purified as described previously [27, 28] with few alterations.

1. 1 M $MgSO_4$: Weigh 24.65 g of $MgSO_4 \cdot 7H_2O$ (reagent grade) into a 100-mL graduated cylinder and add water to 100 mL. Mix and transfer the solution to heat resistant glass bottles and autoclave. Store at 4 °C.

2. 20× NPS solution: pour approx. 800 mL of water into a 1-L graduated cylinder (*see* **Note 1**). Weigh and add chemicals into the cylinder (reagent grade or higher) in the following order: 66 g of $(NH_4)_2SO_4$ (0.5 M final), 136 g of $KH_2PO_4$ (1 M final), and 178 g of $Na_2HPO_4 \cdot 2H_2O$ (1 M final). Add water to 1000 mL, dissolve, transfer the solution to heat resistant glass bottles, and autoclave. Store at 4 °C.
3. 50× 5052 solution: pour approx. 700 mL of water into a 1-L graduated cylinder (*see* **Note 1**). Weigh and add chemicals into the cylinder (reagent grade or higher) in the following order: 250 g of 100 % glycerol, 25 g of glucose, and 100 g of lactose. Add water to 1000 mL, dissolve, transfer the solution to heat resistant glass bottles, and autoclave. Store at 4 °C.
4. Auto-induction medium: pour approx. 1500 mL of deionized water into a 2-L graduated cylinder. Weigh and add chemicals into the cylinder in the following order: 20 g of peptone, 10 g of yeast extract, 4 mL of 1 M $MgSO_4$, 40 mL of 50× 5052 solution, and 100 mL of 20× NPS solution. Add water to 2000 mL, dissolve, transfer the solution to 5000 mL baffled conical flask, and autoclave. After cooling down add 2 mL of 100 mg/mL ampicillin (*see* **Note 2**).
5. Osmotic shock buffer (*see* **Note 3**): pour approx. 500 mL of water into a 1-L graduated cylinder. Weigh and add chemicals into the cylinder: 3.63 g of Tris-HCl (30 mM final), 400 g of sucrose (reagent grade or higher), and 0.58 g of EDTA. Add water to approx. 900 mL, dissolve, and set the pH to 7.2 with HCl. Fill the cylinder to 1000 mL, mix, and store at room temperature.
6. Saturated $MgCl_2$ solution. Store at room temperature.
7. Ice-cold ultrapure water.
8. 1 M HCl and 1 M NaOH. Store at room temperature.
9. 0.45 or 0.2 μm pore size filter.
10. Q-sepharose ion exchange chromatography column (*see* **Note 4**).
11. FPLC system.
12. Buffer A: 20 mM Tris–HCl, pH 7.5. Pour approx. 100 mL of water into a 1-L graduated cylinder and add 2.42 g of Tris-HCl. Fill with water to 900 mL, mix, and adjust pH to 7.4 with HCl. Add water to 1000 mL, filter through 0.2 μm pore size filter, and degas (*see* **Note 5**). Store at 4 °C.
13. Buffer B: 20 mM Tris–HCl, 1 M NaCl, pH 7.5. Pour approx. 100 mL of water into a 1-L graduated cylinder, add 2.42 g of Tris-HCl and 58.44 g of NaCl. Fill with water to 900 mL, mix, and adjust pH to 7.4. Add water to 1000 mL, filter through 0.2 μm pore size filter, and degas (*see* **Note 5**). Store at 4 °C.

14. 30 kDa molecular weight cutoff (MWCO) ultrafiltration discs and stirring cell (Millipore, USA).
15. 30 kDa MWCO dialysis membrane.

### 2.2 Oligomeric αSN Purification Components

1. 10× Phosphate buffered saline (PBS): 20 mM phosphate, 150 mM NaCl, pH 7.4. Pour approx. 700 mL water to a 1-L graduated cylinder and add 14.4 g of $Na_2HPO_4{\cdot}7H_2O$, 2.4 g of $KH_2PO_4$, 80 g of NaCl, and 2 g of KCl. Store at room temperature.
2. 0.2 μm pore size syringe filter and plastic syringe.
3. Preparative Superose 6 gel filtration column (*see* **Note 6**).
4. 30 kDa MWCO conical ultrafiltration unit (Millipore, USA).

### 2.3 Sodium Dodecyl Sulfate Polyacrylamide Gel Electrophoresis (SDS-PAGE)

1. 3.5× gel buffer: pour 100 mL of water into a 250-mL graduated cylinder and add 65.4 g of BisTris. Fill with water up to 200 mL, mix, and adjust pH with HCl to pH 6.6. Add water to 250 mL. Store at room temperature.
2. 10 % sodium dodecyl sulfate (SDS) solution in water.
3. 0.75 mm thick glass gel casting plates and casting chamber.
4. Resolving gel: 15 % (for 12 gels). Pour into a 100-mL graduated cylinder: 17 mL of 3.5× gel buffer, 30 mL of BisTris-Acrylamide, 0.6 mL of 10 % SDS, and 12 mL of water.
5. Stacking gel: 5 % (for 12 gels). Pour into a 100-mL graduated cylinder: 10 mL of 3.5× gel buffer, 6 mL of BisTris-Acrylamide, 0.35 mL of 10 % SDS, 18.3 mL of water and minimal amount of bromophenol blue powder.
6. 10 % ammonium persulfate solution (APS) in water. Store at −20 °C.
7. *N*,*N*,*N*,*N*′-tetramethyl-ethylenediamine (TEMED) (Sigma, USA). Store at 4 °C.
8. 10× SDS-PAGE running buffer: pour approx. 600 mL of water into a 1-L graduated cylinder and add 30 g of Tris-HCl, 144 g of glycine, and 10 g of SDS. Fill with water to 1000 mL and mix. Store at room temperature.
9. 375 mM Tris-HCl solution: pour 50 mL of water and add 4.54 g of Tris-HCl. Fill with water to 90 mL, mix, adjust pH with HCl to 6.8, and add water to 100 mL. Store at room temperature.
10. 6× SDS-PAGE sample buffer: pour 10 mL of 375 mM Tris-HCl solution into a 25-mL graduated cylinder. Add 1.5 g of SDS, 12.5 g of glycerol, 2.25 g of β-mercaptoethanol, and 7.5 mg of bromophenol blue. Fill with water to 25 mL, mix, and aliquot 1 mL per tube. Store at −20 °C.
11. Staining solution: pour 400 mL of water into a 1-L graduated cylinder and add 1 g of Coomassie Brilliant Blue R-250 (CBB),

400 mL of 96 % methanol, and 100 mL of glacial acetic acid. Mix and fill with water to 1000 mL.

12. Destaining solution: pour 400 mL of water into a 1-L graduated cylinder and add 40 mL of 96 % methanol and 40 mL of glacial acetic acid. Mix and fill with water to 1000 mL.

### 2.4 Circular Dichroism (CD) Spectroscopy

1. 1 mm path length quartz cuvette (*see* **Note 7**).
2. 2 % Hellmanex solution in water.
3. 70 % EtOH solution in water.
4. Ultrapure water.

### 2.5 Electron Microscopy (EM)

1. EM grids (*see* **Note 8**).
2. Staining solution (*see* **Note 9**).
3. Ultrapure water.
4. Soft drying paper.

### 2.6 Atomic Force Microscopy (AFM)

1. Mica matrix.
2. 30 kDa MWCO dialysis membrane or buffer exchange unit.
3. Inert gas supply (optional).

### 2.7 Calcein Releases

1. 1,2-dioleoyl-*sn*-glycero-3-[phospho-*rac*-(1-glycerol)] (DOPG) (Avanti Polar Lipids, Alabaster, AL).
2. PBS buffer (20 mM phosphate, 150 mM NaCl, pH 7.4) (*see* Subheading 2.2, **item 1**).
3. Lipid solution: Weigh 5 mg of DOPG (5 mg/mL final concentration), 46 mg of calcein disodium salt (70 mM final concentration) and dissolve in 1 mL PBS buffer.
4. Calcein disodium salt.
5. Liquid nitrogen.
6. Heating block.
7. Thermometer.
8. Extruder.
9. PD-10 desalting column.
10. 2 % Triton X-100.
11. 96-well plate.
12. Crystal clear sealing tape.
13. Fluorescence plate reader.

### 2.8 Fourier-Transform Infrared Spectroscopy (FTIR)

1. Inert gas supply.
2. 70 % EtOH solution in water.
3. Ultrapure water.
4. Dust free drying paper.

## 3 Methods

### 3.1 Monomeric αSN Purification

1. Inoculate prepared auto-induction media with bacterial cell culture (*see* **Note 10**).
2. Grow the cells in a shaking incubator at 37 °C, 120 rpm for 24 h.
3. Harvest cells by centrifugation: 4000 × *g*, 4 °C, 20 min.
4. Resuspend cell pellet in 10 % volume of osmotic shock buffer (100 mL per 1000 mL of cell culture before centrifugation) and incubate in room temperature for 10 min.
5. Centrifuge suspension: 9000 × *g*, 20 °C, 20 min.
6. Discard supernatant and resuspend pellet in ice-cold water. Use 40 mL of water per 1000 mL of cell culture (*see* **Note 11**).
7. Add 40 μL of saturated $MgCl_2$ per 100 mL of cell suspension, mix, and incubate on ice for 3 min.
8. Centrifuge suspension: 9000 × *g*, 4 °C, 30 min.
9. Collect supernatant and titrate it with 1 M HCl to pH 3.5.
10. Incubate with magnet stirring at room temperature for 10 min. The stirring should be gentle to avoid the formation of air bubbles.
11. Centrifuge: 9000 × *g*, 4 °C, 30 min.
12. Collect supernatant and titrate it with 1 M NaOH to pH 7.5 (*see* **Note 12**).
13. Filter protein extract through a pore size filter of 0.45 μm, or lower.
14. Equilibrate the Q-sepharose column with buffer A.
15. Load the protein extract on the column (*see* **Note 13**).
16. Wash the column with 3 column volumes (CV) of 10 % buffer B.
17. Elute bound proteins with constant gradient of buffer B from 10 to 50 % over 8 CV (*see* **Note 14**). Collect necessary fractions. αSN will normally elute around 30 % of buffer B.
18. Analyze fractions using SDS-PAGE and collect the fractions containing αSN (it migrates as a 15 kDa protein).
19. Pass through a 30 kDa MWCO ultrafiltration membrane to remove high molecular weight proteins and aggregates (*see* **Note 15**).
20. Dialyze purified αSN against water overnight at 4 °C using a 30 kDa MWCO dialysis membrane.
21. Determine the protein concentration with a UV–VIS spectrophotometer using a theoretical extinction coefficient at 280 nm of 0.412 $(mg/mL)^{-1}\,cm^{-1}$ (*see* **Note 16**).
22. Lyophilize obtained protein in adequate aliquots (*see* **Note 17**) and store at −20 °C until further analysis.

### 3.2 Oligomeric αSN Purification

1. Prepare 1× PBS solution by diluting 10× PBS with water (*see* **Note 18**).
2. Equilibrate Superose 6 gel filtration column with 1× PBS (*see* **Note 19**).
3. Dissolve the lyophilized monomeric αSN to the final concentration of 12 mg/mL.
4. Filter the solution using a 0.2 μm pore size syringe filter.
5. Aliquot the solution to adequate number of Eppendorf tubes.
6. Incubate the samples in a shaking incubator (*see* **Note 20**) at 37 °C with 900 rpm agitation for 5 h (*see* **Note 21**).
7. Centrifuge the samples at 12,000 × *g*, 10 min, 4 °C to remove insoluble protein aggregates.
8. Transfer the supernatant into the injection syringe. Avoid the uptake of insoluble aggregates from the pellet (*see* **Note 22**).
9. Inject the prepared solution into a sample loop of the chromatography equipment.
10. Inject the sample into the Superose 6 column.
11. Run the sample through the column and collect fractions of interest. The large aggregates will elute in the void volume at around 150 mL. The oligomer will start to appear at 200–210 mL with a peak centroid around 235 mL. Remaining monomer will start to elute around 320 mL. A typical elution profile is presented in Fig. 1.
12. Concentrate oligomer samples using a 30 kDa cutoff conical ultrafiltration unit at 4 °C (optional).
13. Store oligomers at 4 °C (*see* **Note 23**).

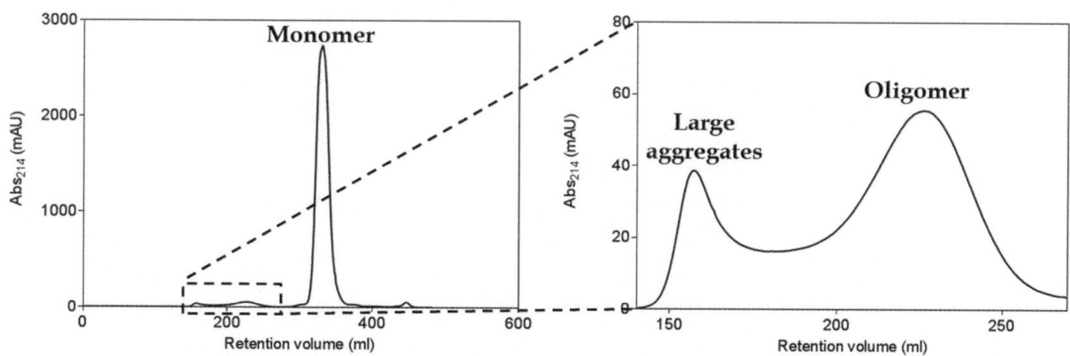


**Fig. 1** Typical elution profile observed during αSN oligomer purification. Three major peaks are present: void volume peak containing large aggregates—starting around 150 mL; oligomer peak—starting around 200 mL; and monomer peak starting around 320 mL. Adapted from Lorenzen, N., Nielsen, S. B., Buell, A. K., Kaspersen, J. D., Arosio, P., Vad, B. S., Paslawski, W., Christiansen, G., Valnickova-Hansen, Z., Andreasen, M., Enghild, J. J., Pedersen, J. S., Dobson, C. M., Knowles, T. J., and Otzen, D. E. (2014) The role of stable α-synuclein oligomers in the molecular events underlying amyloid formation, *J. Am. Chem. Soc.* **136**, 3859–3868. Copyright (2014) American Chemical Society

### 3.3 SDS-PAGE

1. Cast the gels (*see* **Note 24**): assemble glass plates in a casting chamber. Mix the resolving gel solution with 0.6 mL of 10 % APS and 15 μL of TEMED. Pour the solution into the casting chamber and overlay the resolving gel mixture with water or isopropanol. Wait until the gel has polymerized and remove the water/isopropanol layer. Mix the stacking gel solution with 0.35 mL of 10 % APS and 50 μL of TEMED. Pour the solution into the casting chamber and immediately insert a gel comb avoiding creation of air bubbles. Wait until the gel has polymerized.
2. Place the gel in a running chamber and fill it with 1× SDS-PAGE running buffer.
3. Mix the protein sample (*see* **Note 25**) with 6× SDS-PAGE loading buffer in a 5:1 sample/buffer volume ratio.
4. Incubate samples at 95 °C for 5 min and briefly spin them down to collect liquid droplets from the tube walls.
5. Load the sample into the gel and run the gel until the bromophenol blue line (blue line migrating on a gel) will exit the gel. Remember to also load a protein marker (*see* **Note 26**).
6. Remove the gel from in-between glass plates and transfer it to staining solution. Incubate the gel for at least 30 min.
7. Discard the staining solution, wash the gel with water, and transfer it into destaining solution. Change the destaining solution every hour until the gel is destained.
8. Wash the gel with water and scan it for your laboratory notebook.
9. A typical gel image obtained using a 15 % BisTris acrylamide gel is presented in Fig. 2.

### 3.4 CD Spectroscopy

1. Turn on the CD spectrometer (*see* **Note 27**).
2. Clean the cuvette prior to use with a 2 % Hellmanex solution, rinse with water and 70 % EtOH. Remember to dry out remaining liquid before transferring the sample into the cuvette.
3. Transfer αSN oligomer solution to a quartz cuvette (*see* **Note 28**).
4. Record a far-UV wavelength spectra (from 190 to 260 nm) of the oligomer sample and of pure buffer at room temperature (*see* **Note 29**).
5. Subtract the buffer spectrum from the oligomer spectrum and calculate the MRE using the following equation:

$$\mathrm{MRE}\left(\deg \mathrm{cm}^2\,\mathrm{dmol}^{-1}\right) = \frac{\mathrm{Ellipticity}\left(\mathrm{mdeg}\right) \times 10^6}{\mathrm{Pathlength}\left(\mathrm{cm}\right) \times \mathrm{Protein\ concentration}\left(\mu\mathrm{M}\right) \times \mathrm{Number\ of\ residues} \times 10}$$

6. A typical αSN oligomer spectrum is characterized by a single negative peak with a local minimum around 218 nm (Fig. 3).

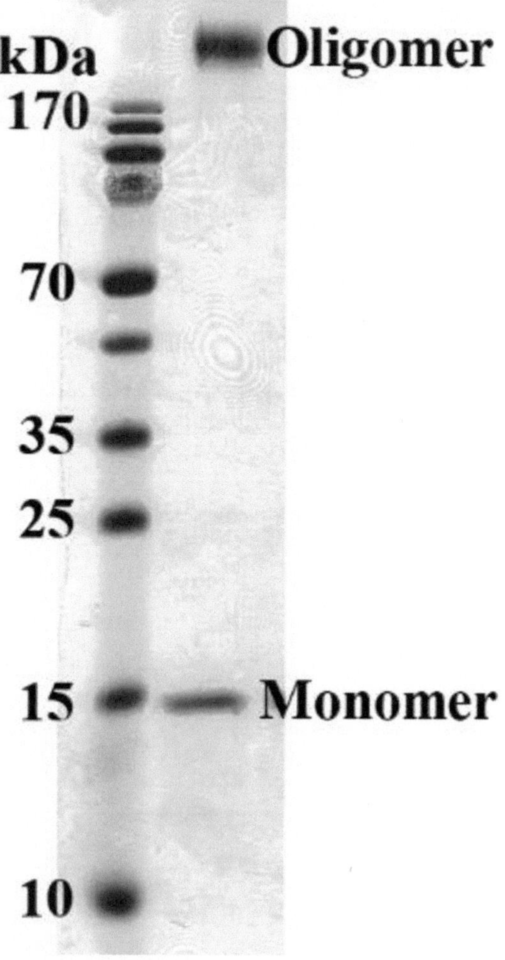


**Fig. 2** Scanned SDS-PAGE gel with αSN oligomer sample. The monomeric band is observed around 15 kDa, while oligomeric αSN migrates only few millimeters into the gel

### 3.5 EM

1. Prepare the EM grids (*see* **Note 30**).
2. Transfer a drop of oligomer solution onto the grid (*see* **Note 31**).
3. Wait for 1 min and dry the sample by gently touching the side of the grid with dust free soft paper.
4. Wash with water and dry as in **step 3**.
5. Pipette a drop of staining solution onto the grid.
6. Wait and dry as in **step 3**.
7. Transfer the grid into the EM instrument and obtain images (*see* **Note 32**).
8. A typical image will contain spherical αSN oligomers with a diameter around 20 nm (Fig. 4) (*see* **Note 33**).

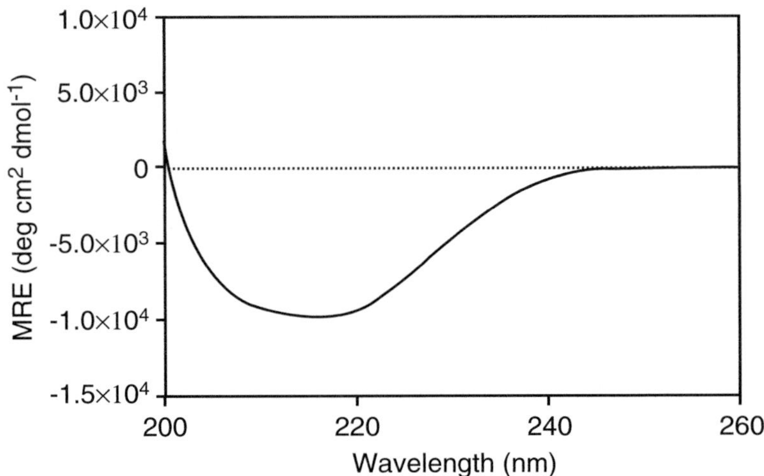


**Fig. 3** Typical CD spectrum of αSN oligomer

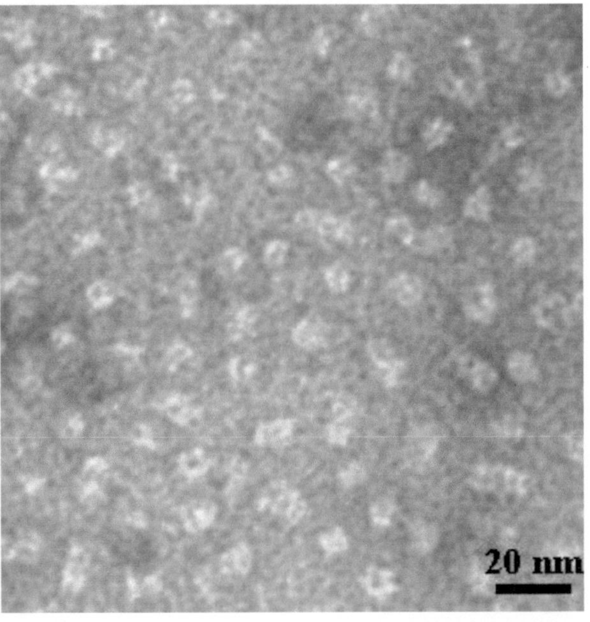


**Fig. 4** EM image showing the characteristic round structure of αSN oligomers. Imaged by Karen Thomsen

### 3.6 AFM

1. Desalt the αSN oligomer solution (*see* **Note 34**).
2. Transfer approx. 30 μL of solution onto the mica matrix.
3. Leave the sample to dry or use a low flow of inert gas to facilitate the process.
4. Image the sample with AFM instrument.
5. Typical image will contain spherical αSN oligomers with a diameter around 20 nm.

### 3.7 Calcein Release Assay

1. Prepare freshly 1 mL of lipid solution.
2. Vortex the sample thoroughly to dissolve the lipid.
3. Freeze-thaw cycles: Freeze the sample in liquid nitrogen for approx. 2 min. Transfer the lipid solution to a water bath of 40 °C. When the lipid solution is completely thawed, repeat the freeze-thaw cycle ten times in total.
4. Extrude the lipid solution 21 times back and forth through a filter with a 100 nm pore size (*see* **Note 35**). It is important to end the extrusion at the alternating side of the filter from where the extrusion was started. Normally this produces a monodisperse vesicle solution with an average diameter of around 100–115 nm, which can be measured with dynamic light scattering.
5. Separate vesicles from free calcein: Thoroughly pre-wash a PD-10 desalting column with PBS buffer. Add the extruded lipid solution to the column and collect fractions of 3–5 drops manually with Eppendorf tubes.
6. Identify the fractions which include vesicles with entrapped calcein and a low degree of free calcein. Pipette 148 μL PBS buffer to the wells of a 96-well plate (one well per collected fraction). Add 2 μL of the collected fraction. Also include a buffer control with 150 μL buffer and no addition of lipid. Seal the 96-well plate with a UV transparent sealing tape to avoid evaporation (*see* **Note 36**).
7. Measure 10–20 cycles of calcein fluorescence with a plate reader (*see* **Note 37**) using an excitation wavelength of 485 nm and an emission wavelength of 520 nm, at 37 °C. In between each measurement the plate is quickly shaken for 2 s.
8. Add 2 μL of 2 % Triton X-100 to each well.
9. Measure calcein fluorescence again according to Subheading 3.7, **step 7**.
10. Pool the fractions which have a low background fluorescence signal and have a 3–5 times (or higher) increase in fluorescence signal upon addition of Triton X-100. (Store the vesicle solution at 4 °C until use.)

11. To determine the purified oligomers potency to permeabilize vesicles, prepare a tenfold dilution series of monomers and oligomers (six solutions at $10^{-1}$–$10^{-6}$ mg/mL) in triplicates at 148 μL in PBS buffer. Remember to include a triplicate with buffer as a control to measure background fluorescence.
12. Add 2 μL of vesicle batch to each well. *Proceed right away to next step.*
13. Measure calcein release for minimum 1 h (for saturation) according to Subheading 3.7, **step 7**.
14. Add 2 μL Triton X-100 to each well.
15. Measure calcein release for minimum 15 min according to Subheading 3.7, **step 7**.
16. Determine the average signal of the calcein release signal before addition of Triton X-100 ($F$) (*see* **Note 38**), when Triton X-100 has been added ($Ft$) and for the buffer control ($F_0$) and calculate the calcein release percentage as follows:

$$\text{Calcein release}\left(\%\right) = \frac{F - F_0}{F_t - F_0} \times 100$$

17. Plot the calcein release percentage (CR%) of monomer and oligomer as Fig. 5 and estimate the concentration needed for 50 % calcein release. Normally we see that the concentration needed for oligomers is ~17 times lower than for monomers (*see* Fig. 5).
18. To apply this setup for inhibitor studies, we refer to [18]. For the comparison of different monomers and oligomers, we refer to [17].

### 3.8 FTIR

1. Turn on the FTIR spectrometer (*see* **Note 39**).
2. Transfer 1 μL of αSN oligomer solution onto the spectrometer crystal plate. In our experience a protein concentration of 1 mg/mL is ideal.
3. Dry out the sample with a low flow of inert gas.
4. Record an absorption spectrum from 1000 to 4000 $cm^{-1}$ (*see* **Note 40**).
5. Perform atmospheric compensation, by subtracting reference spectrum from sample spectrum, and baseline subtraction (*see* **Note 41**).
6. A typical αSN oligomer spectrum is characterized by a peak with a maximum around 1654 $cm^{-1}$ (disordered regions), a peak at 1627 $cm^{-1}$ (β-sheet structure), and a small shoulder at 1695 $cm^{-1}$ (anti-parallel β-sheet structure) (Fig. 6).

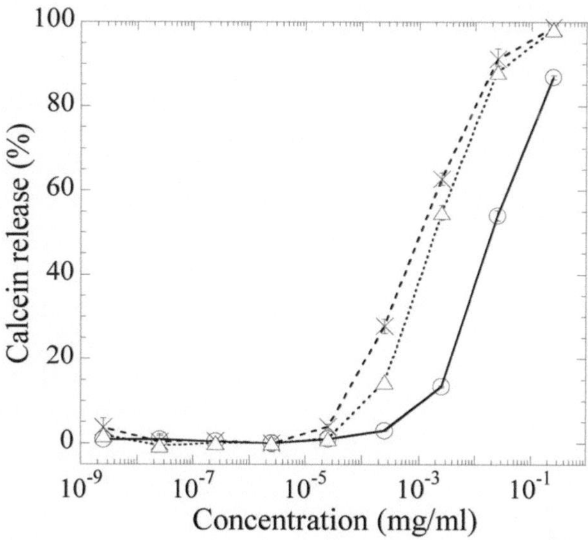


**Fig. 5** Calcein release of DOPG vesicles by oligomers (×), large aggregates (Δ), and monomers (○) (*see* Fig. 1). 100 % represents complete permeabilization of vesicles in the presence of Triton X-100. Data points are averaged triplicates, and standard deviation is given. Adapted with permission from Lorenzen, N., Nielsen, S. B., Buell, A. K., Kaspersen, J. D., Arosio, P., Vad, B. S., Paslawski, W., Christiansen, G., Valnickova-Hansen, Z., Andreasen, M., Enghild, J. J., Pedersen, J. S., Dobson, C. M., Knowles, T. J., and Otzen, D. E. (2014) The role of stable α-synuclein oligomers in the molecular events underlying amyloid formation, *J. Am. Chem. Soc.* **136**, 3859–3868. Copyright (2014) American Chemical Society

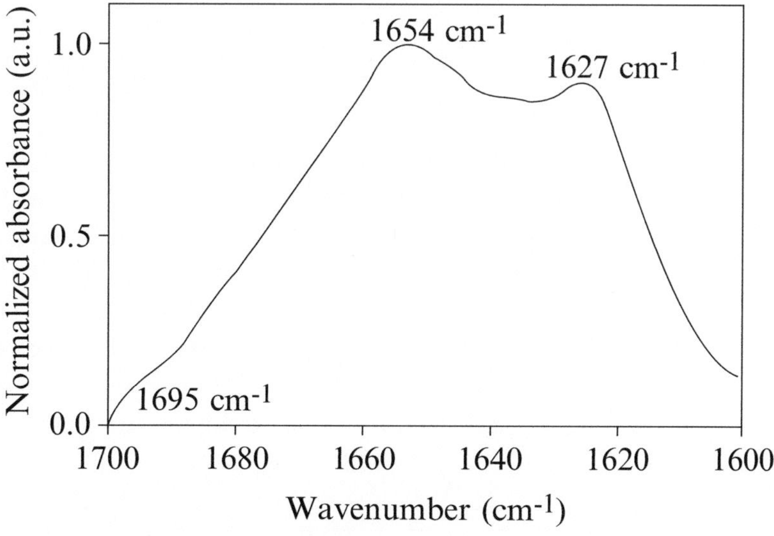


**Fig. 6** Typical FTIR spectra of αSN oligomer. Characteristic peak maxima around 1654, 1627 $cm^{-1}$, and a weak but significant shoulder at 1695 $cm^{-1}$ are observed

## 4 Notes

1. 20× NPS and 50× 5052 might be difficult to dissolve without heating the solution. Therefore a glass cylinder or beaker can be used in case the solutions need to be heated. Alternatively, non-fully dissolved solutions can be transferred into heat resistant bottles and autoclaved. The heating applied during the autoclaving process will dissolve remaining insoluble chemicals.
2. Ampicillin should be added just prior to inoculation of the media with bacterial cells. An alternative is carbenicillin, which is more stable. Also note that the antibiotic used here is dependent on the plasmid's resistance marker.
3. The solution prepared at the day of purification can be kept at room temperature. If the solution is prepared the day before (or longer), store the solution in 4 °C and move it to room temperature on the purification day.
4. We use a Hitrap DEAE FF Q-HP column (GE Healthcare, USA).
5. Degassing may be performed by keeping the bottle with buffer in a water bath sonicator for 10–15 min.
6. We use Superose 6 XK 26/100 (GE Healthcare, USA).
7. 0.1 mm path length cuvettes improve signal-to-noise ratios but are more fragile.
8. We use carbon-coated copper grids—mesh.
9. We use 1 % uranyl acetate solution in water.
10. We assume that you have your own αSN expression vector.
11. In the case of highly dense cell cultures, which will yield higher levels of αSN, it may be necessary to increase the volume of water. Otherwise the concentration of αSN may become high enough to induce aggregation.
12. At this point the samples can be kept frozen (−20 °C) until further analysis.
13. Check the capacity of the used ion exchange column. Be sure not to exceed the maximum capacity of the column, as this will lead to a decreased yield of α-synuclein.
14. To increase the purity of the eluted proteins, a more shallow gradient can be used to improve separation.
15. This step can be skipped for fractions which do not contain higher molecular weight bands as analyzed with SDS-PAGE.
16. We have compared the extinction coefficients of monomeric and oligomeric αSN by total amino acid hydrolysis and found them to have the same value.

17. We lyophilize protein by flash freezing aliquoted samples in liquid nitrogen and subsequently remove water by sublimation in a vacuum lyophilizer.
18. Usually around 1.5–2 L of buffer is needed for the whole experiment (including sample preparation, column equilibration, and sample run).
19. Depending on laboratory guidelines for column storage, the column may need to be pre-equilibrated with water prior to equilibration with 1× PBS. For oligomer purification on the column, it is often convenient to begin column equilibration 1 day prior to oligomer purification due to the large size of the column.
20. We use a tube shaker adapted for 1.5-mL tubes with adjustable shaking and temperature.
21. We observe that the 5 h time point gives good yield of oligomers with none or only small number of fibrillar aggregates.
22. Loading of insoluble aggregates on the SEC column may clog filters and compromise the analytical quality of the column. Furthermore, a high fraction of larger aggregates eluting in the void volume can lead to poor separation and overlapping of the void volume peak and the oligomer peak.
23. Based on our observations, αSN oligomers can be safely stored at 4 °C for up to 1 week. Longer storage may lead to oligomer clustering and formation of amorphous aggregates.
24. We use self-cast 15 % BisTris Acrylamide SDS-PAGE gels. Commercially available 4–15 % BisTris Acrylamide gels are also suitable to detect αSN oligomers. Any other SDS-PAGE gel systems might also be appropriate, but might need to be optimized to obtain good quality results. For the use of Pore-limit gel electrophoresis we refer to [15].
25. When using CBB staining solution, the concentration of αSN oligomers should be above 0.3 mg/mL to obtain clearly visible band on an SDS-PAGE gel.
26. Preferably use a prestained protein marker, which will give a better estimation on how far proteins have migrated in the gel. In that case let the gel run until the 10 kDa MW band will reach the bottom of the gel. This allows the αSN oligomers to enter the gel without losing the monomeric αSN band.
27. Remember that most CD spectrometers must be purged with inert gas such as nitrogen before turning on the instrument lamp.
28. Depending on your instrument use 0.1–0.4 mg/mL αSN oligomer.
29. Low protein concentration or sample purity might result in a low signal from 190 to 200 nm. In this case record the spectrum

only from 200 to 260 nm. For each samples at least three spectra should be accumulated, averaged, and a wavelength step of maximum 1 nm should be used.

30. Depending on used grids, carbon coating and charging of grids might be necessary.
31. Optimal results are observed with αSN oligomer concentrations between 0.1 and 0.5 mg/mL. Higher concentrations may lead to overcrowding and impair image quality. The volume of added sample should be just enough to cover the grid surface.
32. 72,000× to 90,000× magnification works well for the oligomers.
33. Sometimes a dark spot in the center of the oligomer structure can be observed. This is a sample drying artifact and shouldn't be misinterpreted as a hole in the αSN oligomer structure.
34. Remaining salt can be removed by dialysis or any buffer exchange system. Alternatively, sample might be diluted with water if the concentration of αSN oligomers is high. Any remained salt will crystallize on the surface of the mica and subsequently result in poor quality of obtained images.
35. We use a mini extruder (Avanti Polar Lipids, Alabaster, AL, USA).
36. We use crystal clear sealing tape (Hampton Research, Aliso Viejo, CA, USA).
37. We use a Genios pro plate reader from (Tecan, Mänersdorf, Switzerland). For this instrument a fluorescence gain of 20–40 is ideal for this calcein release setup; however, this is instrument dependent.
38. Make sure to only average data points from the saturated phase. An example of raw data is provided in Fig. 7.

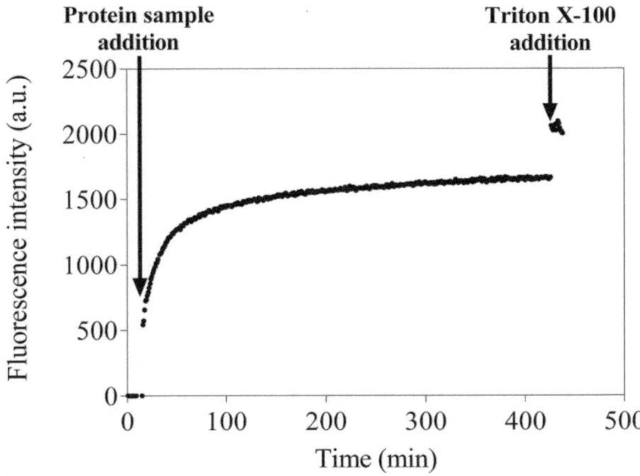


**Fig. 7** Typical raw data obtained for 0.05 mg/mL αSN oligomer solution and DOPG vesicles during a calcein release experiment

39. Remember that most FTIR spectrometers must be purged with inert gas such as nitrogen prior to measurement, due to the high absorption of water.
40. The resolution of obtained spectra should not be lower than 2 $cm^{-1}$ and depending on the instrument at least 50 accumulations should be averaged to obtain acceptable spectral quality.
41. Both atmospheric compensation and baseline subtraction can be usually done by the FTIR spectrometer software. The atmospheric compensation is performed by subtracting a reference spectrum (blank spectrum without any sample) from the sample spectrum. Baseline correction can be done using any mathematical software which allows the user to design the most appropriate baseline function. This baseline function also needs to be subtracted from the sample spectrum.

## Acknowledgements

We are grateful for support from the Danish Research Foundation (inSPIN) and the Michael J. Fox Foundation.

### References

1. Giehm L, Svergun DI, Otzen DE et al (2011) Low-resolution structure of a vesicle disrupting α-synuclein oligomer that accumulates during fibrillation. Proc Natl Acad Sci U S A 108: 3246–3251
2. Lashuel HA, Hartley D, Petre BM et al (2002) Neurodegenerative disease: amyloid pores from pathogenic mutations. Nature 418:291
3. Conway KA, Lee SJ, Rochet JC et al (2000) Acceleration of oligomerization, not fibrillization, is a shared property of both alpha-synuclein mutations linked to early-onset Parkinson's disease: implications for pathogenesis and therapy. Proc Natl Acad Sci U S A 97:571–576
4. Winner B, Jappelli R, Maji SK et al (2011) In vivo demonstration that alpha-synuclein oligomers are toxic. Proc Natl Acad Sci U S A 108: 4194–4199
5. Gurry T, Ullman O, Fisher CK et al (2013) The dynamic structure of alpha-synuclein multimers. J Am Chem Soc 135:3865–3872
6. Mysling S, Betzer C, Jensen PH et al (2013) Characterizing the dynamics of alpha-synuclein oligomers using hydrogen/deuterium exchange monitored by mass spectrometry. Biochemistry 52:9097
7. Celej MS, Sarroukh R, Goormaghtigh E et al (2012) Toxic prefibrillar alpha-synuclein amyloid oligomers adopt a distinctive antiparallel beta-sheet structure. Biochem J 443:719–726
8. Hong DP, Han S, Fink AL et al (2011) Characterization of the non-fibrillar alpha-synuclein oligomers. Protein Pept Lett 18: 230–240
9. Emadi S, Kasturirangan S, Wang MS et al (2009) Detecting morphologically distinct oligomeric forms of alpha-synuclein. J Biol Chem 284:11048–11058
10. van Rooijen BD, van Leijenhorst-Groener KA, Claessens MM et al (2009) Tryptophan fluorescence reveals structural features of alpha-synuclein oligomers. J Mol Biol 394:826–833
11. Ehrnhoefer DE, Bieschke J, Boeddrich A et al (2008) EGCG redirects amyloidogenic polypeptides into unstructured, off-pathway oligomers. Nat Struct Mol Biol 15:558–566
12. Hong DP, Fink AL, Uversky VN (2008) Structural characteristics of α-synuclein oligomers stabilized by the flavonoid baicalein. J Mol Biol 383:214–223
13. Apetri MM, Maiti NC, Zagorski MG et al (2006) Secondary structure of alpha-synuclein oligomers: characterization by Raman and atomic force microscopy. J Mol Biol 355:63–71
14. Cremades N, Cohen SI, Deas E et al (2012) Direct observation of the interconversion of normal and toxic forms of α-synuclein. Cell 149:1048–1059
15. Lorenzen N, Nielsen SB, Buell AK et al (2014) The role of stable alpha-synuclein oligomers in

the molecular events underlying amyloid formation. J Am Chem Soc 136:3859–3868

16. Paslawski W, Mysling S, Thomsen K et al (2014) Co-existence of two different α-synuclein oligomers with different core structures determined by hydrogen/deuterium exchange mass spectrometry. Angew Chem Int Ed Engl 53:7560
17. Lorenzen N, Lemminger L, Pedersen JN et al (2013) The N-terminus of α-synuclein is essential for both monomeric and oligomeric interactions with membranes. FEBS Lett 588: 497–502
18. Lorenzen N, Nielsen SB, Yoshimura Y et al (2014) How epigallocatechin gallate can inhibit α-synuclein oligomer toxicity in vitro. J Biol Chem 289:21299, jbc.M114.554667
19. van Rooijen BD, Claessens MM, Subramaniam V (2010) Membrane permeabilization by oligomeric alpha-synuclein: in search of the mechanism. PLoS One 5:e14292
20. Stockl M, Claessens MMAE, Subramaniam V (2012) Kinetic measurements give new insights into lipid membrane permeabilization by α-synuclein oligomers. Mol Biosyst 8:338–345
21. Stockl MT, Zijlstra N, Subramaniam V (2013) Alpha-synuclein oligomers: an amyloid pore? Insights into mechanisms of alpha-synuclein oligomer-lipid interactions. Mol Neurobiol 47:613–621
22. van Rooijen BD, Claessens MM, Subramaniam V (2010) Membrane interactions of oligomeric alpha-synuclein: potential role in Parkinson's disease. Curr Protein Pept Sci 11:334–342
23. Zijlstra N, Blum C, Segers-Nolten IM et al (2012) Molecular composition of sub-stoichiometrically labeled alpha-synuclein oligomers determined by single-molecule photobleaching. Angew Chem Int Ed Engl 51: 8821–8824
24. Paslawski W, Andreasen M, Nielsen SB et al (2014) High stability and cooperative unfolding of cytotoxic α-synuclein oligomers. Biochemistry 53:6252
25. Volles MJ, Lee SJ, Rochet JC et al (2001) Vesicle permeabilization by protofibrillar a-synuclein: implications for the pathogenesis and treatment of Parkinson's disease. Biochemistry 40: 7812–7819
26. Studier FW (2005) Protein production by auto-induction in high density shaking cultures. Protein Expr Purif 41:207–234
27. Huang C, Ren G, Zhou H et al (2005) A new method for purification of recombinant human alpha-synuclein in Escherichia coli. Protein Expr Purif 42:173–177
28. Giehm L, Lorenzen N, Otzen DE (2011) Assays or alpha-synuclein aggregation. Methods 53: 295–305

Chapter 10

# Fluorescence Methods for Unraveling Oligomeric Amyloid Intermediates

**Niels Zijlstra, Nathalie Schilderink, and Vinod Subramaniam**

## Abstract

Amyloid oligomers are considered to be the relevant toxic species in many amyloid diseases and much research effort has been devoted to fully characterize these oligomers. Despite their importance, oligomers have proven to be difficult to characterize structurally. Information on their aggregation number is scarce, largely because standard techniques struggle to provide reliable results. In this chapter, we present two different methods that reproducibly yield fluorescently labeled α-Synuclein oligomers. We then discuss a new approach, combining single-molecule photobleaching and sub-stoichiometric fluorescent labeling, that we have developed to determine the aggregation number of supramolecular protein assemblies.

**Key words** Single molecule, Oligomer, α-Synuclein, Sub-stoichiometric, Photobleaching, Supramolecular assembly

## 1 Introduction

During the last 15 years, we have witnessed a major shift in the research focus to understand the cause of amyloid diseases such as Parkinson's, Alzheimer's, or Huntington's disease. The attention has shifted from the fully mature amyloid fibrils to the nanometer-sized aggregation intermediates called oligomers as the cytotoxic species that are at the basis of these diseases. There have been an increasing number of reports in the literature discussing the structure, composition, and the role of amyloid oligomers in disease [1–4]. In our laboratory, we focus on α-Synuclein (αSyn) oligomers, potentially key players in Parkinson's disease [5–7].

Ever since the first realization that the αSyn oligomers are cytotoxic and might play an important role in Parkinson's disease, much effort has been devoted to: (1) fully characterize these oligomers in terms of structure, morphology, and aggregation number [8–11] and (2) obtain detailed information on the formation process of these oligomers [4, 12]. One would expect that all the research effort combined would quickly provide detailed biophysical

David Eliezer (ed.), *Protein Amyloid Aggregation: Methods and Protocols*, Methods in Molecular Biology, vol. 1345, DOI 10.1007/978-1-4939-2978-8_10, 

insights into the αSyn oligomers. Instead, it became abundantly clear that, depending on the preparation protocol used, these oligomers exhibit a significant degree of structural and morphological heterogeneity, making it extremely challenging to identify a specific cytotoxic type of oligomer and to fully characterize these [12–14]. We, in particular, have extensively characterized the membrane binding and permeabilization capabilities of a specific, isolated, oligomeric species [15–17], and have focused on understanding the molecular composition and structural characteristics of this oligomeric species [10, 18].

Information on the aggregation numbers of the different oligomers is lacking, simply because standard techniques struggle to provide reliable results, since they either need suitable reference samples or need to determine the molecular weight of the oligomers first, which is very difficult for the typically unstable and heterogeneous oligomeric amyloid aggregates. Hence, a more direct approach is needed that avoids these problems.

We therefore developed a new approach to determine the aggregation number of protein aggregates that combines single-molecule photobleaching and sub-stoichiometric fluorescent labeling [18]. By counting the number of discrete photobleaching steps in the intensity time traces for a large number of distinct oligomers and by applying a statistical analysis on the histogram of bleaching steps, the aggregation number can be determined.

This approach allows us to directly study the aggregation number of αSyn oligomers, and large macromolecular protein assemblies in general, without the need to determine the molecular mass of the oligomers first, the need to compare it with a reference sample, or the need for a high spatial resolution. This approach is therefore very suitable for the sensitive detection of subtle changes in the aggregation number and makes a systematic study of the influence of the aggregation conditions on the aggregation number of the oligomers formed possible. Additionally, it allows us to detect a possible distribution in the number of monomers per oligomer.

First, we will discuss how to prepare fluorescently labeled αSyn monomers. Second, we will focus on two methods that reproducibly yield fluorescently labeled αSyn oligomers suitable for the combination of single-molecule photobleaching and sub-stoichiometric labeling. The first method is based on high protein concentrations and long incubation times [18], while the second method is based on short incubation times and the presence of the neurotransmitter dopamine [19]. Third, we will present the custom-built, single-molecule sensitive, optical microscope that we used to study the αSyn oligomers. We will highlight a few important aspects that need to be considered carefully for single-molecule detection. Subsequently, we will discuss our newly developed single-molecule photobleaching approach discussed above, which we have recently applied to the oligomers formed under both conditions mentioned above [18, 19]. Here, we will present some of the results obtained for the dopamine-induced αSyn oligomers [19].

## 2 Materials

αSyn is recombinantly expressed in *E. coli* BL21 DE3 cells using a pT7-7 vector, containing DNA encoded for αSyn (*see* **Note 1**). Single-cysteine mutants were engineered with a cysteine at position 140 of the amino acid sequence using the QuickChange Site-Directed Mutagenesis kit (Stratagene, USA). Expression and purification is performed as described previously, with minor alterations [20].

### 2.1 Expression and Purification of Monomeric αSyn

All solutions and buffers are prepared in 0.2 μm filtered ultrapure water and using the highest purity grade chemicals available. All chemicals are from Sigma-Aldrich unless mentioned otherwise.

To prevent the forming of disulfide bonds during purification of cysteine mutants, 1 mM freshly prepared DTT was added to all buffers.

1. Culture medium: Weigh 25 g of LB-Broth high salt. Add water to 1 L, dissolve, and autoclave. Store at room temperature. Transfer to a sterile 2 L baffled conical flask and add 1 mL of 100 mg/mL ampicillin (*see* **Note 2**).
2. 1 M IPTG: Weigh 2.38 g Isopropyl β-D-1-thiogalactopyranoside. Dissolve in water to a total volume of 10 mL. Store in 1 mL aliquots at −20 °C.
3. 0.5 M EDTA, pH 8.0: Weigh 18.6 g EDTA-$Na_2 \cdot 2H_2O$ (Molecular Weight 372.24). Add 80 mL water and adjust to pH 8.0 with 10 M NaOH while stirring (*see* **Note 3**). Add water to 100 mL and autoclave. Store at room temperature.
4. 1 M Tris–HCl, pH 8.0: Weigh 121.1 g Tris. Dissolve in 900 mL water and adjust pH to 8.0 with HCl. Add water to 1000 mL and autoclave. Store at room temperature.
5. 1 M DTT: Prepare a 154 mg/mL DL-Dithiothreitol solution in water just prior to use.
6. Lysis buffer: 10 mM Tris–HCl, 1 mM EDTA, 1 mM PMSF, pH 8.0: Take 500 μl of 1 M Tris–HCl, pH 8.0, and add 100 μl of 0.5 M EDTA, pH 8.0. Add ice-cold water to 49.5 mL. Add 500 μl of 100 mM PMSF just prior to use.
7. Tip sonicator.
8. Streptomycin sulfate.
9. Ammonium sulfate.
10. 0.2 μm pore size filter.

11. FPLC system.
12. Anion exchange column (*see* **Note 4**).
13. Buffer A: 10 mM Tris–HCl, pH 7.4. Weigh 1.21 g of Tris. Dissolve in 900 mL water and adjust to pH 7.4 with HCl. Add water to 1000 mL, filter through 0.2 μm pore size filter, and degas (*see* **Note 5**).
14. Buffer B: 10 mM Tris–HCl, 1 M NaCl, pH 7.4: Weigh 1.21 g of Tris and 58.44 g of NaCl. Dissolve in 900 mL water and adjust to pH 7.4 with HCl. Add water to 1000 mL, filter through 0.2 μm pore size filter, and degas.
15. Desalting column (*see* **Note 6**).

### 2.2 Fluorescent Labeling of Monomeric αSyn

1. Alexa Fluor 647 C2 Maleimide (Life Technologies, Invitrogen, Carlsbad, CA) (*see* **Note 7**).
2. 1 M DTT: Weigh 15.4 mg and dissolve in 100 μL water.
3. Buffer: 10 mM Tris–HCl, pH 7.4. Weigh 1.21 g of Tris. Dissolve in 900 mL water and adjust to pH 7.4 with HCl. Add water to 1000 mL and filter through 0.2 μm pore size filter.
4. Zebaspin desalting column 7K MWCO (Pierce, Thermo Scientific, USA).

### 2.3 Preparation and Purification of (Fluorescently Labeled) αSyn Oligomers

1. Vacuum evaporator.
2. 10 mM Phosphate buffer, pH 7.4: take 3.8 mL of 0.5 M $NaH_2PO_4$ and 16.2 mL of 0.5 M $Na_2HPO_4$, both not adjusted to pH. Add water to make a total volume of 1000 mL.
3. 200 mM Dopamine: make a 37.9 mg/mL solution in water of dopamine hydrochloride (*see* **Note 8**).
4. 0.22 μm Spin-x centrifuge filter (Corning, USA).
5. Size exclusion gel filtration column (*see* **Note 9**).
6. Eluent: 10 mM Tris–HCl, 150 mM NaCl, pH 7.4: Weigh 1.21 g of Tris and 8.76 g of NaCl. Dissolve in 900 mL water and adjust to pH 7.4 with HCl. Add water to 1000 mL, filter through 0.2 μm pore size filter, and degas.
7. 10 kDa MWCO Vivaspin concentrator (GE Healthcare, USA).

### 2.4 Single-Molecule Photobleaching

1. Spectroscopically very pure methanol (Methanol Uvasol, Merck Millipore, Germany).
2. Spectroscopically very pure water (Chromasolv Plus for HPLC, Sigma-Aldrich, Germany).
3. Glass Coverslips, thickness #1.5, 25 mm diameter (EMS Diasum, USA).
4. Ozone Cleaner (UV/Ozone Procleaner Plus, Bioforce, USA).
5. Spin coater (WS-400-6NPP, Laurell Technologies, USA).

6. Single-molecule sensitive optical microscope with sample scanning capability (*see* **Note 10**).
7. Emission filters: Long-pass filter (Razoredge, 664 nm, Semrock, USA) and band-pass filter (Brightline, 708/75 nm, Semrock, USA) (*see* **Note 11**).

## 3 Methods

### *3.1 Expression and Purification of Monomeric αSyn*

1. Inoculate 20 mL culture medium with bacteria cell culture and grow cells O/N in the shaking incubator at 37 °C, 150 rpm.
2. Transfer the O/N culture to 1 L of fresh culture medium.
3. Induce expression when $OD_{600}$ reaches between 0.6 and 0.8 with 1 mM IPTG (*see* **Note 12**), grow for another 3.5 h at 37 °C, 150 rpm.
4. Harvest cells by centrifugation: 4000 × *g*, 4 °C, 10 min (*see* **Note 13**).
5. Cell lysis: Add 50 mL ice-cold lysis buffer, stir for up to 1 h at 4 °C to completely resuspend the pellet, sonicate for 2 min using the tip sonicator while keeping the lysate on ice.
6. Centrifuge lysate: 9000 × *g*, 4 °C, 30 min.
7. Collect supernatant, add 0.5 g streptomycin sulfate (*see* **Note 14**), stir for 15 min, 4 °C.
8. Centrifuge lysate: 9000 × *g*, 4 °C, 30 min.
9. Collect supernatant, add 15 g ammonium sulfate, stir for 1 h, 4 °C (*see* **Note 15**).
10. Centrifuge lysate: 9000 × *g*, 4 °C, 30 min.
11. Carefully remove supernatant. At this point the precipitated protein pellet can be stored at −20 °C for several weeks.
12. Re-dissolve the protein pellet in 50 mL Buffer A and filter through a 0.2 μm pore size filter.
13. Equilibrate the Q-sepharose column with Buffer A.
14. Load the protein extract on the column.
15. Wash the column with 3 column volumes of Buffer A.
16. Elute bound proteins with a linear gradient of Buffer B from 0 to 50 % over 20 column volumes. Collect protein containing fractions. αSyn will elute at around 30 % of Buffer B.
17. Determine the purity by SDS-PAGE and Coomassie staining.
18. Equilibrate the desalting column with Buffer A.
19. Load the purified αSyn protein on the desalting column.
20. Elute the protein with Buffer A and collect the protein fractions.

21. Determine the concentration: Measure the absorbance at 276 nm using an extinction coefficient of 5600 $M^{-1}$ $cm^{-1}$. For each cysteine present in the protein add another 145 $M^{-1}$ $cm^{-1}$ [21].
22. Store purified protein in aliquots at a concentration of 250 μM in 10 mM Tris–HCl, pH 7.4 at −80 °C. 1 mM of freshly prepared DTT was added as required for the cysteine mutant.

### 3.2 Fluorescent Labeling of αSyn Monomers

1. Take 500 μl of 250 μM αSyn 140C in 10 mM Tris–HCl, 1 mM DTT, pH 7.4.
2. Add sixfold molar excess of freshly prepared DTT, incubate for 30 min at room temperature (*see* **Note 16**).
3. Remove excess DTT using a desalting column (*see* **Note 17**).
4. Add a threefold molar excess of the fluorescent dye Alexa Fluor 647 (*see* **Note 18**).
5. Incubate for 1 h in the dark at room temperature.
6. Remove free label by using two consecutive desalting columns.
7. Determine the concentration of both the fluorescently labeled αSyn and the unlabeled αSyn present in the sample by measuring the absorbance spectrum. The protein concentration was determined from the absorbance at 276 nm using an extinction coefficient of 5745 $M^{-1}$ $cm^{-1}$ [21], and the Alexa Fluor 647 concentration from the absorbance at 650 nm using an extinction coefficient of 239,000 $M^{-1}$ $cm^{-1}$ (*see* **Note 19**).
8. Determine the labeling efficiency (*see* **Note 20**).

### 3.3 Preparation and Purification of (Fluorescently Labeled) αSyn Oligomers

#### 3.3.1 Protocol Based on a High Protein Concentration and Long Incubation Time [18]

1. Prepare a mixture of protein with the desired ratio between αSyn wild-type and fluorescently labeled αSyn (*see* **Note 21**).
2. Completely dry the αSyn in a vacuum evaporator (*see* **Note 22**).
3. Dissolve the dried αSyn in water at a final protein concentration of 1 mM.
4. Incubate for 18 h in a shaking incubator at room temperature at 1250 rpm in the dark.
5. Equilibrate the size exclusion column with 10 mM Tris–HCl, pH 7.4, 150 mM NaCl (*see* **Note 23**).
6. Transfer the sample to 37 °C, incubate for 2 h without shaking in the dark.
7. Filter the solution over a 0.22 μm spin filter for 15 min at 15,000 × *g* to remove larger aggregates.
8. Inject the sample into the size exclusion column (*see* **Note 24**).
9. Separate the oligomers from monomers with 10 mM Tris–HCl, pH 7.4, 150 mM NaCl as eluent. Collect fractions of interest identified by the absorbance at 276 nm and the dye absorbance

at 650 nm. The oligomer fraction will appear around 8 mL and the monomer fraction will follow eluting at around 12 mL.

*OPTIONAL STEPS* (*see* **Note 25**):

10. Concentrate the oligomer sample with a Vivaspin 10 kDa concentrator.
11. Determine the concentration: Measure the absorbance at 276 nm using an extinction coefficient of 5600 $M^{-1}$ $cm^{-1}$ (*see* **Note 26**).
12. Store the oligomers at 4 °C (*see* **Note 27**).

#### *3.3.2 Dopamine-Induced Oligomers [19]*

1. Equilibrate the size exclusion column with 10 mM Tris–HCl, pH 7.4, 50 mM NaCl (*see* **Note 23**).
2. Prepare a mixture of protein with the desired ratio of αSyn wild-type and fluorescently labeled αSyn (*see* **Note 21**).
3. Divide in four vials each containing 500 μL and completely dry αSyn in a vacuum evaporator.
4. Add 890 μL of 10 mM phosphate buffer, pH 7.4, to each vial to obtain a final protein concentration of 140 μM.
5. Add 0.89 μL of 200 mM dopamine to make a final concentration of 200 μM dopamine.
6. Incubate for 3 h at 37 °C in the dark (*see* **Note 28**).
7. Filter the solution over a 0.22 μm spin filter for 15 min at 15,000 ×*g* to remove larger aggregates.
8. Inject the sample into the size exclusion column (*see* **Note 24**).
9. Separate the oligomers from monomers using 10 mM Tris–HCl, pH 7.4, 50 mM NaCl as eluent. Collect fractions of interest identified by the protein absorbance at 276 nm and the dye absorbance at 650 nm. The oligomer fraction will appear around 8 mL and the monomer fraction will follow eluting at around 12 mL.

*OPTIONAL STEPS* (*see* **Note 25**):

10. Concentrate the oligomer sample with a Vivaspin 10 kDa concentrator.
11. Determine the concentration using a BCA Protein Assay (*see* **Note 29**).
12. Store the oligomers at 4 °C (*see* **Note 30**).

### 3.4 Typical Single-Molecule Sensitive Optical Microscope

There are many ways to realize a single-molecule sensitive optical microscope. For the measurement procedure presented in this chapter, we used a custom-built apparatus (*see* Fig. 1) (for more details, *see* [22]). The main components are indicated in the figure, but in general high quality components should be used. Both the counting hardware and the Symphotime software used

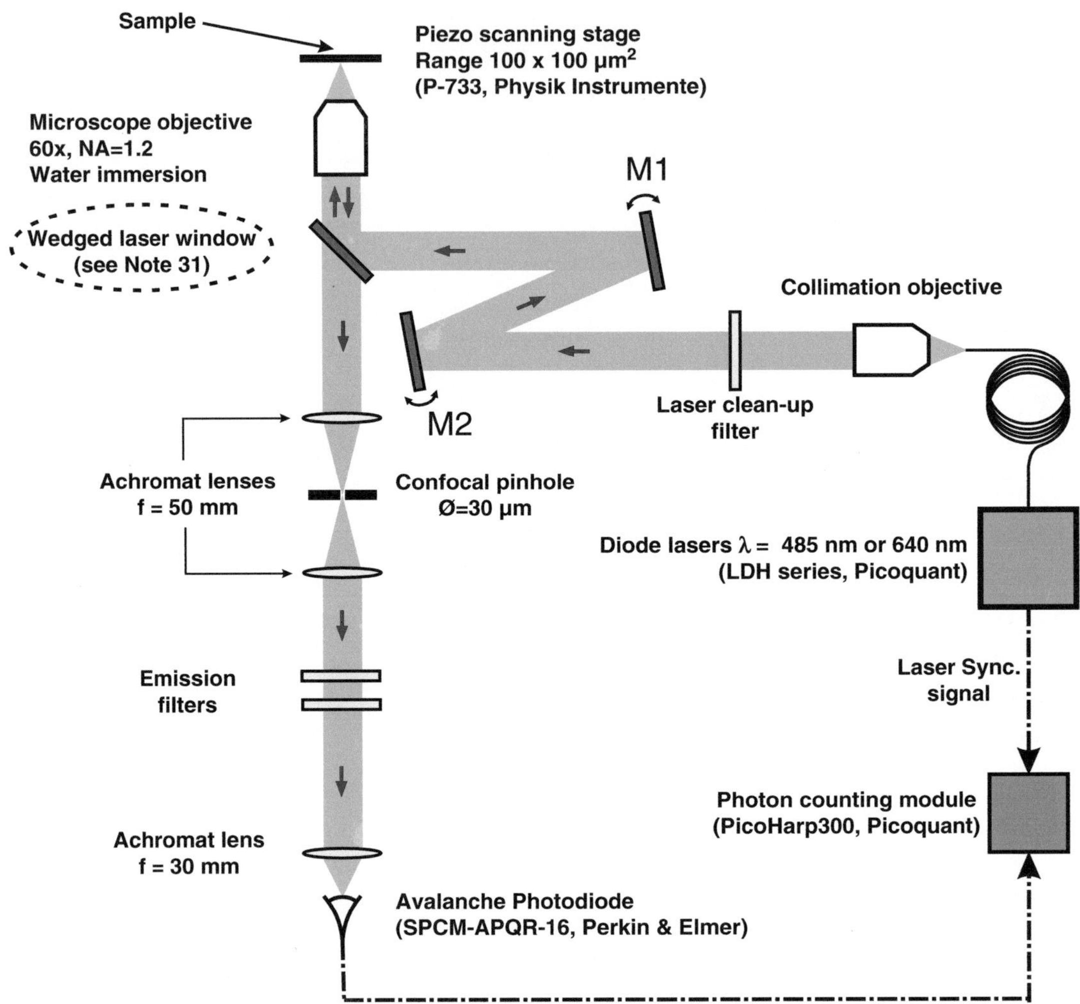


**Fig. 1** Schematic of the custom-built confocal microscope. *Gray paths* and *dash-dotted lines* denote optical and electrical signals, respectively. *Arrows* indicate the direction of the signals. An epi-illumination setup was used, i.e., illumination and emission collection through the same microscope objective. The emission light is spatially filtered by a single confocal pinhole before it is focused onto a single-photon counting avalanche photodiode

to control the setup were supplied by Picoquant (Germany). The Symphotime software is also used for the data visualization, although this could also be done in an external program, such as Matlab.

#### *3.4.1 Important Aspects Concerning the Optical Microscope*

1. It is very important to accurately collimate the laser beam after the collimation objective to ensure a full overlap between the excitation and detection volume. Therefore, the fiber output was mounted on a 5-axis platform that allows for a precise beam collimation and a high quality beam profile. The collimation can be checked by using, for example, a testing telescope (FR500/65/14.7, Möller-Wedel Optical, Germany).

2. To maintain the high quality Gaussian beam profile, a microscope objective was used for collimation instead of an achromat lens. Additionally, a microscope objective is typically better corrected for chromatic aberrations as compared to a lens, therefore allowing to maintain the precise beam collimation even when switching between different excitation wavelengths.
3. Make sure that the excitation beam goes straight and on-axis through the microscope objective. An easy way to do this is by replacing the microscope objective with a tube that has two diaphragms: one at the beginning and one at the end of the tube. By using the two adjustable mirrors M1 and M2, the position and angle of the excitation beam can be adjusted to go perfectly straight and on-axis through the two diaphragms and hence the microscope objective.
4. Minimize the number of interfaces in the detection path: every interface the emission light encounters on its way to the detector will introduce additional Fresnel losses and will hence decrease the fluorescence signal detected. The total collection efficiency of a typical single-molecule sensitive setup is inherently limited to about 8 %.
5. A confocal pinhole is not absolutely necessary, but it will significantly reduce the background signal. The lower the background intensity level, the easier it will be to determine whether the signal reached this level after the photobleaching time trace was recorded (*see* also **Note 39**).

### 3.5 Single-Molecule Photobleaching

#### 3.5.1 Clean Substrates

For single-molecule spectroscopy, it is very important to minimize the number of fluorescent contaminations on the sample substrate, since even the smallest fluorescent contamination can already be confused with a single or even a few fluorophores. Starting with a clean substrate is therefore essential. There are many different methods to obtain clean substrates. In our laboratory, we used the following protocol (*see* **Note 32**):

1. Rinse the glass coverslips with spectroscopically very pure Methanol to remove large particles and let it dry passively.
2. Place the coverslips in the UV/Ozone cleaner for at least 1 h to oxidize the contaminations on the surface and hence remove their florescence.

#### 3.5.2 Sample Preparation and Quality Control

To study the oligomers, they need to be immobilized and spatially separated, that is, the oligomers should not overlap within the diffraction limit of the microscope. We realized this as follows:

1. Dilute the fluorescently labeled oligomers to about 1 nM concentrations using the spectroscopically very pure water. Make sure you have at least about 100 μL of solution.

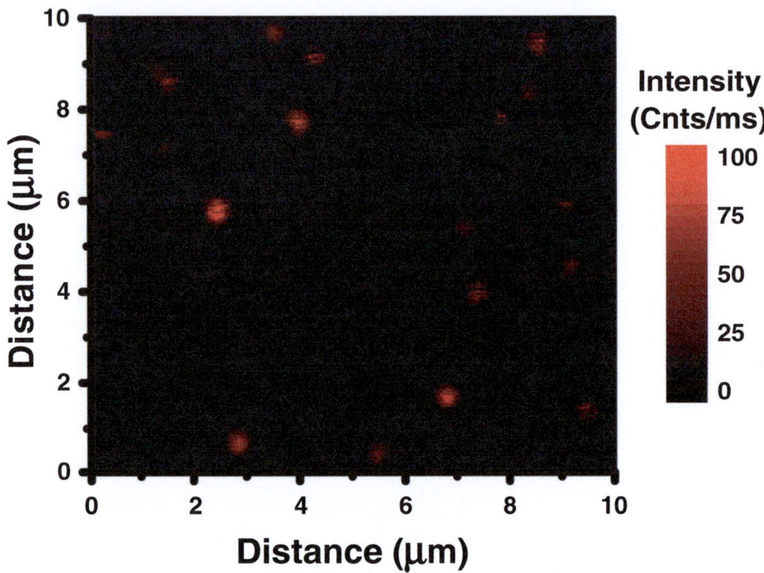


**Fig. 2** Typical area scan of fluorescently labeled αSyn oligomers spin coated on top of a clean coverslip recorded using an excitation power of ~750 W/cm$^2$ at 640 nm excitation wavelength (figure taken from [19]). The scan clearly shows well-separated fluorescence spots. The differences in intensity between the spots are on the one hand the result of different numbers of fluorescently labeled monomers incorporated into the oligomer due to the stochastic nature of aggregation and on the other hand due to the different dipole orientations of the fluorophores

2. Place the cleaned coverslip onto the spin coater. Make sure you turn on the vacuum at this point to immobilize the coverslip.
3. Add a drop of about 100 μL of the oligomer solution on top of the coverslip.
4. Directly after adding the drop, start the spin coating. Spin coat for 10 s at 6000 rpm (*see* **Note 33**).
5. Place the sample on the microscope and make an area scan to confirm that the oligomers are not overlapping within the diffraction limit of the microscope. A typical area scan is shown in Fig. 2, showing spatially well-separated oligomers (*see* **Note 34**).

#### 3.5.3 Measurement Procedure

In the next sections, turning the laser off just means that the laser light must not reach the sample. The laser does not have to be physically turned off; inserting a beam block into the beam is also sufficient.

1. Turn the laser off and reduce the laser power to 50–100 W/cm$^2$ (*see* **Note 35**).
2. Move to a new area on the sample as compared to the area used in the previous step.

3. Turn the laser on and make an initial area scan using short integration times to locate the individual oligomers (*see* **Note 36**). Turn the laser off.
4. Increase the laser power to about 750 W/cm$^2$ (*see* **Note 37**).
5. Position an empty area in the focus of the microscope objective, turn the laser on, and record a time trace to determine the background intensity level.
6. Turn the laser off.
7. Localize an oligomer in the focus of the microscope objective.
8. Start recording the intensity time trace.
9. Turn the laser on after the recording has been started (*see* **Note 38**).
10. After the recording is finished, turn the laser off again and do a quick visual inspection of the time trace to determine if it will be suitable for analysis (*see* **Note 39**). Typical time traces are shown in Fig. 3 (*see* **Note 40**).

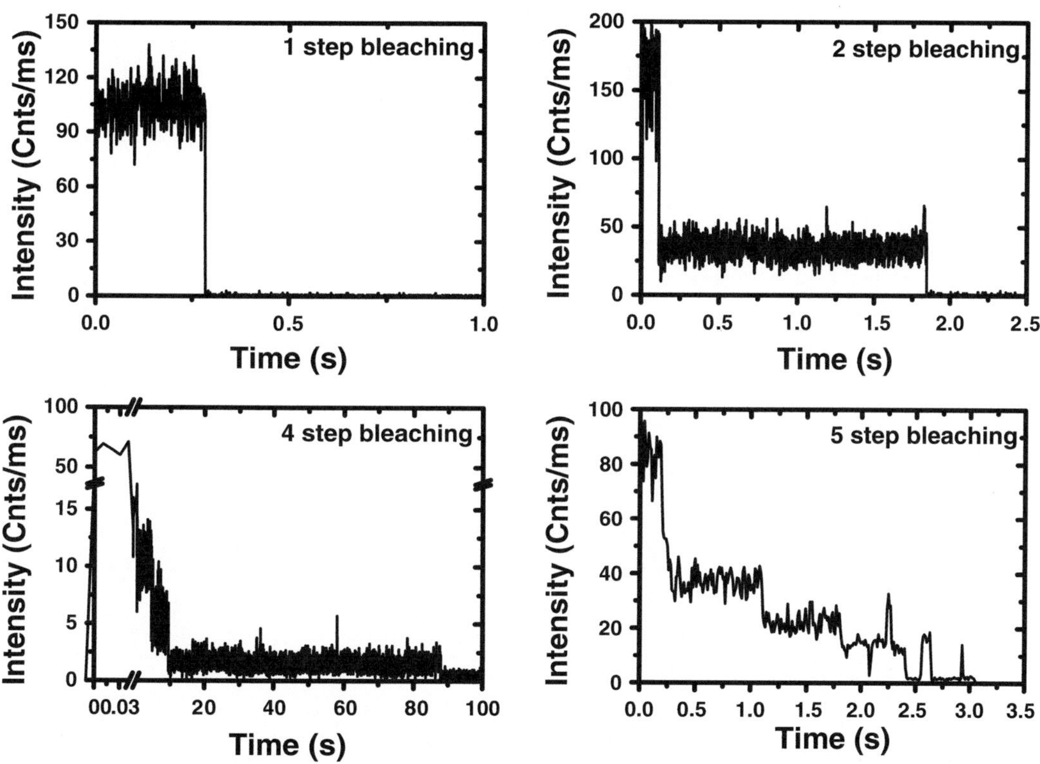


**Fig. 3** Four typical time traces for dopamine-induced αSyn oligomers fluorescently labeled with Alexa fluor 647 (figure taken from [19]). The time traces clearly show discrete photobleaching steps. The time binning is optimized for each individual time trace to visualize the bleaching steps best. The intensity is background subtracted

11. Locate the next oligomer into the focus of the microscope objective (*see* **Note 41**).
12. Repeat **steps 8–10** until time traces for about 120 oligomers are recorded (*see* **Note 42**).

#### 3.5.4 Analyzing Time Traces and Interpreting the Results

1. Go through all time traces and determine the number of bleaching steps per oligomer (*see* **Note 43**).
2. Make a histogram of all the number of bleaching steps found for the individual oligomers (*see* **Note 44**).
3. Check the peak position of the histogram. Depending on the position, it might be necessary to repeat the measurement with a different label density (*see* **Note 45**).
4. Fit the histogram of bleaching steps with the appropriate number of species to determine the aggregation number(s) (*see* **Notes 46** and **47**).
5. Repeat the entire procedure for at least two different label densities and do one label density twice. All label densities should give the same aggregation number if you are working within the optimal range of label densities. A typical series of histograms and fits for dopamine-induced αSyn oligomers are shown in Fig. 4 (*see* **Note 48**).

## 4 Notes

1. We assume you have your own αSyn expression vectors.
2. The antibiotic used here, ampicillin, is based on the expression vector's selection criteria. Alternatively, carbenicillin can be used, which is more resistant to degradation than ampicillin. Always add the antibiotic just prior to use.
3. EDTA is not easy to dissolve; it will require the solution to be at pH 8.0. Stir vigorously while adjusting the pH to dissolve all the powder.
4. We use a ResourceQ column (GE Healthcare, USA). Check the capacity of your column, since exceeding the maximal capacity of the column will result in loss of protein.
5. Degassing can be performed by sonicating the buffer solution in a water bath sonicator for about 10 min.
6. We use a HiPrep 26/16 Desalting column (GE Healthcare, USA). For small volumes a PD10 desalting column can be used (GE Healthcare, USA).
7. Alternatively, Alexa Fluor 488 C5 maleimide can be used (Life Technologies, Invitrogen, Carlsbad, CA), or similar dyes from Atto-Tec (Germany). Always dissolve the dye in anhydrous DMSO to prevent hydrolysis and store aliquots at −80 °C.

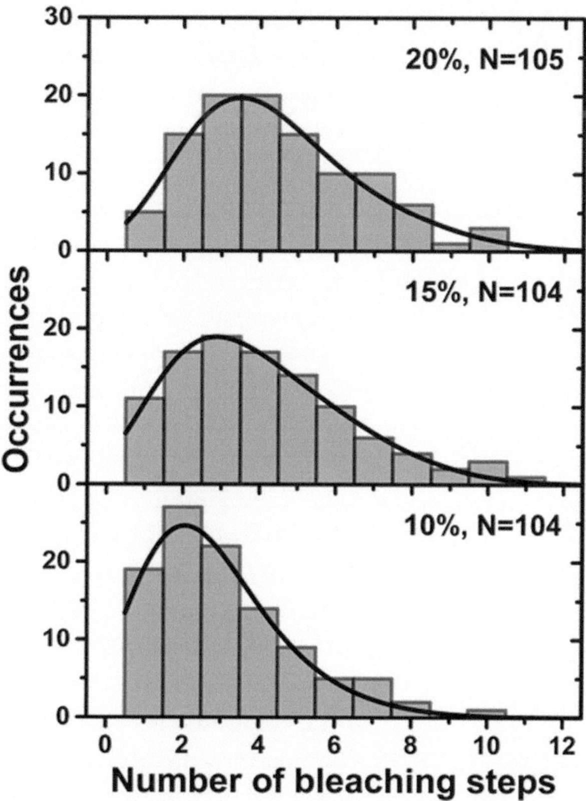


**Fig. 4** Histograms of bleaching steps for dopamine-induced αSyn oligomers with a 20, 15, and 10 % label density (figure taken from [19]). Each histogram is built from bleaching traces of at least 100 distinct oligomers (*N*). The histograms are fitted with a combination of two binomial distributions (*solid black lines*), according to Eq. 1. The mean number of monomers per oligomer is determined as 17 ± 5 and 31 ± 6 for the 20 % label density, 18 ± 5 and 36 ± 4 for the 15 % label density, and 23 ± 5 and 44 ± 14 for the 10 % label density

8. Dopamine solutions should be kept in the dark at all times to prevent light-induced degradation. Do not use the solution anymore if it is discolored in any way.
9. We use the Superdex200 10/300GL size exclusion column (GE Healthcare, USA).
10. A sample scanning approach is used, since it is essential to move the sample with a high repeatability to guarantee that the same single molecule is located in the detection volume again after the initial area scan used to localize them. A high repeatability, typically less than a few nanometers, is provided by piezo scanning stages. Beam scanning approaches do typically not provide this high repeatability.
11. The emission filters should be optimized for each experiment and depend on the excitation wavelength and fluorophores used.

The filters used to suppress the remaining laser light should offer at least an optical density of 6.

12. This will take approximately 3 h. At $OD_{600}$ between 0.6 and 0.8 the bacteria are in log phase of their growth and are most viable.
13. At this point the bacteria pellet can be stored at −20 °C.
14. After adding streptomycin sulfate, the nucleic acids will bind via electrostatic interactions and the complex precipitates. Do not incubate for longer than 15 min, since also protein can be precipitated by streptomycin sulfate.
15. Slowly add the ammonium sulfate to the solution, while stirring at 4 °C. Adding it all at once will result in wrongly precipitated proteins, as the initial concentration will be too high.
16. The cysteines need to be reduced prior to labeling, since maleimides do not react with disulfides.
17. We use 2 mL Zeba spin desalting columns 7K MWCO (Pierce, Thermo Scientific, USA). As a buffer, 10 mM Tris–HCl pH 7.4 is used, following manufacturer's protocol.
18. A twofold excess of the dye is in principle sufficient for labeling, but a threefold excess will increase the labeling efficiency.
19. To measure the absorbance spectrum, it is very important to use a spectrophotometer with a high sensitivity and reproducibility. We therefore advise to use a Shimadzu UV-2401PC spectrophotometer or equivalent.

    To determine the protein concentration, it is important to correct for the contribution of the absorbance of the fluorescent dye at 276 nm. For Alexa Fluor 647, this contribution is about 5 % of the total absorbance at 650 nm. The total protein concentration can therefore be calculated using:

$$[\alpha\mathrm{Syn}] = \frac{A_{276} - (0.05 \times A_{650})}{\varepsilon}$$

    where [αSyn] is the total protein concentration, $A_{276}$ is the absorbance at 276 nm, $A_{650}$ is the absorbance at 650 nm, and $\varepsilon$ is the extinction coefficient, which is in this case 5745 $M^{-1}$ $cm^{-1}$ [21].

    Furthermore, the concentration of labeled αSyn is assumed to be equal to the dye concentration, since we removed all free dye from the solution, while the concentration of unlabeled protein is the difference between the total protein concentration and the concentration of labeled αSyn.
20. The labeling efficiency is the ratio between the concentration of labeled protein and the total protein concentration. It is a measure of the quality of the labeling procedure. Generally, a labeling efficiency of >90 % is achieved.

21. We start with 2 mL of 250 μM protein concentration with a 7.5–30 % label density. When calculating how much wild-type protein to add to the labeled protein, do not forget to take the unlabeled protein determined in **Note 19** into account.
22. We divide the protein solution in several vials, each containing 250 μl. It takes 4–5 h to completely dry the samples.
23. Depending on the storage conditions, the column must first be equilibrated with water. It is convenient to start with column equilibration 1 day prior to oligomer purification due to the low flow rate used and therefore long equilibration time.
24. Make sure that the total volume of the oligomer solution is matching the volume the size exclusion column can accept. If necessary, concentrate the sample using a Vivaspin 10 kDa concentrator to the appropriate volume.
25. For the single-molecule photobleaching experiments mentioned in this chapter, it is not necessary to do **steps 10** and **11**.

    The oligomer yield will be no more than 5 % of the total amount of protein started with. Increasing the amount of starting material is possible, but the total sample volume needs to match the volume that the size exclusion column can accept (*see* also **Note 24**).
26. Depending on the fraction of labeled protein present in the sample, a different extinction coefficient might be more appropriate. If the sample mainly consists of wild-type protein, as is the case for the low label densities used here, the extinction coefficient of 5600 $M^{-1}$ $cm^{-1}$ gives a good estimate of the protein concentration. Of course, also in this case there will be a contribution of the fluorescent label on the protein absorbance that needs to be taken into account (*see* also **Note 19**).
27. The oligomers are stable for about 4 weeks when stored at 4 °C.
28. For the fluorescent dye as well as for the dopamine used, it is required to keep the sample in the dark.
29. Dopamine shows absorption at 280 nm; a standard protein absorption measurement is therefore not possible. To quantify the dopamine-induced oligomers, a BCA protein assay can be used.
30. The dopamine-induced oligomers are stable for about 2 weeks when stored at 4 °C.
31. A wedged glass plate is used to deflect the ghost beam (e.g., BSF10-A, Thorlabs, Germany). Alternatively, a dichroic mirror could be used. Make sure that the dichroic mirror is sufficiently thick (at least 3 mm) to ensure maintaining the high quality beam profile.
32. This method was chosen because it offers an easy method to clean many coverslips simultaneously without the need to use

aggressive chemicals. Alternative methods that we explored and also result in clean substrates are prolonged soaking in 65 % Nitric acid (at least 2 days), cleaning with a Piranha solution, or oxygen plasma cleaning.

33. It is important to start the spin coating as quickly as possible after the drop is added. Waiting before starting the spin coating might result in much more densely packed oligomers in the center of the sample where the drop was deposited due to pre-drying.
34. An easy way to determine whether the oligomers are indeed well separated and not, for example, clustered, is by determining the spot size and shape. If the spots are larger than the diffraction limit or if they are elongated in one direction, it is likely that there is more than one oligomer present. Typically, we aim for about 10–15 oligomers per $10 \times 10\ \mu m^2$ area, which will ensure spatially separated oligomers.
35. Low excitation powers are necessary to prevent photobleaching of the dyes before recording of the actual photobleaching time trace starts. We typically use about 50–100 $W/cm^2$, but this of course depends on the fluorophores used. In general, the lower the excitation power, the better.
36. Similar to **Note 34**, short integration times are necessary to prevent photobleaching of the dyes before the actual measurements start. We typically use 2 ms integration time.
37. This power needs to be optimized for the specific microscope used to ensure time traces that can be analyzed while at the same time all fluorophores are bleached within a reasonable time. We optimized the powers for recording times of about 180 s.
38. It is important not to turn on the laser before the recording is started, because the first fluorophores usually photobleach very quickly. Usually it takes some time before the actual recording starts and the first bleaching events might therefore be missed.
39. The first important check is whether the time trace levels off at the background intensity level, since this shows that all fluorophores are indeed bleached. If this is not the case, the trace is not useable. Furthermore, the oligomer might contain too many fluorescent labels and as a consequence, the time trace will look like an exponentially decaying curve, which is not analyzable. Additionally, focus drift or heavy blinking could render the time trace not useable.
40. The discrete bleaching steps have different amplitudes since the oligomers are immobilized on a surface. The immobilization will restrict the rotational freedom of the fluorescent dyes and hence, depending on their dipole orientation, the fluorescent dyes will show different intensity drops.

41. It is important not to spend too much measuring time at one area, since there is the possibility of focus drift. Focus drift will make it much more difficult to analyze the time traces since the collected signal will significantly decrease resulting in difficulty identifying the individual bleaching steps. Re-focusing without making a new area scan is not recommended, since the sample only has to move about 100 nm to completely lose the position of the oligomers. Of course, the total time you can spend on one area depends on the stability of the microscope used.
42. Since sub-stoichiometric labeling always results in a distribution in the number of bleaching steps, it is important to analyze a statistically relevant number of distinct oligomers, that is, more than 100 distinct oligomers. In our experience, about 10 % of the time traces is not analyzable (even after the initial quick visual inspection of **Note 39**), due to heavy blinking, focus drift, or the presence of too many fluorophores. It is therefore recommended to record about 120 traces in total. As mentioned in **Note 33**, typically there are 10–15 oligomers per area, so about ten different areas need to be studied. To avoid that you look at sample dependent fluorescence contaminations, it is advisable to use at least three different samples.
43. Not all time traces will be analyzable immediately, and you probably have to fine-tune the time binning and zoom. The optimal time binning does not have to be the same for the entire time trace, that is, some time windows might require a different time binning. Make sure you do not count intensity levels twice. Be careful with determining steps in the beginning of the time trace; the very fast bleaching events will be at the beginning and are easy to miss when using an incorrect time binning.
44. Make sure at least 100 distinct oligomers were analyzed. *See* **Note 41** for the reason for at least 100 distinct oligomers.
45. If the oligomers are too small for the label density used, the histogram will have a large peak at one bleaching step. To obtain more reliable results, the label density should be increased. On the other hand, if the aggregates are too large for this label density, there will be many bleaching traces that cannot be analyzed due to the many bleaching steps (*see* **Note 38**). In this case, the peak in the histogram will be at large number of bleaching steps (9 or 10). The label density should now be decreased for more reliable results. In general, multiple runs with different label densities are needed to obtain sufficient insight into the system to determine whether the optimal label density is used.
46. We use OriginPro 9 to fit the histograms of bleaching steps, but of course other programs can be used. The stochastic incorporation of labeled monomers in the oligomers is described by a classical Bernoulli process in which there is no preference

for either labeled or unlabeled monomers. The histograms should therefore be fitted with a (combination of) binomial distribution(s), given by:

$$\sum_{i=1}^{s} A_i \binom{n_i}{k} p^k (1-p)^{n_i-k} \qquad (1)$$

where $A_i$ is the total number of analyzed oligomers for species $i$ with aggregation number $n_i$, $s$ is the total number of species, $k$ the number of fluorescent labels, and $p$ is the label density.

The mean values of the two binomial distributions give the mean number of monomers per oligomer for all species.

47. The appropriate number of species can be determined using the fit quality. If the fit quality reduces after adding more species, the histogram is being overfitted. The fit quality is characterized by the reduced chi-squared parameter $\chi_{\text{red}}^2$:

$$\chi_{\text{red}}^2 = \frac{\sum_{i=1}^{n} \left(y_i - y_{i,\text{fit}}\right)^2}{\text{df}}$$

where $y_i$ is the experimental value, $y_{i,\text{fit}}$ is the fit value, $n$ is the total number of data points, and df are the degrees of freedom, that is, the total number of data points minus the number of fit parameters.

48. The numbers of monomers that form an oligomer are, within the error bars, identical for the 20 and 15 % label densities. For the 10 % label density, we observe deviations that can be explained by the low label density. In this case, more than 15 % of the oligomers do not contain a fluorescent label and are hence not included in the histogram of bleaching steps. This will result in the observed overestimation of the aggregation number. This severe overestimation of the aggregation number of the smaller species also results in an overestimation of the number of monomers in the larger species.

## Acknowledgements

This work was financially supported by the Nederlandse Organisatie voor Wetenschappelijk Onderzoek (NWO) through the NWO-CW TOP program number 700.58.302. We further acknowledge support from the Stichting Internationaal Parkinson Fonds. NS is supported by a research program of the Foundation for Fundamental Research on Matter (FOM) entitled "A Single Molecule View on Protein Aggregation." NZ is also supported by NanoNextNL, a consortium of the Government of the Netherlands and 130 partners.

## References


1. Volles MJ, Lansbury PT (2003) Zeroing in on the pathogenic form of α-synuclein and its mechanism of neurotoxicity in Parkinson's disease. Biochemistry 42(26):7871–7878
2. Uversky VN (2010) Mysterious oligomerization of the amyloidogenic proteins. FEBS J 277(14): 2940–2953
3. Demuro A et al (2005) Calcium dysregulation and membrane disruption as a ubiquitous neurotoxic mechanism of soluble amyloid oligomers. J Biol Chem 280(17): 17294–17300
4. Lorenzen N et al (2014) The role of stable α-synuclein oligomers in the molecular events underlying amyloid formation. J Am Chem Soc 136:3859–3868
5. Lashuel HA et al (2002) Neurodegenerative disease - amyloid pores from pathogenic mutations. Nature 418:291
6. Caughey B, Lansbury PT (2003) Protofibrils, pores, fibrils, and neurodegeneration: separating the responsible protein aggregates from the innocent bystanders. Annu Rev Neurosci 26: 267–298
7. Winner B et al (2011) In vivo demonstration that alpha-synuclein oligomers are toxic. Proc Natl Acad Sci U S A 108:4194–4199
8. Kim H-y et al (2009) Structural properties of pore-forming oligomers of alpha-synuclein. J Am Chem Soc 131:17482–17489
9. Hong D-P, Fink AL, Uversky VN (2008) Structural characteristics of alpha-synuclein oligomers stabilized by the flavonoid baicalein. J Mol Biol 383:214–223
10. van Rooijen BD et al (2009) Tryptophan fluorescence reveals structural features of alpha-synuclein oligomers. J Mol Biol 394: 826–833
11. Giehm L et al (2011) Low-resolution structure of a vesicle disrupting α-synuclein oligomer that accumulates during fibrillation. Proc Natl Acad Sci 108(8):3246–3251
12. Cremades N et al (2012) Direct observation of the interconversion of normal and toxic forms of α-synuclein. Cell 149:1048–1059
13. Emadi S et al (2009) Detecting morphologically distinct oligomeric forms of α-synuclein. J Biol Chem 284(17):11048–11058
14. Apetri MM et al (2006) Secondary structure of alpha-synuclein oligomers: characterization by Raman and atomic force microscopy. J Mol Biol 355:63–71
15. van Rooijen BD, Claessens MMAE, Subramaniam V (2009) Lipid bilayer disruption by oligomeric α-synuclein depends on bilayer charge and accessibility of the hydrophobic core. Biochim Biophys Acta Biomembr 1788(6):1271–1278
16. Stöckl M, Zijlstra N, Subramaniam V (2013) α-Synuclein oligomers: an amyloid pore? Mol Neurobiol 47(2):613–621
17. Stefanovic AND et al (2014) α-Synuclein oligomers distinctively permeabilize complex model membranes. FEBS J 281(12):2838–2850
18. Zijlstra N et al (2012) Molecular composition of sub-stoichiometrically labeled α-synuclein oligomers determined by single-molecule photobleaching. Angew Chem Int Ed 51(35): 8821–8824
19. Zijlstra N et al (2014) Elucidating the aggregation number of dopamine-induced α-synuclein oligomeric assemblies. Biophys J 106(2):440–446
20. van Raaij ME, Segers-Nolten IMJ, Subramaniam V (2006) Quantitative morphological analysis reveals ultrastructural diversity of amyloid fibrils from α-synuclein mutants. Biophys J 91(11):L96–L98
21. Pace CN et al (1995) How to measure and predict the molar absorption coefficient of a protein. Protein Sci 4(11):2411–2423
22. Zijlstra N (2014) Parkinson's disease in the spotlight: unraveling nanoscale α-Synuclein oligomers using ultrasensitive single-molecule spectroscopy. University of Twente, Enschede

# Part IV

## Fibrils

# Chapter 11

# Preparation of Amyloid Fibrils for Magic-Angle Spinning Solid-State NMR Spectroscopy

Marcus D. Tuttle, Joseph M. Courtney, Alexander M. Barclay, and Chad M. Rienstra

## Abstract

Solid-state NMR spectroscopy (SSNMR) is an established and invaluable tool for the study of amyloid fibril structure with atomic-level detail. Optimization of the homogeneity and concentration of fibrils enhances the resolution and sensitivity of SSNMR spectra. Here, we present a fibrillization and fibril processing protocol, starting from purified monomeric α-synuclein, that enables the collection of high-resolution SSNMR spectra suitable for site-specific structural analysis. This protocol does not rely on any special features of α-synuclein and should be generalizable to any other amyloid protein.

**Key words** Fibrillization, α-synuclein, Amyloid fibrils, Solid-state NMR, Size exclusion chromatography

## 1 Introduction

Characterization of amyloid fibrils via magic-angle spinning (MAS) solid-state NMR (SSNMR) spectroscopy has led to previously inaccessible insights into the structure of proteins including amyloid-β [1–3], Het-S [4, 5], and α-synuclein [6–8]. Obtaining high-resolution data through the proper production of fibrils from monomeric protein solution is paramount to the investigation of the structure, function, and pathology of amyloids. Many biophysical studies rely on assays performed directly on aliquots taken from the fibrillization mixture [9] or on sample production techniques with insufficient yields to perform NMR spectroscopy. However, in MAS SSNMR, spectra of the unrefined fibrillization mixture can exhibit broad spectral features arising from heterogeneities in the fibril size or polymorphism, as well as background signal from residual, soluble monomer [10]. The following protocol maximizes the fraction of protein incorporated into mature fibrils and describes a washing procedure to reduce ionic strength and

David Eliezer (ed.), *Protein Amyloid Aggregation: Methods and Protocols*, Methods in Molecular Biology, vol. 1345, DOI 10.1007/978-1-4939-2978-8_11, © Springer Science+Business Media New York 2016

remove remaining monomer and small protofibrillar aggregates, greatly improving the homogeneity of the sample as well as the performance of the NMR instrument, thereby resulting in higher quality NMR data.

The use of this procedure has consistently produced α-synuclein samples of homogenous morphology that give high-resolution spectra (<0.5 ppm $^{13}C$ linewidths in uniformly $^{13}C$,$^{15}N$ labeled samples and <0.3 ppm $^{13}C$ linewidths for sparsely labeled samples [7, 11]) that facilitate the assignment of chemical shifts for the fibrillar cores of wild-type α-synuclein and the early-onset related mutants: A30P, E46K, and A53T [7, 12, 13].

## 2 Materials

All solutions are made with ultrapure water deionized to a resistivity of 18.2 MOhm, filtered through a 0.22 μm filter, and stored at room temperature until use, unless otherwise noted. Reagents for buffers and solutions are analytical grade. Follow all applicable waste disposal regulations when disposing of waste materials. Although the safety of handling recombinant amyloid proteins has not been assessed in all cases, we recommend using appropriate personal protective equipment to avoid direct contact whenever possible [9].

### 2.1 Size Exclusion Chromatography

1. Purified α-synuclein monomer in solution [10, 14] (*see* **Note 1**).
2. Amicon stirred cell concentrator 200 mL, Model 8200 (Part Number part 5123) (Millipore).
3. 3 kDa molecular weight cut-off membrane for Amicon stirred cell concentrator.
4. Amicon Ultra-15 Centrifugal Filter Unit with Ultracel-3 membrane, 3 kDa NMWL (Part Number UFC900308) (EMD Millipore).
5. Eppendorf Centrifuge 5810R equipped with swinging bucket rotor A-4-62.
6. 0.22 μm PES syringe tip filter.
7. 3 L Size exclusion buffer: 100 mM sodium phosphate, 0.1 mM EDTA, 0.01 % w/w $NaN_3$, pH 7.4, stored at 4 °C, sterile filtered and degassed.
8. 2 L Ultrapure water, stored at 4 °C, sterile filtered and degassed.
9. GE Healthcare's HiPrep 26/60 Sephacryl S-200 High Resolution gel filtration column, 320 mL column volume.
10. GE Healthcare's ÄKTAprime plus FPLC equipped with fraction collector, UV detector, and chart recorder or recording software.
11. 50 fraction collection tubes of at least 5 mL each.
12. 5 mL sample loop.

### 2.2 Final Concentration

1. Amicon stirred cell concentrator 200 mL, Model 8200 (Part Number part 5123) (EMD Millipore).
2. 3 kDa molecular weight cut-off membrane for Amicon stirred cell concentrator.
3. Pre-fibrillization Dilution Buffer: 0.1 mM EDTA, 0.01 % $NaN_3$ in $H_2O$.
4. Amicon Ultra-15 Centrifugal Filter Unit with Ultracel-3 membrane, 3 kDa NMWL (Part Number UFC900308) (EMD Millipore).
5. Eppendorf Centrifuge 5810R equipped with swinging bucket rotor A-4-62.

### 2.3 Seeding and Fibrillization

1. Fibrillization Buffer: 50 mM sodium phosphate, 0.1 mM EDTA, 0.01 % $NaN_3$, pH 7.4.
2. Beckman Coulter Tubes, with Snap-On Cap, Polypropylene, 1.5 mL, 11 × 38 mm, Natural (Part Number 357448) (*see* **Note 2**).
3. 0.22 μm PES syringe tip filter.
4. Preformed α-synuclein fibrils (*see* **Note 3**).
5. Parafilm.
6. New Brunswick Scientific Co., C-24 Classic Benchtop Incubator Shaker.

### 2.4 Fibril Harvest

1. Handheld pellet pestle grinder (Part Number Z359971-1EA) (Sigma-Aldrich).
2. Pellet pestle, autoclaved (Part Number Z359947) (Sigma-Aldrich).
3. Beckman TLA-100.3 ultracentrifuge rotor with adapters (Part Number 364701).
4. An air manifold with nylon tubing.

### 2.5 Rotor Packing

1. Rotor Sleeve, 3.2 mm, 22 μL (Part Number MSPA005053) (Agilent Technologies).
2. Packing Tool Kit, 3.2 mm, 22 μL (Part Number MSPA000272) (Agilent Technologies).
3. 3.2 mm, 22 μL, Drive Cap, Torlon, CS (Part Number MP4140-001) (Revolution NMR).
4. Half Top Cap, 3.2 mm, 22 μL (Part Number MP4603-001) (Revolution NMR).
5. Two Spacers, Custom Made, Kel-F, one 0.078″ diameter, 0.073″ length, one 0.078″ diameter, 0.153″ length. All measurements within +/−0.001″ tolerance.
6. Rubber discs, Custom Made, 0.078″ diameter, 0.063″ length.
7. Disposable aluminum dishes (Part Number Z154857-1PAK) (Sigma-Aldrich).

## 3 Methods

Carry out all procedures at room temperature unless otherwise specified.

### 3.1 Size Exclusion Chromatography

Size exclusion chromatography is performed in a cold box or cold room at 4 °C.

1. Equilibrate the 26/60 Sephacryl S-200 High Resolution gel filtration column with at least 3 column volumes of size exclusion buffer using a flow rate of 0.5 mL/min with the ÄKTAprime.
2. Using the Amicon stirred cell concentrator with a 3 kDa molecular weight cut-off filter, bring the total volume of purified α-synuclein in solution to ~5 mL (*see* **Note 4**).
3. Transfer the concentrate to an Amicon Ultra-15 centrifugal Filter Unit, rinse the stirred cell membrane and concentrator with ~5 mL of size exclusion buffer, and add the rinsate to the same centrifugal filter unit (*see* **Note 5**).
4. Concentrate the sample down to ~5.5 mL by centrifugation in the swinging bucket rotor A-4-62 at 4000 rpm (3217 ×*g*). Spin time will vary with monomer concentration.
5. Draw the concentrate into a 10 mL syringe and pass through a 0.22 μm filter into a sterile plastic tube. Draw the filtered solution into a second 10 mL syringe and tap out all bubbles just prior to injection into the FPLC sample loop.
6. Run the sizing column at a flow rate of 1.3 mL/min after injecting the sample into the sample loop as follows: 0–5 mL, inject valve set to LOAD; 5.1–25 mL, inject valve set to INJECT; 25.1–980 mL, inject valve set to LOAD. 6.5 mL fractions are collected over 40.1–320 mL.
7. Combine Fractions containing the protein (*see* **Notes 6** and **7**).

### 3.2 Final Concentration

1. Add the combined fractions to the stirred cell concentrator with a 3 kDa cut-off membrane and add enough pre-fibrillization dilution buffer to double the total volume, bringing the phosphate buffer concentration to 50 mM (*see* **Note 4**).
2. Take an $A_{280}$ to determine the protein concentration. We use an experimentally determined extinction coefficient equal to 5165 mM -1 cm -1, corresponding to 2.8 mg/mL. Literature values may vary. The target concentration for fibrillization is 15 mg/mL.
3. Concentrate the solution down until you have less than 15 mL remaining or you have reached 15 mg/mL concentration of the monomer. Transfer the concentrate to an Amicon Ultra-15 centrifugal Filter Unit, rinse the stirred cell membrane and concentrator with ~5 mL of diluted size exclusion buffer, and add the rinsate to the same centrifugal filter unit (*see* **Note 5**).

4. Concentrate further with the Amicon centrifugal concentrator to a concentration slightly above 15 mg/mL as measured by $A_{280}$ and add the solution to a 25 mL conical centrifuge tube.
5. Use an appropriate volume of Fibrillization Buffer to rinse the Amicon concentrator such that, when added to the monomer solution, it reaches a final monomer concentration to 15 mg/mL (*see* **Note 8**).

### 3.3 Seeding and Fibrillization

1. Measure the total volume of concentrated α-synuclein monomer. Gather enough Beckman Coulter ultracentrifuge tubes to divide the solution into 0.5 mL fractions in each tube.
2. Using a syringe and long needle, collect the monomer solution into the syringe.
3. Remove the needle and attach the 0.22 μm syringe tip filter.
4. Divide the solution by 0.5 mL into separate microfuge tubes. If there is less than 100 μL remaining, split it between the other tubes, otherwise add it to a new tube.
5. Combine the preformed fibrils and enough of the Fibrillization buffer to bring the concentration to ~15 mg/mL (*see* **Note 9**).
6. Bath sonicate the preformed fibril mixture for 30 s or until the fibrils appear to be well dispersed.
7. Add 25 μL of the sonicated mixture to each microfuge tube to seed the fibril growth at a 5 % seeding level. If a tube has a volume other than 0.5 mL, adjust the volume of seed solution to compensate.
8. Close and tightly parafilm each tube. Vortex each for 30 s.
9. Add the fibrils to the Benchtop Incubator Shaker set to 37 °C at 200 rpm.
10. Within 24–72 h the solution should begin forming a gel and may become cloudy.
11. Allow to fibrillize for 3 weeks. This time frame shows a convergence of SSNMR chemical shifts (Fig. 1) within the first week and a maximized fibril yield by the end of the 3-week incubation.

### 3.4 Fibril Harvest

1. Collect the microfuge tubes from the incubator and remove all parafilm.
2. Ultracentrifuge at 4 °C and 55,000 rpm (163,202 ×*g*) for 1–2 h in a TLA-100.3 rotor with adapters for the microfuge tubes. Decant the supernatant and save for analysis (*see* **Note 10**).
3. Add 100 μL of sterile filtered ultrapure water to each tube and use the pellet pestle grinder to resuspend the fibrils.
4. Combine three tubes into one and rinse out each empty microfuge tube with an additional 100 μL of sterile filtered ultrapure water and add to the combined suspension.

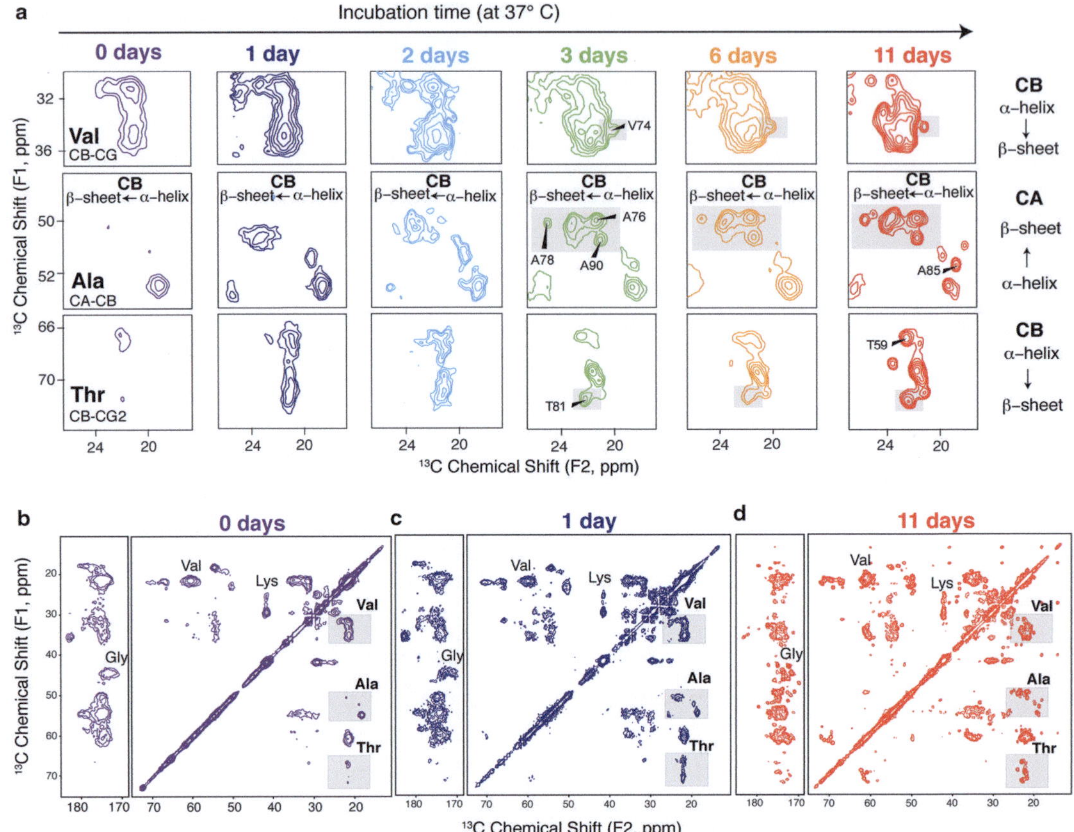


**Fig. 1** Transition from $\alpha$-helix to $\beta$-sheet evidenced by the changes in chemical shifts of $\alpha$-synuclein incubated in the presence of phospholipid. Note that although the protocol described here does not include phospholipids, a similar time dependence of spectral quality has been observed for $\alpha$-syn fibrillized following this protocol. Differing fibrillization times have been reported for amyloid fibrils of different proteins. (**a**) Expanded regions of $^{13}C$–$^{13}C$ 2D spectra of $\alpha$-synuclein incubated in the presence of phospholipid for 0, 1, 2, 3, 6, and 11 days. All contours were drawn at ~6$\sigma$. Full $^{13}C$–$^{13}C$ 2D spectra of samples incubated for (**b**) 0, (**c**) 1, and (**d**) 11 days. Expanded regions shown in part (**a**) are highlighted in *gray*. Reprinted with permission from G. Comellas, et al. *J. Am. Chem. Soc.* **134**, 5090–5099 (2012). Copyright 2012 American Chemical Society

5. Ultracentrifuge again at 4 °C and 55,000 rpm (163,202 × *g*) for 1–2 h in a TLA-100.3 rotor with adapters for the microfuge tubes. Decant and save the supernatant.
6. Repeat **steps 3–5** for an additional wash (*see* **Note 11**).
7. Arrange the air manifold to dry the fibrils over $N_2$ (Fig. 2). Be sure to poke a hole in the caps of the microfuge tubes with a syringe needle for air release and to include the water bubbler to maintain a safe pressure.
8. Turn on the nitrogen gas to a very low flow and a pressure of 1–2 p.s.i.

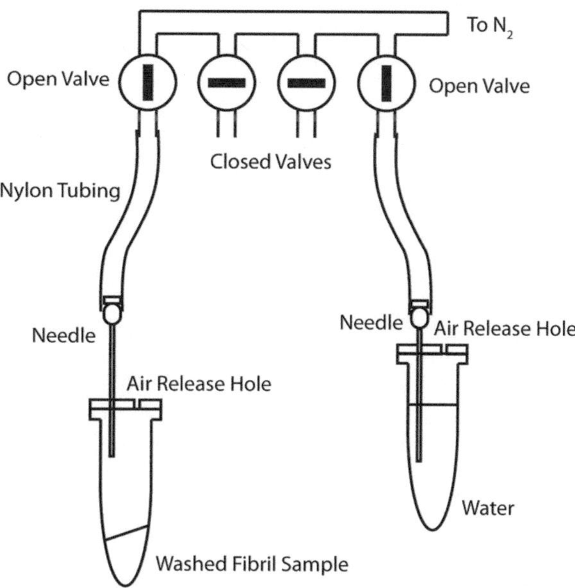


**Fig. 2** Diagram of the drying apparatus. An air manifold is attached to an $N_2$ cylinder and nylon tubing to dry the fibril sample. The water bubbler is a safety feature to keep the manifold from over-pressurizing. The valve to the bubbler should never be shut off

9. Dry the fibrils under $N_2$ for at least 2–3 h and up to overnight until the mass stops changing. The resulting dry fibrils should look like a thin clear sheet of mica. Dry fibrils can be stored at 4 °C until ready to pack into a SSNMR Rotor.

### 3.5 SSNMR Rotor Packing

1. Collect the Rotor Sleeve, drive cap, top cap, both Kel-F spacers, and rubber discs. Wash them thoroughly, first with water, then ethanol, and a second wash with water. Allow to dry. Record mass of all rotor components (*see* **Note 12**).
2. Insert one of the rubber discs into the rotor sleeve (*see* **Note 13**).
3. Lightly start inserting the top cap into the same end of the rotor as the rubber disc using finger pressure. The drive tip should enter the sleeve ~3/4 of the way using finger pressure.
4. Put the rotor into the collet chuck from the packing tool kit and attach the screw press assembly. Finish inserting the top cap into the sleeve using the screw press ensuring that there is no gap between the rotor sleeve and drive cap (*see* **Note 14**).
5. From the opposite end of the rotor, insert the shorter Kel-F spacer and push it to the opposite end so that it is pressed against the rubber disc. Re-record the mass of all rotor components.
6. Mass the partially assembled rotor in a disposable aluminum weighboat to determine the starting mass. Slowly add dried fibrils to the rotor until you have added up to ~12 mg. Record the total mass of the protein added (*see* **Notes 15** and **16**).

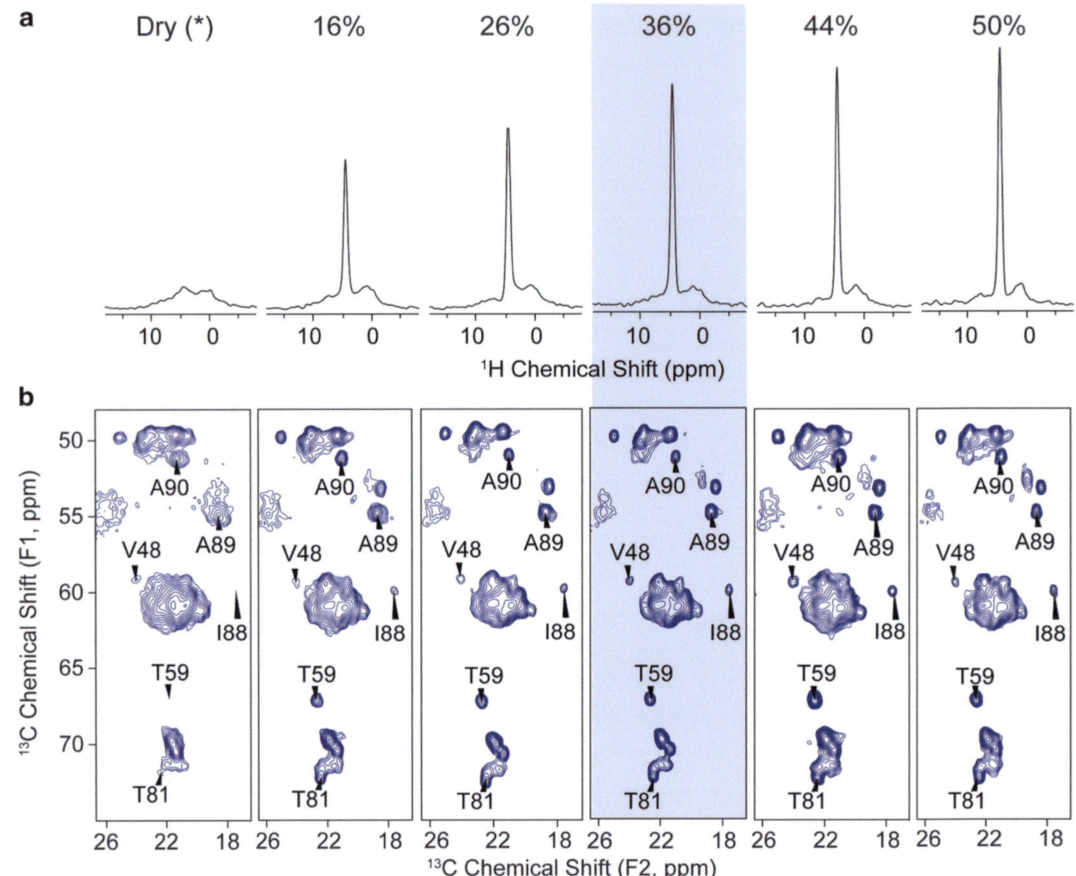


**Fig. 3** Optimizing hydration of α-synuclein fibril samples for the solid-state NMR experiments (mass of water/total mass). (**a**) $^{1}H$ 1D spectra of U-$^{13}C$,$^{15}N$ α-synuclein fibrils at different hydration levels (dry, 16, 26, 36, 44, and 50 %) and (**b**) $^{13}C$–$^{13}C$ 2D spectra with 50-ms DARR mixing. All spectra were acquired under identical conditions at 600-MHz ($^{1}H$ frequency) and 13.3-kHz MAS rate. Dry fibrils correspond to fibrils dried by $N_2$ gas until no changes in mass were observed, as previously described. Spectra marked in *blue* correspond to the optimal conditions selected for the multidimensional NMR experiments. Reprinted from Comellas et al., Structured regions of α-synuclein fibrils include the early-onset Parkinson's disease mutation sites. *J. Mol. Biol.* 411, 881–895 (2011), with permission from Elsevier

7. Rehydrate the dried fibrils by adding ultrapure water to 50 % by weight of the total protein added in the previous step. This water is vital for obtaining high-resolution spectra (Fig. 3). Record the amount of water added (*see* **Note 17**).
8. Spin the rotor for 30 s in the swinging arm rotor at 5000 rpm (3217 × *g*) to help distribute the water. This can be done by placing the rotor inside of a microfuge tube placed within a 50 mL conical centrifuge tube.
9. Place the long Kel-F spacer into the top of the rotor, followed by the rubber disc and finally with the drive cap (*see* **Note 18**).

10. Similar to inserting the top cap, use the collet chuck and screw press to insert the drive cap completely. Add a mark to the top cap using a black sharpie for the tachometer. Record the final mass of the rotor.

## 4 Notes

1. We have stored purified α-synuclein monomer at ~0.7 mg/mL in 50 mM sodium phosphate, 1 mM EDTA, 0.01 % $NaN_3$, pH 7.4 at 4 °C for periods of several months with no observable adverse effects. We typically start a fibril prep with a total volume of 75–100 mL, corresponding to approximately 50–70 mg of pure protein.
2. It is important that the microfuge tubes be ultracentrifuge compatible.
3. Preformed fibrils can be prepared using this protocol without the addition of seeds during the fibrillization stage by spontaneous nucleation.
4. Running ~10 mL of buffer through the stirred cell concentration apparatus prior to adding the protein solution will ensure that it was correctly assembled and not leaking, preventing the loss of material.
5. We typically save the stirred cell membrane in submerged size exclusion buffer to later quantify protein loss due to adhesion to the membrane surface. The loss is usually minimal if the rinse is properly performed.
6. The sizing column trace may have a large shoulder that runs at a larger size than the monomer peak (45 kDa, *see* **Note 7**). This can indicate the presence of an aggregate in the monomer solution that can lead to inconsistent fibrillization results.
7. We have observed that in the case of α-synuclein, the monomer elutes under these conditions at a volume corresponding to a 45 kDa globular protein.
8. After reaching this point, if seeding (*see* **Note 9**), make all attempts to move to fibrillization as quickly as possible to prevent the possibility of undesired spontaneous aggregates forming in the solution.
9. Seeding is optional. In our experience this protocol will produce consistent α-synuclein fibrils from batch to batch regardless of whether preformed fibril seeds were added or not. We suggest seeding to see an increase in fibril yield and guard against any possible undesired-pathway aggregates forming.
10. There should be a very clear gelatin-like pellet after ultracentrifugation. Sometimes the pellet is hazy or has apparent layers, but these are removed during the washing steps.

11. The washing steps remove any remaining monomer and/or protofibrils from the fibrils and enhance the quality of the data of the fibrils.
12. It is possible that improperly sized parts can cause instabilities in rotor spinning. We find it beneficial to test the rotor packed with baking soda to ensure it spins properly prior to packing the NMR sample.
13. If the rubber disc is not placed evenly into the rotor sleeve, it can introduce spinning instabilities.
14. The screw press assembly applies a constant and symmetric force to the drive tip, which should ensure that you do not add torque that could cause the rotor to crack.
15. It helps to crush the dried fibrils lightly with a small spatula prior to packing.
16. The crushed fibrils can cling to surfaces through static electricity. We recommend using the aluminum weigh boats to circumvent the fibrils sticking.
17. It can be beneficial to add the water in stages during the packing to ensure even distribution of the water.
18. Sometimes water squeezes out of the rotor during this step, so adding 50 % w/w water ensures you reach a minimum of ~36 % hydration in the enclosed rotor.

## Acknowledgements

This work was supported by NIH grants R01-GM073770 and P50-NS053488. M.D.T. and A.M.B. were supported by NIH Training Grant PHS 5T32 GM008276 and J.M.C. was supported by a National Science Foundation Graduate Research Fellowship.

### References

1. Petkova AT, Yau WM, Tycko R (2006) Experimental constraints on quaternary structure in Alzheimer's β-amyloid fibrils. Biochemistry 45:498–512
2. Lu J-X, Qiang W, Yau W-M et al (2013) Molecular structure of β-amyloid fibrils in Alzheimer's disease brain tissue. Cell 154: 1257–1268
3. Petkova AT, Ishii Y, Balbach JJ et al (2002) A structural model for Alzheimer's β-amyloid fibrils based on experimental constraints from solid state NMR. Proc Natl Acad Sci USA 99:16742–16747
4. Wasmer C, Lange A, Van Melckebeke H et al (2008) Amyloid fibrils of the HET-s(218-289) prion form a solenoid with a triangular hydrophobic core. Science 319:1523–1526
5. Van Melckebeke H, Wasmer C, Lange A et al (2010) Atomic-resolution three-dimensional structure of HET-s(218-289) amyloid fibrils by solid-state NMR spectroscopy. J Am Chem Soc 132:13765–13775
6. Comellas G, Lemkau LR, Zhou DH et al (2012) Structural intermediates during α-synuclein fibrillogenesis on phospholipid vesicles. J Am Chem Soc 134:5090–5099
7. Comellas G, Lemkau LR, Nieuwkoop AJ et al (2011) Structured regions of α-synuclein fibrils include the early-onset Parkinson's disease mutation sites. J Mol Biol 411:881–895

8. Bousset L, Pieri L, Ruiz-Arlandis G et al (2013) Structural and functional characterization of two α-synuclein strains. Nature Commun 4:2575
9. Volpicelli-Daley LA, Luk KC, Lee VMY (2014) Addition of exogenous α-synuclein preformed fibrils to primary neuronal cultures to seed recruitment of endogenous α-synuclein to Lewy body and Lewy neurite-like aggregates. Nat Protocols 9:2135–2146
10. Kloepper KD, Hartman KL, Ladror DT et al (2007) Solid-state NMR spectroscopy reveals that water is nonessential to the core structure of α-synuclein fibrils. J Phys Chem B 111: 13353–13356
11. Castellani F, van Rossum B, Diehl A et al (2002) Structure of a protein determined by solid-state magic-angle- spinning NMR spectroscopy. Nature 420:98–102
12. Lemkau LR, Comellas G, Kloepper KD et al (2012) Mutant protein A30P α-synuclein adopts wild-type fibril structure, despite slower fibrillation kinetics. J Biol Chem 287: 11526–11532
13. Lemkau LR, Comellas G, Lee SW et al (2013) Site-specific perturbations of α-synuclein fibril structure by the Parkinson's disease associated mutations A53T and E46K. PloS one 8:e49750
14. Kloepper KD, Woods WS, Winter KA et al (2006) Preparation of α-synuclein fibrils for solid-state NMR: expression, purification, and incubation of wild-type and mutant forms. Protein Expr Purif 48:112–117

# Chapter 12

# Spin Labeling and Characterization of Tau Fibrils Using Electron Paramagnetic Resonance (EPR)

**Virginia Meyer and Martin Margittai**

## Abstract

Template-assisted propagation of Tau fibrils is essential for the spreading of Tau pathology in Alzheimer's disease. In this process, small seeds of fibrils recruit Tau monomers onto their ends. The physical properties of the fibrils play an important role in their propagation. Here, we describe two different electron paramagnetic resonance (EPR) techniques that have provided crucial insights into the structure of Tau fibrils. Both techniques rely on the site-directed introduction of one or two spin labels into the protein monomer. Continuous-wave (CW) EPR provides information on which amino acid residues are contained in the fibril core and how they are stacked along the long fibril axis. Double electron–electron resonance (DEER) determines distances between two spin labels within a single protein and hence provides insights into their spatial arrangement in the fibril cross section. Because of the long distance range accessible to DEER (~2–5 nm) populations of distinct fibril conformers can be differentiated.

**Key words** Tau fibril, Amyloid, Spin labeling, EPR spectroscopy, DEER, Protein structure, Alzheimer's disease

## 1 Introduction

Fibrils composed of the microtubule-associated protein Tau are a pathological hallmark of Alzheimer's disease and other neurodegenerative disorders [1, 2]. The primary location of these fibrils is the interior of nerve cells. Short Tau fibrils can transfer between nerve cells [3] and then recruit new Tau monomers onto their ends [4]. This mechanism of transfer and growth appears to be central to the spreading of Tau pathology [5, 6]. Similar mechanisms of propagation are thought to exist for other amyloid fibrils [7]. Given their central role in pathology, understanding the structure of Tau fibrils is essential. This is a challenging task, as the fibrils are large and heterogeneous.

One of the methods that has provided invaluable structural insights in the past few years is site-directed spin labeling combined with electron paramagnetic resonance (EPR) spectroscopy. This

David Eliezer (ed.), *Protein Amyloid Aggregation: Methods and Protocols*, Methods in Molecular Biology, vol. 1345,
DOI 10.1007/978-1-4939-2978-8_12, 

technique is applicable to proteins of any size and structure [8]. A key feature is the introduction of one or two small spin labels into the protein backbone. The most commonly used spin label is (1-oxyl-2,2,5,5-tetramethyl-Δ3-pyrroline-3-methyl)methanethiosulfonate [9], referred to as MTSL. It possesses a small side chain volume and causes only minimal structural perturbation. The label attaches selectively to cysteine residues, so a first step in the overall procedure is the removal of all native cysteines in the protein. This is achieved by site-directed mutagenesis. The introduction of cysteines at specific positions follows. After the proteins have been labeled, their structural properties can be investigated using two different EPR techniques: (a) continuous-wave (CW) or (b) double electron–electron resonance (DEER).

The first CW EPR study on amyloid fibrils was published in 2002 [10] and involved the structural characterization of the Aβ peptide. The study of other amyloid fibrils by EPR, including those of Tau [11], α-synuclein [12], IAPP [13], Sup-NM [14], Ure2p1-89-M [15], β2-microglobulin [16], and the human prion protein, PrP90-231 [17] followed. CW EPR spectra provide information on the mobility of spin labels and their inter-spin distances. Regions in the protein that are highly mobile can be clearly distinguished from those that are immobilized. Systematic scanning of the protein sequence, one spin-labeled residue at a time, thus allows delineating the core regions of amyloid fibrils. A common structural feature of many amyloid fibrils is the parallel, in-register arrangement of β-strands [18]. As a consequence, spin labels from neighboring proteins in the fibril are stacked on top of each other. This results in spin exchange and the collapse of the typical hyperfine structure of the EPR spectrum. Longer distances between spin labels (~1.0–2.0 nm) result in dipolar interactions that cause spectral broadening and can also be resolved by CW EPR [12]. Short-distance exchange interactions and dipolar interactions between labels along the fibril axis can be suppressed by diluting the labeled proteins with an excess of unlabeled proteins prior to fibril formation. Intramolecular spin interactions of doubly labeled proteins are unaffected by such dilutions.

DEER spectroscopy offers an additional handle on amyloid fibrils, as it allows distance measurements in the ~2–5 nm range. The technique has provided long-range distance constraints for amyloid fibrils composed of IAPP [19], α-synuclein [20, 21], and Tau [22, 23]. These distances are between two spin labels in the same protein and hence offer insights into the packing of β-sheets. CW and DEER measurements are complementary: combined, they provide information on the stacking along the long axis (spin exchange), the extent of the fibril core (mobility), and the spatial relationship between residues in the cross section of the fibril (DEER or CW distance information). Here, we focus on the spin labeling and EPR characterization of Tau fibrils. A similar approach can be chosen to study other amyloid fibrils.

## 2 Materials

### 2.1 Production of Spin-Labeled Tau Protein

#### 2.1.1 Transformation and Expression

1. QuikChange site-directed mutagenesis kit (Agilent).
2. Tau pET28b plasmid DNA (*see* **Note 1**).
3. BL21 (DE3) *Escherichia coli.*
4. NZY medium: To NZY broth (autoclave 5 g NaCl, 5 g yeast extract, 10 g NZ amine in 1 L deionized water at pH 7.5) add sterile filtered $MgCl_2$ to 12.5 mM, $MgSO_4$ to 12.5 mM and glucose to 20 mM final concentrations.
5. LB-agar plates: Autoclave 20 g LB broth and 15 g LB agar in 1 L deionized water. Allow solution to cool to 50–60 °C and add 50 mg kanamycin. Mix well and pour into 100 × 15 mm petri dishes, filling each approximately ¼″.
6. LB media with kanamycin: Autoclave 30 g LB broth in 1.5 L deionized water. Allow to cool and add 30 mg (20 μg/mL) kanamycin.
7. Extraction buffer: 20 mM PIPES pH 6.5, 500 mM NaCl, 1 mM ethylenediaminetetraacetic acid (EDTA), 50 mM β-mercaptoethanol.

#### 2.1.2 Purification

1. Microtip sonicator, D100 series (Fisher Scientific).
2. ÄKTA FPLC protein purification system (GE Healthcare) (*see* **Note 2**).
3. Mono S cation exchange column (GE Healthcare).
4. Superdex 200 gel filtration column (GE Healthcare).
5. SDS-PAGE gel electrophoresis equipment.
6. 0.45 μm Acrodisc PSF syringe filters and 0.45 μm nylon membrane filters (*see* **Note 3**).
7. 55 % w/v ammonium sulfate: Dissolve the ammonium sulfate fully by rocking or shaking for at least 1 h at 25 °C.
8. DTT-water: 2 mM dithiothreitol (DTT) in nanopure water.
9. Low-salt (LS) buffer: 10 mM PIPES pH 6.5, 50 mM NaCl, 1 mM EDTA, 2 mM DTT.
10. High-salt (HS) buffer: 10 mM PIPES pH 6.5, 1000 mM NaCl, 1 mM EDTA, 2 mM DTT.
11. Gel filtration (GF) buffer: 20 mM Tris-HCl pH 7.4, 100 mM NaCl, 1 mM EDTA, 2 mM DTT.

#### 2.1.3 Spin Labeling

1. Spin label stock solution: (1-oxyl-2,2,5,5-tetramethyl-Δ3-pyrroline-3-methyl) methanethiosulfonate (MTSL spin label, Toronto Research Chemicals) dissolved in DMSO (40 mg/mL) (*see* **Note 4**).
2. PD-10 column (GE Healthcare).

3. Protein buffer (PB): 10 mM HEPES pH 7.4, 100 mM NaCl, 1 mM $NaN_3$.
4. Bicinchoninic acid (BCA) assay kit (Pierce).

### 2.2 Characterization of Tau Fibrils by CW EPR and DEER

#### 2.2.1 Seeding and Fibril Assembly

1. Heparin (Celsus, average MW = 5000) (*see* **Note 5**).
2. Microtip sonicator.
3. Microstir bar.

EPR Sample Preparation

1. 1.6 mm outer diameter (o.d.) quartz Q-band EPR sample tube.
2. 4 mm o.d. quartz X-band EPR sample tube.
3. 0.84 mm o.d. borosilicate capillary.
4. Critoseal.
5. 1.5 and 15 mL centrifuge tubes.
6. 83 mm gel-loading pipette tip.
7. 20 gauge syringe.

#### 2.2.2 Instrumentation

1. Bruker EMXplus X-band CW EPR spectrometer.
   (a) ER 4119HS resonator.
2. Bruker ELEXSYS E580 pulse spectrometer with Q-band DEER components.
   (a) ER5107D2 dielectric resonator.
   (b) E580-400U ELDOR unit.
   (c) SuperQ-FT bridge.
   (d) PatternJet.
   (e) CF935 cryostat (Oxford).

## 3 Methods

### 3.1 Production of Spin-Labeled Tau Protein

#### 3.1.1 Transformation and Expression

1. Incubate 900 ng plasmid DNA with 17 μL BL21 (DE3) competent *E. coli* cells on ice for 20 min. Heat shock cells for 50 s at 42 °C and incubate on ice for an additional 2 min. Add 800 μL NZY medium and incubate at 37 °C for 1 h while shaking at 200 rpm.
2. Plate 50 μL of cells onto kanamycin-containing LB-agar plates, and incubate overnight at 37 °C.
3. Prepare expression starter cultures by introducing a single colony from the transformation plate into 50 mL LB media with kanamycin. Incubate at 37 °C with shaking at 200 rpm for 16–17 h.
4. Transfer 15 mL starter culture into 1.5 L fresh LB media with kanamycin. Incubate at 37 °C with shaking at 200 rpm until

the optical density of the culture reaches 0.8 at 600 nm, approximately 3 h.

5. Induce protein expression by adding isopropyl-ß-D-thiogalactopyranoside to a final concentration of 1 mM. Allow expression to continue for 3.5 h by incubating cells at 37 °C with shaking at 200 rpm.
6. Centrifuge cells at 3000 × *g* for 20 min. Each 1.5 L expression flask is divided between two 1 L centrifuge bottles. Remove supernatant and resuspend bacterial pellet with 9.5 mL extraction buffer. Store cells at −80 °C until purification.

### *3.1.2 Purification*

1. Thaw cells at 80 °C for 30 min. At this temperature, most bacterial proteins precipitate, leaving natively unfolded Tau in solution. Cool cells at least 5 min before sonication.
2. Sonicate cells for 1 min using a microtip probe set to approximately 50 % total power. Power settings differ between sonicators, so check the manual for the most appropriate setting.
3. Centrifuge cells at 15,000 × *g* for 30 min at 4 °C.
4. Transfer the supernatant into 55 % w/v ammonium sulfate to precipitate Tau protein. The pellet from **step 3**, which contains cellular debris, can be discarded.
5. Collect Tau protein by centrifugation at 15,000 × *g* for 10 min at 25 °C. Discard the supernatant, and dissolve the pellet in 8 mL DTT-water.
6. Sonicate the mixture using a microtip set to 50 % power for 40 s. Filter the sample immediately through a 0.45 μm syringe filter. Dilute the sample with additional DTT-water to approximately 100 mL (more water may be added later if the conductivity of the sample is too high).
7. Equilibrate the Mono S cation-exchange column with several column volumes of LS buffer. Inject the sample onto the column at 3 mL/min. Protein elutes at approximately 25 mS, so if the sample conductivity is above 16 mS, it is advisable to add additional DTT-water to the sample.
8. Elute protein from the column by steadily increasing the concentration of HS buffer, thereby running a linear salt gradient (50–1000 mM NaCl). Collect eluate in 3 mL fractions.
9. Combine the three fractions containing the highest concentration of protein as determined by SDS-PAGE gel electrophoresis. Add 5 mM DTT (*see* **Note 6**) to the sample and store at −80 °C until gel filtration purification.
10. Thaw the sample for 10 min at 40 °C. Inject the sample onto an equilibrated Superdex 200 gel filtration column with GF buffer. Protein elutes over time according to size and shape.

Because Tau is unfolded, it elutes faster than globular proteins of similar molecular weight. Collect eluate in 5 mL fractions.

11. Combine fractions containing pure protein and add 5 mM DTT. Using a threefold volumetric excess of cold acetone (truncated Tau) or a twofold excess of cold methanol (full-length Tau), precipitate protein overnight at 4 °C (*see* **Note 7**).
12. Centrifuge samples at 15,000 × $g$ and transfer protein pellets into 1.5 mL tubes. Store pellets in fresh 2 mM DTT-acetone at −80 °C.

#### 3.1.3 Spin Labeling

1. Dissolve purified protein pellet in 200 μL 8 M guanidine HCl (*see* **Note 8**).
2. Add a tenfold molar excess of spin label, using spin label stock solution, to the dissolved pellet of the single- or double-cysteine mutant, and incubate for 1 h at 25 °C. The first time protein is labeled, the initial concentration should be determined by BCA assay to ensure use of an appropriate amount of spin label. Addition of MTSL for subsequent experiments can be estimated from the expected protein concentration.
3. Equilibrate PD-10 columns with three column volumes of PB. Load the 200 μL samples of monomerized wild-type Tau and spin-labeled Tau mutant onto separate PD-10 columns. Add 1.8 mL PB to each column and allow to flow through. Collect 2 mL eluate from each column.
4. Determine protein concentrations for the wild-type Tau and the spin-labeled Tau using a BCA assay. The PD-10 eluate forms a concentration gradient in the collection tube and should be mixed thoroughly for accurate BCA concentration measurement.

### 3.2 Characterization of Tau Fibrils by CW EPR

To determine short-range spin exchange interactions, fibrils are formed from fully spin-labeled Tau monomers. To resolve side-chain mobility, fibrils are assembled from spin-diluted proteins.

#### 3.2.1 Fibril Assembly

1. 30 μM of spin-labeled Tau or 60 μM of spin-diluted Tau (3 μM spin-labeled Tau mixed with 57 μM wild-type Tau) (*see* **Note 9**) in PB are placed into 2 mL tubes. To facilitate aggregation, heparin is added at a protein:heparin molar ratio of 4:1. The volumes are adjusted to 1.8 mL and a small microstir bar (2 × 8 mm) is added to each tube.
2. Incubate samples for 3 days stirring at 25 °C (*see* **Note 10**).
3. To verify the fibrillar nature of the sample, small aliquots may be taken off for electron microscopic analysis.

*3.2.2 CW Measurement and Analysis*

1. In order to separate fibrils from unincorporated monomers, centrifuge all samples for 30 min at 100,000 × *g*. To remove residual monomers, wash once with PB and repeat centrifugation step.
2. Depending on pellet size add 10–20 μL PB. Mix the pellet with buffer thoroughly and gently using the pipet tip.
3. Transfer the sample into a round borosilicate capillary (0.6 mm inner diameter, 0.84 mm o.d., 10 mm length, two open ends) by either suction or capillary flow.
4. Seal one end of the capillary with Critoseal.
5. Place capillary into X-band spectrometer fitted with an ER 4119HS resonator, and tune resonator to achieve maximal EPR signal [24].
6. During the CW measurement at X-band, the microwave frequency remains constant at 9.5 GHz while the field is swept. The typical sweep width for Tau fibrils is 150 G. This width ensures that the full breadth of the spectrum is recorded. The microwave power is chosen to avoid saturation effects. Whereas the modulation frequency is set at 100 kHz, the modulation amplitude is optimized for best signal-to-noise ratio (SNR). This is typically achieved at amplitudes of 1–3 G. Overmodulation, which results in spectral distortion, should be avoided. The number of scans may be varied depending on the SNR of the sample.
7. The CW spectra are depicted as first derivatives, and normalization to the same number of spins is achieved by double integration. For the MTSL label, three hyperfine lines are observed due to coupling between the single electron and the $^{14}N$ nucleus. The central linewidth and the spacing between the outer peaks provide information on the mobility of the label [25]. A spin label that is attached to the intrinsically disordered Tau monomer will produce an EPR spectrum with small central linewidth and narrow spacing between the outer lines (Fig. 1a). These spectral features reflect the high degree of mobility of the label. The central linewidth and the spacing between outer peaks will increase significantly when the monomer becomes incorporated into the fibril and the spin label is immobilized (Fig. 1b). In order to observe the outer lines, however, fibrils have to be formed from a mixture of labeled and unlabeled monomers. Fibrils formed from only labeled monomers will result in the stacking of spin labels along the fibril axis, because of the parallel, in-register arrangement of ß-strands [18]. This causes spin exchange between spin labels resulting in the loss of hyperfine structure. As a consequence a single-line EPR spectrum is observed (Fig. 1c). The spectrum reflects a crystal-like order of the fibrils along the long axis. Amorphous or disordered structures can be excluded.

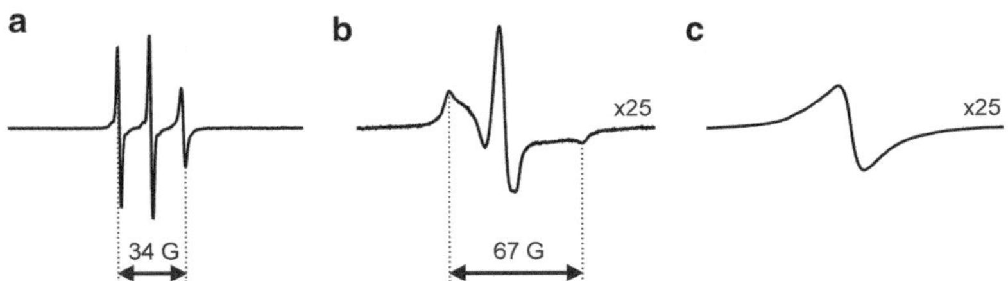


**Fig. 1** CW line shape analysis. A truncated form of Tau (K18), which contains the amyloid-forming core, was labeled at a single cysteine in position 280. The monomeric form produces a spectrum with three sharp lines (**a**). Fibrils formed from a mixture of 5 % spin-labeled Tau and 95 % wild-type Tau result in a broadened spectrum (**b**). Fibrils formed from spin-labeled Tau only produce a single-line spectrum due to spin exchange between stacked labels in the parallel, in-register β-strands (**c**). The distances between outer peaks are indicated for (**a**) and (**b**). To compare line shapes the spectra in (**b**) and (**c**) are amplified 25-fold. The scan width for all spectra is 150 G

### 3.3 Characterization of Tau Fibrils by DEER

#### 3.3.1 Seeded Fibril Assembly

Like other amyloid proteins, assembly of Tau fibrils relies on self-nucleation of monomer, imparting a relatively long lag phase prior to fibril elongation. It is convenient to introduce seeds to overcome this lag phase [26], which provide the structural template which the monomer adopts.

1. Initial fibrils are formed by stirring a mixture of 25 μM Tau protein and 50 μM heparin cofactor at a total volume of 1500 μL in PB for 3 days at room temperature. Stirring is most effective in a 2 mL tube with a relatively flat bottom using a microstir bar (*see* **Note 11**).
2. To create seeds, sonicate 500 μL of initial fibrils using a tip sonicator for 20 s at 50 % power. These seeds can be added directly to Tau monomer to induce templated fibril growth, or seed preparation can continue over several cycles (*see* **Note 12**).
3. Fibril assembly: Combine the following in a 2 mL tube to a total volume of 1.8 mL with PB:
   (a) 49 μM wild-type Tau (98 %).
   (b) 1 μM spin-labeled Tau (2 %) (*see* **Note 13**).
   (c) 5 % seeds (based on monomer equivalents).
   (d) 12.5 μM heparin.
4. Allow fibrils to elongate overnight at 37 °C.

#### 3.3.2 DEER Sample Preparation

1. Centrifuge 2 % spin-labeled fibrils at 100,000 ×*g* for 30 min at 10 °C. Wash pellet with 1 mL fresh PB and repeat centrifugation for 20 min. The pellet may be transparent.
2. Remove all excess buffer from the pellet. Depending on the size of the pellet, add 10–30 μL fresh PB. Mix the pellet with

buffer thoroughly using the pipette tip. Aspiration is not recommended as the pellets can be particularly viscous.

3. Using an 83 mm long gel-loading tip, add the pellet and buffer mixture directly to a 1.6 mm o.d. quartz Q-band (34 GHz) EPR tube, sealed at one end. Support the 1.6 mm o.d. tube within a 4 mm o.d. X-band EPR tube and place this in a 15 mL tube. Centrifuge at 2000 × *g* for 1 min to compress the fibrils into the bottom of the Q-band tube. Add additional fibrils using the gel-loading tip and centrifuge, repeating this process until the sample in the bottom of the tube is longer than the active space of the resonator (~8 mm for a Bruker Q-band ER5107-D2 dielectric resonator).
4. Following centrifugation, the buffer may have formed a layer above the fibrils in the tube, which can be removed using a 20-gauge or smaller syringe needle. Once the buffer is removed, centrifuge the sample a final time at 2000 × *g* for 45 min to further compress the fibrils. This step ensures that the highest number of spins possible are present in the resonator active space, enhancing the SNR of the DEER measurement.
5. It is recommended to record a room-temperature CW spectrum prior to freezing the sample for DEER measurement. This can be performed at X-band (9.5 GHz) by placing the 4 mm o.d. tube containing the 1.6 mm Q-band tube directly into the resonator, or CW can be measured at Q-band. This step provides a check that the spin-labeled monomer has been incorporated into the fibrils and that the exchange interaction is negligible.
6. Following the final 45 min centrifugation, additional residual buffer should be removed and the sample immediately flash frozen in the Q-band tube using liquid $N_2$. The sample can be stored at −80 °C until DEER measurement (*see* **Note 14**).

#### 3.3.3 DEER Acquisition and Analysis

DEER measurements are described for data collection at Q-band (34 GHz) on a Bruker ELEXSYS E580 system using an ER5107D2 dielectric resonator. The resonator is fully overcoupled ($Q \sim 500$) for pulse experiments to reduce resonator ringdown. The E580 can be equipped with an Oxford CF935 cryostat for low temperature measurements using either liquid $N_2$ (≥80 K) or liquid He (20–80 K). The second microwave source required for DEER measurement was an E580-400U ELDOR unit, along with a SuperQ-FT bridge capable of dual frequencies. The bridge is equipped with a 1 W amplifier, and microwave pulses are formed using a Bruker PatternJet. Detailed information on DEER methodology for the study of proteins can be found in [27].

1. The DEER sample should be maintained in liquid $N_2$ after removal from −80 °C and inserted into the pre-cooled Q-band resonator at or below 80 K.

2. It is important to collect spin-spin ($T_2$) and spin–lattice ($T_1$) relaxation times prior to beginning a DEER experiment. These values provide information regarding the available time window for the DEER measurement ($T_2$) as well as the necessary shot repetition time (SRT) of the instrument (SRT = $1.2 \times T_1$).
3. Create a new DEER experiment and record a field-swept echo-detected spectrum (Fig. 2a) to find the field position corresponding to the maximum echo amplitude. Enter this value as the field position of the DEER pump pulse and the frequency at which the field-swept echo-detected spectrum was recorded as the pump/ELDOR frequency ($\nu_2$ in Fig. 2a).
4. In the tuning panel, adjust the operating frequency to 37 MHz below the ELDOR frequency. This is the frequency at which modulation of the spin echo will be observed ($\nu_1$ in Fig. 2a) (*see* **Note 15**). The effect of this frequency difference on the field position is shown in Fig. 2a.
5. The following parameters are used for most Tau fibril DEER measurements. Corresponding times are shown in Fig. 2b. Some parameters require modification when studying different protein systems. Adjust the $\pi/2$ observer pulse at $\nu_1$ to obtain the maximum echo height. The typical range for this pulse is from 36 to 46 ns. The $\pi$ pump pulse at $\nu_2$ can also be adjusted to obtain maximal echo modulation, but is typically set to 40 ns. Eight-step phase cycling is used to remove unwanted echoes. $\tau_1$ and $\tau_2$ are held constant for each experiment at 200 ns and 2500 ns, respectively. The relative echo height at $\tau_2 = 2500$ ns is shown in Fig. 2c (*see* **Note 16**). The pump pulse starts at $T = 100$ ns and is stepped in 8-ns increments. The SRT is set to 1.2 times the $T_1$ measured by inversion recovery. Using a 1 W amplifier, the pulse power at both frequencies should be 0 dB. For higher power amplifiers, the pulse power may need to be adjusted to avoid spin saturation. Depending on signal strength, spectra may be averaged for several hours to a week.
6. There are several programs available for use in MATLAB designed to fit dipolar evolution oscillations and obtain a distance distribution. The programs differ in background subtraction algorithm and data fit models. DeerAnalysis [28] and DEFit [29] require user-defined background selection, whereas GLADD [30] determines the most probable background subtraction and distance distribution simultaneously, using no a priori background correction. Both DEFit and GLADD fit dipolar evolution curves to single-, double-, or triple-Gaussian functions. Additional models are available in DeerAnalysis, including Tikhonov regularization, which does not limit the distance distribution to a defined number of peaks or peak shapes. For this reason, Tikhonov regularization is beneficial for studying changes in peak ensembles corresponding to mixtures

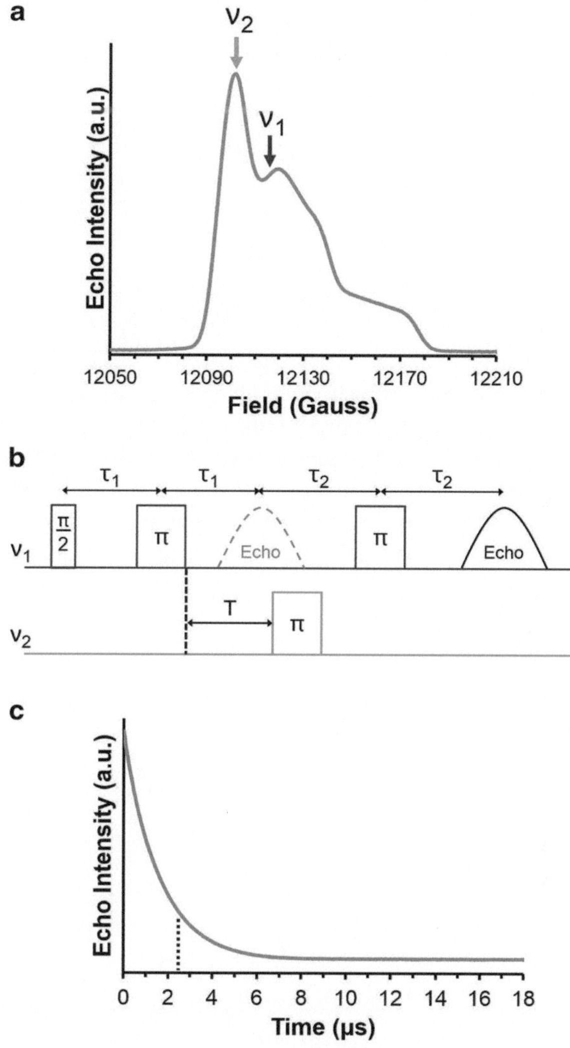


**Fig. 2** Q-band field-swept echo-detected spectrum of spin-labeled Tau fibrils at 80 K (**a**), dual-frequency DEER pulse sequence (**b**), and Q-band spin-echo decay of spin-labeled Tau fibrils at 80 K (**c**). The frequencies, $\nu_1$ and $\nu_2$, indicated by *arrows* in (**a**) correspond to the separation of pulse frequencies in the sequence (**b**). An echo forms from the two initial pulses at the observer frequency, $\nu_1$, which is refocused by a third pulse at the same frequency. A pulse at the pump frequency, $\nu_2$, perturbs the spin system, inducing oscillation of the observed echo intensity. The available time delay between pulses is dictated by the $T_2$ relaxation time, which is measured by echo decay (**c**). The *dotted line* in (**c**) corresponds to the echo intensity for a pulse delay time, $\tau_2$, of 2.5 μs

of protein structures, as limitation to a set number of Gaussians does not provide a way to model the multiple components. Fig. 3 shows distance distributions obtained from fitting the same DEER data to single- and double-Gaussian distributions compared to Tikhonov regularization.

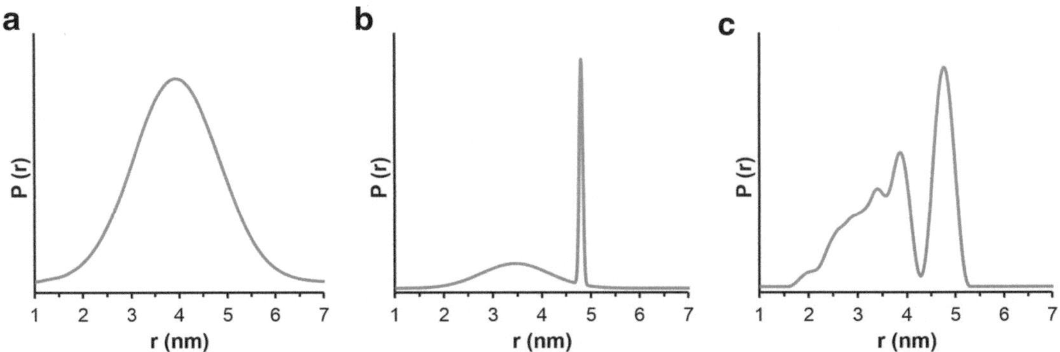


**Fig. 3** Distance distributions resulting from fitting a dipolar oscillation curve of Tau to a single-Gaussian distribution (**a**), a double-Gaussian distribution (**b**), and Tikhonov regularization (**c**). The Gaussian distributions constrain the number of oscillation frequencies to one (**a**) or two (**b**), which is not well suited for a system containing multiple components or mixtures of protein conformations. The distance distribution calculated from Tikhonov regularization is a compromise between smoothness of the oscillation fit function and sharpness of the resulting distance peaks. This model is highly beneficial for studying amyloid proteins, where conformational heterogeneity may exist

## 4 Notes

1. Truncated Tau (K18) that contains the amyloid-forming repeat region (amino acid residues 244–372) was cloned into pET28b via the Nde1 and Xho1 cleavage sites. In this construct the native cysteines at positions 291 and 322 were replaced by serines. For convenience this construct is referred to as wild-type. Single and double cysteines were introduced by site-directed mutagenesis.
2. The purifications are carried out at room temperature as Tau is heat stable.
3. To prevent clogging of the ion exchange column, it is essential to filter all buffers and protein samples.
4. The spin label is connected to the protein via a disulfide linkage. In order to avoid loss of label after attachment, reducing agents should not be added.
5. The heparin used here is a low-molecular-weight form derived from porcine intestinal mucosa. Its primary function is to facilitate aggregation of positively charged Tau. Based on the heterogeneous nature of heparin, experiments that are compared with each other should be performed using the same heparin batch.
6. As the protein contains either one or two cysteines, the protein needs to be in a reducing environment at all times. Oxidation damage of the cysteines will reduce the labeling efficiency.
7. Since Tau is intrinsically disordered the protein can be precipitated by organic solvents. After monomerization in denaturant, the protein does not have to refold.

8. Depending on pellet size it may be necessary to combine multiple pellets to achieve a higher protein concentration. Other amyloid proteins such as Aβ, α-synuclein, or Sup35NM, which can be obtained as lyophilized powder or may be precipitated by organic solvent after purification, should also be taken up by guanidine HCl. This ensures that the proteins are monomeric.
9. Alternatively, the spin-labeled protein may be mixed with a nonparamagnetic label such as [1-acetyl-2,2,5,5-tetramethyl-Δ3-pyrroline-3-methyl]methanethiosulfonate. Also, to increase the signal strength in the CW measurements of fibrils, the percentage of spin label may be increased to about 25 %. Although this causes some spectral broadening due to spin interactions, the line shape will still provide information on the mobility of the label.
10. Stirring of the sample greatly accelerates aggregation. The formation of fibrils composed of full-length Tau, however, will require longer incubation times (typically 8–12 days).
11. Since Tau fibrils exist as conformational ensembles, the particular reaction conditions (temperature, stirring speed, volume, tube geometry, etc.) could influence the structural composition.
12. The use of multiple seeding steps can result in conformational selection. To perform multiple seeding steps, incubate 10 % seeds with 25 μM fresh Tau monomer and 50 μM heparin in PB at 37 °C for 1 h. Form new seeds from these fibrils through sonication. Repeat the cycles of incubation (with fresh monomer and heparin) and sonication until the desired number of seeding steps has been achieved.
13. In order to measure distances between spin labels in the same Tau protein, interactions with spin labels in neighboring Tau proteins have to be suppressed. This is achieved by forming the fibrils with a large excess of unlabeled wild-type protein.
14. For fibrillar samples, removal of residual buffer prevents the sample tube from breakage upon freezing. Proteins in solution require addition of a cryoprotectant to the sample, such as 30 % sucrose or 24 % glycerol, to prevent sample expansion and tube breakage.
15. The 37 MHz difference in $\nu_1$ and $\nu_2$ is specific to nitroxide spin labels at Q-band and should be optimized if different spin systems are used at different microwave frequencies.
16. The distance information is limited to less than 5 nm, because of the short spin echo memory dephasing times ($T_M$) for the nitroxides in the fibrils. High concentrations of protons in the vicinity of the spin label and rotating methyl groups are known to decrease $T_M$ [31]. Spin labels in the hydrophobic core of the fibril are thus expected to have shortened $T_M$. The distance range could be increased by using fully deuterated samples as this will increase $T_M$.

## Acknowledgement

This work was supported by National Institute of Neurological Disorders and Stroke Grant R01NS076619.

### References

1. Ballatore C, Lee VM, Trojanowski JQ (2007) Tau-mediated neurodegeneration in Alzheimer's disease and related disorders. Nat Rev Neurosci 8:663–672
2. Spillantini MG, Goedert M (2013) Tau pathology and neurodegeneration. Lancet Neurol 12:609–622
3. Wu JW, Herman M, Liu L et al (2013) Small misfolded Tau species are internalized via bulk endocytosis and anterogradely and retrogradely transported in neurons. J Biol Chem 288:1856–1870
4. Kfoury N, Holmes BB, Jiang H et al (2012) Trans-cellular propagation of tau aggregation by fibrillar species. J Biol Chem 287:19440–19451
5. de Calignon A, Polydoro M, Suarez-Calvet M et al (2012) Propagation of tau pathology in a model of early Alzheimer's disease. Neuron 73:685–697
6. Liu L, Drouet V, Wu JW et al (2012) Trans-synaptic spread of tau pathology in vivo. PLoS One 7:e31302
7. Soto C (2012) Transmissible proteins: expanding the prion heresy. Cell 149:968–977
8. Hubbell WL, Lopez CJ, Altenbach C et al (2013) Technological advances in site-directed spin labeling of proteins. Curr Opin Struct Biol 23:725–733
9. Berliner LJ, Grunwald J, Hankovszky HO et al (1982) A novel reversible thiol-specific spin label: papain active site labeling and inhibition. Anal Biochem 119:450–455
10. Torok M, Milton S, Kayed R et al (2002) Structural and dynamic features of Alzheimer's Abeta peptide in amyloid fibrils studied by site-directed spin labeling. J Biol Chem 277:40810–40815
11. Margittai M, Langen R (2004) Template-assisted filament growth by parallel stacking of tau. Proc Natl Acad Sci U S A 101:10278–10283
12. Chen M, Margittai M, Chen J et al (2007) Investigation of alpha-synuclein fibril structure by site-directed spin labeling. J Biol Chem 282:24970–24979
13. Jayasinghe SA, Langen R (2004) Identifying structural features of fibrillar islet amyloid polypeptide using site-directed spin labeling. J Biol Chem 279:48420–48425
14. Tanaka M, Chien P, Yonekura K et al (2005) Mechanism of cross-species prion transmission: an infectious conformation compatible with two highly divergent yeast prion proteins. Cell 121:49–62
15. Ngo S, Gu L, Guo Z (2011) Hierarchical organization in the amyloid core of yeast prion protein Ure2. J Biol Chem 286:29691–29699
16. Ladner CL, Chen M, Smith DP et al (2010) Stacked sets of parallel, in-register beta-strands of beta2-microglobulin in amyloid fibrils revealed by site-directed spin labeling and chemical labeling. J Biol Chem 285:17137–17147
17. Cobb NJ, Sonnichsen FD, McHaourab H et al (2007) Molecular architecture of human prion protein amyloid: a parallel, in-register beta-structure. Proc Natl Acad Sci U S A 104:18946–18951
18. Margittai M, Langen R (2008) Fibrils with parallel in-register structure constitute a major class of amyloid fibrils: molecular insights from electron paramagnetic resonance spectroscopy. Q Rev Biophys 41:265–297
19. Bedrood S, Li Y, Isas JM et al (2012) Fibril structure of human islet amyloid polypeptide. J Biol Chem 287:5235–5241
20. Karyagina I, Becker S, Giller K et al (2011) Electron paramagnetic resonance spectroscopy measures the distance between the external beta-strands of folded alpha-synuclein in amyloid fibrils. Biophys J 101:L1–L3
21. Pornsuwan S, Giller K, Riedel D et al (2013) Long-range distances in amyloid fibrils of alpha-Synuclein from PELDOR spectroscopy. Angew Chem Int Ed Engl 52:10290–10294
22. Siddiqua A, Luo Y, Meyer V et al (2012) Conformational basis for asymmetric seeding barrier in filaments of three- and four-repeat tau. J Am Chem Soc 134:10271–10278
23. Meyer V, Dinkel PD, Luo Y et al (2014) Single mutations in tau modulate the populations of

fibril conformers through seed selection. Angew Chem Int Ed Engl 53:1590–1593

24. Eaton GR, Eaton SS, Barr DP, Weber RT (2010) Quantitative EPR. Springer, Wien
25. Margittai M, Langen R (2006) Spin labeling analysis of amyloids and other protein aggregates. Methods Enzymol 413:122–139
26. Friedhoff P, von Bergen M, Mandelkow EM et al (1998) A nucleated assembly mechanism of Alzheimer paired helical filaments. Proc Natl Acad Sci U S A 95:15712–15717
27. Jeschke G (2012) DEER distance measurements on proteins. Annu Rev Phys Chem 63:419–446
28. Jeschke G, Chechik V, Ionita P et al (2006) DeerAnalysis2006: a comprehensive software package for analyzing pulsed ELDOR data. Appl Magn Reson 30:473–498
29. Sen KI, Logan TM, Fajer PG (2007) Protein dynamics and monomer-monomer interactions in AntR activation by electron paramagnetic resonance and double electron-electron resonance. Biochemistry 46:11639–11649
30. Brandon S, Beth AH, Hustedt EJ (2012) The global analysis of DEER data. J Magn Reson 218:93–104
31. Huber M, Lindgren M, Hammarstrom P et al (2001) Phase memory relaxation times of spin labels in human carbonic anhydrase II: pulsed EPR to determine spin label location. Biophys Chem 94:245–256

# Chapter 13

# Preparation of Crystalline Samples of Amyloid Fibrils and Oligomers


## Asher Moshe, Meytal Landau, and David Eisenberg



## Abstract

The molecular structures of amyloid fibers and oligomers are required in order to understand and control their formation. Yet, their partially disordered and polymorphic nature hinders structural analyses. Fortunately, short segments from amyloid proteins, which exhibit similar biophysical properties to the full-length proteins, also form fibrils and oligomers and their atomic structures can be determined. Here we describe experimental procedures used to assess fiber-forming capabilities of amyloid peptide segments and their crystallization.



**Key words** Microcrystals, Amyloid-like peptides, Microcrystallography, Cross-β spine, Steric zipper, Cylindrin


## 1 Introduction

Dozens of different proteins, in all kingdoms of life, form amyloid fibrils; yet no obvious sequence motif is directly associated with this phenomena [1]. Nevertheless, the fibrils display similar biophysical characteristics, including distinctive dye-binding and X-ray diffraction properties [1], and they have a common "cross-β spine" structure [2, 3]. Although the spine structure is common among amyloid fibrils, it is the details of side chain interactions that impart the fibril structure and physiochemical properties. Knowledge of the spine structure can be elucidated only through high resolution structural studies, which then serve as the basis for drug design and engineering. Determining the molecular structure of a full fibril is so challenging that we must often settle for learning the structure of the adhesive protein segments that form the spine of the fibril.

The structures of amyloid fibril spines determined to date tend to be similar. They are formed by short segments of fibril-forming proteins (as short as tetra-peptides); they are themselves well-ordered amyloid-like fibrils; and they are normally built from pairs of closely mating β-sheets. They display many properties in


David Eliezer (ed.), *Protein Amyloid Aggregation: Methods and Protocols*, Methods in Molecular Biology, vol. 1345, DOI 10.1007/978-1-4939-2978-8_13, © Springer Science+Business Media New York 2016

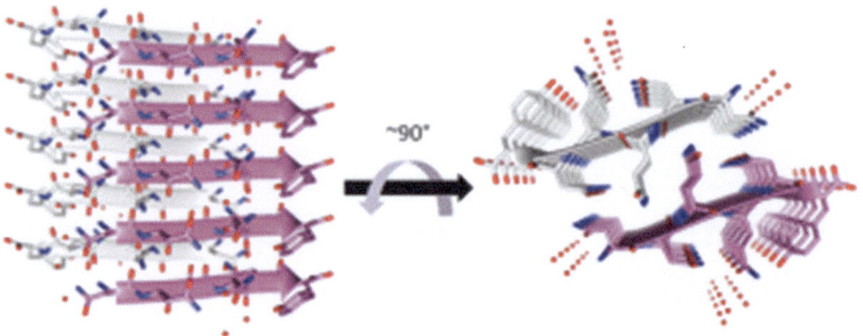


**Fig. 1** Structure of the NNQQNY segment forming a steric zipper structure. The NNQQNY segments [7] of the yeast prion protein Sup35 are packed as pairs of β-sheets forming the basic unit of the fiber, namely the steric zipper. In the *left panel*, the view is perpendicular to the fiber axis; the β-strands run horizontally. In the *right panel*, the view is down the fiber axis. Here five layers of β-strands are depicted; actual fibers contain probably more than 50,000 layers. NNQQNY is shown as *sticks* with non-carbon atoms colored by atom type. The β-sheets are composed of parallel strands (*cartoon arrows*), with the carbons of the two β-sheets colored *white* or *pink*. The oxygen atoms of water molecules are shown as *red spheres*; the dry steric zipper interface is devoid of water

common with fibrils of their parent proteins, and they illuminate the conversion of their parent proteins to fibril form [1, 4]. Because the spines determine properties of the full fibrils, the Eisenberg lab had determined fully objective atomic models of the common β-spine structure of fibril-forming segments (4–8 residues) using X-ray microcrystallography [5–7] (Fig. 1). The particularly small size of these peptide crystals (often no more than 1 μm in cross section) requires special handling and X-ray diffraction data collection [8] (Fig. 2). The techniques are described below. To date, the structures of over 100 amyloid-like segments from 12 disease-associated proteins were determined by the Eisenberg lab, nearly all showing the dual-β-sheet spine pattern ([5, 7–17] and unpublished results). This previously unobserved structural motif, termed steric zipper (Fig. 1), illuminates the stability of amyloid fibrils, their self-seeding characteristic, and their tendency to form polymorphic structures [18–20].

As for amyloid fibrils, the structures of small amyloid oligomers present severe challenges, owing to their transient and polymorphic character. In 2012 the Eisenberg lab determined the atomic structure of an oligomeric complex from a segment of the amyloid-forming protein αB-crystalline that fits some operational definitions of amyloid oligomers [21] An 11-residue segment was identified as amyloidogenic using computational tools [22], yet its crystal structure revealed an astonishingly different structure from the previously characterized steric zippers [21]: a cylindrical barrel of six identical antiparallel beta strands. This small amyloid oligomer was named cylindrin [21] (Fig. 3).

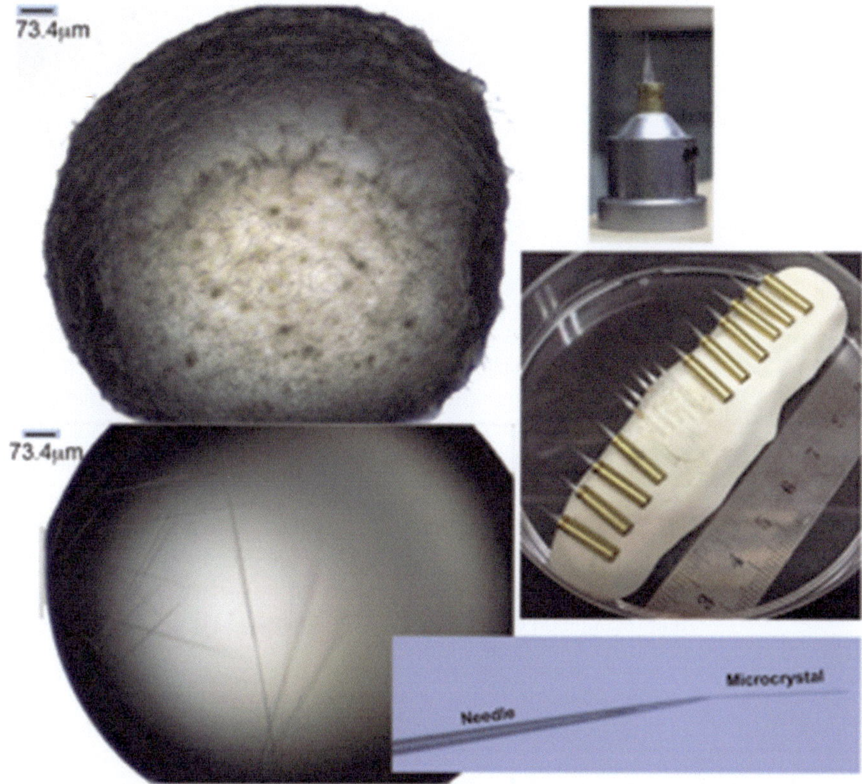


**Fig. 2** Mounting microcrystals of amyloid-like peptides. Drops of ~100 nl are occupied by numerous short microcrystals (*top left*) or by few very long microcrystals (*bottom left*); scale bars are indicated. The sharpened needle capillaries and the glued brass pins (described in Subheading 3.6, **step 1**) are situated on clay stored in a petri dish (*middle right panel*; ruler is marked in centimeters). The needles are used to mount a single microcrystal (*bottom right panel* is a close-up on a microcrystal situated on the tip of the needle). The brass pin holding the microcrystal is inserted into the magnetic crystal mount (*top right panel*) designed to position the brass pin onto a goniometer head for X-ray data collection

## 2 Materials

Prepare all solutions using double deionized water and analytical grade reagents. Preform all experiments at room temperature (unless indicated otherwise). Diligently follow all waste disposal regulations when disposing waste materials, especially uranyl acetate. We do not add sodium azide to the reagents. Because of their seeding properties, amyloid segments may pose dangers to health, so safety precautions should be taken in working with them (*see* **Note 1**).

### 2.1 Amyloid-Like Peptides

1. The peptides are custom synthesized with >98 % purity. Peptides should be stored in lyophilized form at −20 °C. It is better to equilibrate the peptides to room temperature in a

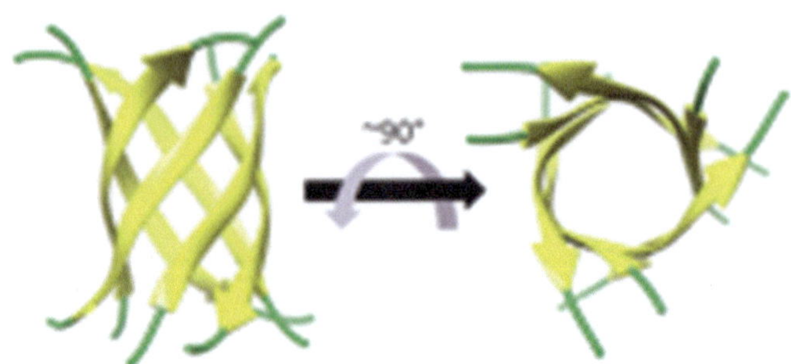


**Fig. 3** The cylindrin crystal structure of a toxic oligomer [21]. Ribbon representation of the cylindrin crystal structure composed of six segments. Each segment is composed of 11 amino acid residues from alphaB crystallin of sequence KVKVLGDVIEV. The segments form beta-strands (*yellow arrows*) that assemble into a barrel-like structure named cylindrin. In the *left panel*, the view is perpendicular to the barrel axis. In the *right panel*, the view is down the barrel axis. The height of the cylindrin is 22 Å. The inner dimension of the cylindrin, around the waist from Cα to Cα, is 12 Å, and at the splayed ends is 22 Å. The cylindrin oligomeric complex exhibits properties of other amyloid oligomers: beta-sheet-rich structure, cytotoxicity, and recognition by an anti-oligomer antibody [21]

desiccator prior to opening and weighing. *See* **Note 1** for safety measures.

2. For fibrillation assays use synthesized peptides with capped termini, namely acetylation of the N-terminus and amidation of the C-terminus.
3. For crystallization experiments leave the termini uncapped.
4. 0.22 μm Ultrafree-MC centrifugal filter device (AMICON, Bedford, MA, USA).

### 2.2 Peptide Fibrillation Assays

All solutions are made using standard lab protocols. References indicate published results in which the outlined conditions have been successfully used for fibrillation assays of amyloid-like peptides.

1. 25 mM TRIS pH 8.5, 150 mM sodium chloride [21].
2. 25 mM TRIS pH 8.5, 150 mM sodium chloride, 10 % dimethyl sulfoxide (DMSO) [21].
3. 150 mM HEPES pH 7.4, 150 mM sodium chloride [15].
4. Phosphate-buffered saline (PBS) [15, 21].
5. 150 mM sodium chloride, 50 mM phosphate, pH 2.5 [5, 8].
6. 1 mM EDTA, 25 mM potassium phosphate, pH 7.0 [23].
7. 50 mM glycine buffer, pH 2.5 [8].
8. 10 mM CAPS pH 11.0, 150 mM sodium chloride, 1 mM EDTA [21].

**2.3 Transmission Electron Microscopy**

1. Copper transmission electron microscopy (TEM) grids, 400 mesh.
2. Poly-lysine solution: 0.1 % w/v poly-lysine in water (CAS Number 25988-63-0).

**2.4 ThT Fluorescence**

1. ThT buffer: 10–40 μM ThT in 10 mM Tris pH 8.0 or PBS.

**2.5 Crystallization Experiments**

1. Borosilicate glass calibrated 5 μl micropipettes (VWR).
2. Capillary cutting stone (Hampton Research HR4-334).
3. Brass pins (Hampton Research HR4-661).
4. Adjustable, magnetic, crystal mount (Hampton Research HR8-028).
5. Standard goniometer key (Hampton Research HR4-659).
6. Dual-Thickness MicroLoops™ LD, SPINE/18 mm length (e.g., Mitegen LLC #M5-L18SP-A1LD).
7. Goniometer cap bases suited for the relevant beam-line (e.g., Hampton Research HR8-112 with suited vials HR4-904).

## 3 Methods

Fibril-formation propensities of segments of amyloid proteins are predicted using computational methods, for example ZipperDB [22], Tango [24], Waltz [25], and Zyggregator [26]. Segments forming amyloid fibers are typically 4–6 residues. Segments forming oligomers are longer, typically about 11 residues [21].

**3.1 Amyloid-Like Peptide Preparation for Fibrillation Assays**

1. Dissolve 1 mM *water-soluble capped peptides* in water. Dissolve 0.5 mM *water-insoluble capped peptides* in 10–60 % DMSO (*see* **Note 2**).
2. Spin filter the solution through a 0.22 μm centrifugal filter device.
3. Prepare the final peptide solutions to 300–500 μl and incubate at 37 °C (*see* **Note 3**).

**3.2 Fiber Formation Assessed by Transmission Electron Microscopy**

1. Incubate the dissolved peptide for 1–7 days (*see* **Note 4**).
2. Charge 400 mesh copper TEM grids (*see* **Note 5**) by either:
   (a) High-voltage, alternating current, glow-discharge immediately before use.
   (b) Apply 5 μl poly-lysine solution on the grid and allow to adhere for 1 min. Drain off excessive fluid using a cellulose filter paper. Wash the grid by applying 5 μl distilled water (careful not to let the grid dry) and drain off excessive fluid

using a cellulose filter paper. Repeat this procedure three times. These charged grids can be stored for future usage.

3. Add 5 μl of sample, and allow to adhere for 4 min. Drain off excessive fluid using a cellulose filter paper.
4. Wash the grid by applying 5 μl distilled water (careful not to let the grid dry) and drain off excessive fluid using a cellulose filter paper. Repeat this procedure twice.
5. Apply 5 μl 1–2.5 % uranyl acetate and allow to adhere for 2 min. Drain off excessive fluid using a cellulose filter paper (*see* **Note 6**).
6. Allow grids to dry in a desiccator for at least 24 h before imaging.
7. Examine specimens in a transmission electron microscope at an accelerating voltage of 75–200 kV.

### 3.3 Fiber Formation Assessed by Thioflavin T (ThT) Fluorescence

1. Incubate freshly prepared peptide samples described in Subheading 3.1. with ThT buffer (*see* **Note 7**).
2. Monitor fibril formation at 37 °C by ThT fluorescence at 444 nm excitation and 482 nm emission wavelengths [5, 8, 10]. Experiments are usually performed using 200 μl sample volume in black, 96-well, optical bottom NUNC plates (Fisher Scientific) [10]. Data should be collected in triplicate. Shaking is needed before taking readouts for homogeneity (*see* **Note 8**).
3. Monitor ThT signal until the curve reaches a plateau (*see* **Note 9**).

### 3.4 Amyloid-Like Peptide Preparation for Crystallization Experiments

1. Dissolve *water-soluble peptides* in water to a final concentration of 10 mg/ml. Dissolve *water-insoluble peptides* in 100 % DMSO to a final concentration of 100 mM and dilute to around 10 mM (*see* **Note 10**).
2. Filter the solution through a 0.1 μm Ultrafree-MC centrifugal filter device (AMICON, Bedford, MA, USA).

### 3.5 Crystallization Screens

1. Perform all crystallizations using the hanging drop/vapor diffusion method at room temperature and store the plates at 18–20 °C.
2. Set 96-well plates with crystallization screens (e.g., using the TTP LabTech Mosquito nanodispenser robot) (*see* **Note 11**). Drops are a mixture of 100 nl peptide solution and 100 nl reservoir solution in three different ratios: 1:1, 2:1, and 1:2.

### 3.6 Crystal Mounting

1. Crystal mounting is the retrieval and transfer of a single crystal from its growth solution into a suitable mounting tool that can be used for X-ray diffraction data collection (*see* **Note 12**).
2. Using a micropipette puller, wet 5 μl borosilicate glass-calibrated micropipettes. Depending on the specification of the

micropipette puller instrument, adjust the heat and pull velocity to obtain sharp needles with a tip width of about 10–20 μm.

3. Cut the sharpened capillaries using a capillary cutting stone to about 1.3 cm.
4. Glue the sharpened capillaries into 12.8 mm long brass pins. Leave about 5–6 mm of the sharp end of the capillary sticking out of the brass pin (Fig. 2).
5. At the synchrotron, mount the brass pins into adjustable, magnetic, crystal mount. You will need a standard goniometer key to adjust the screws (*see* **Note 13**). The Adjustable crystal mount is a magnetic component designed to position the brass pin with the glass capillary onto a goniometer head for X-ray data collection.
6. For crystals of peptides that are eight residues or longer (potential fibers or oligomers), you need to flash freeze and store the crystals. Mount the crystals on very small cryogenic loops depending on the crystal size (*see* **Note 14**).
7. Flash freeze and store the crystals in liquid nitrogen.

### 3.7 X-Ray Diffraction Data Collection for Microcrystals

1. For the microcrystals, X-ray diffraction data is collected in 5° wedges using a 5 μm beam size (*see* **Note 15**).
2. The crystals are cryo-cooled (100 K) for data collection. Transmission intensity should be varied according to the quality of diffraction.

## 4 Notes

1. Safety measures when handling amyloid-like peptides:
   When handling the peptides in the lyophilized form, use personal protective equipment including a respirator protection mask. Carry out all peptide weighing in an analytical balance and keep the doors partially closed. Open the vials only inside the balance. Before weighting and dissolving the peptides centrifuge the vial to minimize powder leftovers on the cap and sides of the vial. Add the dissolving solution quickly on the side on the vial (not directly into the powder) and close the lid. Vortex the vial.
2. There is a large range of peptide concentrations and buffers used in fibrillation assays, and optimal conditions for monitoring fibrillation vary among samples. Here we offer starting conditions along with optional variations used successfully in our past experiments.

   *For water-soluble peptides* start with dissolving peptides to 1–6 mM in water [8, 23]. Optionally, peptide can be diluted to a final concentration of as low as 20 μM in any of the solutions listed in Subheading 2.2.

*Water-insoluble peptides* can be dissolved using the following:

(a) Dissolve 100 mM peptide in 100 % DMSO and dilute with water or any of the solutions listed in Subheading 2.2 to a final concentration of 10 μM–0.5 mM [15]. Avoid using DMSO with peptides containing cysteine or methionine to prevent side-chain oxidation. In such cases, dissolve the peptide in any of the solutions listed in Subheading 2.2 or using the below conditions (b).

(b) Dissolve 1 mM peptide in 100 % hexafluoroisopropanol (HFIP), and then dilute to 10 μM in 20 mM sodium acetate pH 6.5 (1 % HFIP final) [9, 11].

3. Incubating the peptides can also be done at room temperature. Constant shaking at 300–1000 rpm (1.9–20 × *g*) is optional [10].
4. Take into account that some peptides take a longer time to fibrillate, even months.
5. Careful handling of the grid is needed to protect the support film. Hold the grid at the perimeter with the tweezers pointing in an angle.
6. Make sure not to leave excess uranyl acetate on the grid, as it might interfere with imaging. Make sure to properly dispose tips and filter papers contaminated with uranyl acetate.
7. In general, short amyloid-like peptides (4–6 residues) do not necessarily display the common ThT fluorescence signal, even when fiber formation is confirmed via electron microscopy.
8. Constant shaking at 300–1000 rpm (1.9–20 × *g*) is optional [10].
9. Fibrillation time varies tremendously between fiber-forming peptides. While some form fiber immediately, others can take days and even weeks to form.
10. Depending on the solubility of the peptides, final concentration and dissolving buffer vary. Dissolve *water-soluble peptides* in water to a final concentration of 1–40 mg/ml (start with 10 mg/ml) [7–9, 11, 15–17, 23]. Dissolve *water insoluble peptides* in low or high pH buffers [9], DMSO [15, 17], ethanol [17], 15 % acetonitrile in 15 mM Bis-Tris [14], or 10–20 mM lithium hydroxide.
11. Common crystallization screens used for amyloid-like peptides include: Index, Crystal Screen, PEG/Ion, and SaltRX by Hampton Research, Wizard by Emerald Biosystems, as well as ComPAS, JCSG+, and PACT by Qiagen.
12. Most protein crystals require flash freezing in liquid nitrogen and cryo-protection. Yet, due to the low solvent content of the amyloid peptide microcrystals, they can be stored at room temperature. In addition, due to the especially small size of the peptide microcrystals, they are mounted on glass capillaries,

and not on the commonly used cryogenic loop. This enables better centering of the crystal on the goniometer for X-ray diffraction data collection [7, 8, 10, 11, 14, 15, 17].

13. Mounting the brass pins (with glued needle capillaries) into the adjustable crystal mount will result in an overall length of 18 mm standard at nearly all synchrotron beam lines in the world. The pins are super-glued into goniomegter cap bases suited for the relevant beam line (e.g., Hampton Research HR8-112 with suited vials HR4-904).
14. Generally it is recommended to use a loop that is slightly smaller than the crystal (for example 20 μm loop for a 25 μm crystal; Robert Thorne (Mitegen), personal communication). For most amyloid peptide microcrystals, Dual-Thickness MicroLoops™ LD, SPINE/18 mm length (e.g., Mitegen LLC #M5-L18SP-A1LD) are suited.
15. Due to the especially small size of the amyloid-like peptide microcrystals, a Microfocus Beamline (5 μm beam size) is required: for example, NE-CAT 24-ID-E of the Advanced Photon Source (APS), Argonne National Laboratory, USA; ID13 and ID23-2 at the European Synchrotron Radiation Facility (ESRF), Grenoble, France; or X06SA at the Swiss Light Source (SLS), Villigen, Switzerland.

## Acknowledgements

We thank our coworkers for their contributions to development of these methods, and NIH (AG029430 & SG04812), DOE (DE-FC02-02ER63421), and NSF (MCB-0958111) for support. DE and ML thank the U.S.-Israel Binational Science Foundation (BSF). ML thanks the Alon Fellowship from the Israeli Council for Higher Education, the I-CORE Program of the Planning and Budgeting Committee and The Israel Science Foundation, Center of Excellence in Integrated Structural Cell Biology; Grant No 1775/12, the Support for training and career development of researchers (Marie Curie) CIG, Seventh framework program, Single Benefi ciary, the J. and A. Taub Biological Research Fund, and the David and Inez Mayers Career Advancement Chair in Life Sciences.

### References

1. Gazit E (2005) Mechanisms of amyloid fibril self-assembly and inhibition. Model short peptides as a key research tool. FEBS J 272: 5971–5978
2. Sipe JD, Cohen AS (2000) Review: history of the amyloid fibril. J Struct Biol 130:88–98
3. Sunde M, Serpell LC, Bartlam M et al (1997) Common core structure of amyloid fibrils by synchrotron X-ray diffraction. J Mol Biol 273: 729–739
4. Balbirnie M, Grothe R, Eisenberg DS (2001) An amyloid-forming peptide from the yeast

prion Sup35 reveals a dehydrated beta-sheet structure for amyloid. Proc Natl Acad Sci U S A 98:2375–2380

5. Ivanova MI, Thompson MJ, Eisenberg D (2006) A systematic screen of beta(2)-microglobulin and insulin for amyloid-like segments. Proc Natl Acad Sci U S A 103:4079–4082
6. Thompson MJ, Sievers SA, Karanicolas J et al (2006) The 3D profile method for identifying fibril-forming segments of proteins. Proc Natl Acad Sci U S A 103:4074–4078
7. Nelson R, Sawaya MR, Balbirnie M et al (2005) Structure of the cross-beta spine of amyloid-like fibrils. Nature 435:773–778
8. Sawaya MR, Sambashivan S, Nelson R et al (2007) Atomic structures of amyloid cross-beta spines reveal varied steric zippers. Nature 447: 453–457
9. Wiltzius JJ, Sievers SA, Sawaya MR et al (2008) Atomic structure of the cross-beta spine of islet amyloid polypeptide (amylin). Protein Sci 17: 1467–1474
10. Ivanova MI, Sievers SA, Sawaya MR et al (2009) Molecular basis for insulin fibril assembly. Proc Natl Acad Sci U S A 106:18990–18995
11. Wiltzius JJ, Landau M, Nelson R et al (2009) Molecular mechanisms for protein-encoded inheritance. Nat Struct Mol Biol 16:973–978
12. Wiltzius JJ, Sievers SA, Sawaya MR et al (2009) Atomic structures of IAPP (amylin) fusions suggest a mechanism for fibrillation and the role of insulin in the process. Protein Sci 18:1521–1530
13. Laganowsky A, Benesch JL, Landau M et al (2010) Crystal structures of truncated alphaA and alphaB crystallins reveal structural mechanisms of polydispersity important for eye lens function. Protein Sci 19:1031–1043
14. Apostol MI, Wiltzius JJ, Sawaya MR et al (2011) Atomic structures suggest determinants of transmission barriers in mammalian prion disease. Biochemistry 50:2456–2463
15. Colletier JP, Laganowsky A, Landau M et al (2011) Molecular basis for amyloid-beta polymorphism. Proc Natl Acad Sci U S A 108:16938–16943
16. Liu C, Zhao M, Jiang L et al (2012) Out-of-register beta-sheets suggest a pathway to toxic amyloid aggregates. Proc Natl Acad Sci U S A 109:20913–20918
17. Landau M, Sawaya MR, Faull KF et al (2011) Towards a pharmacophore for amyloid. PLoS Biol 9:e1001080
18. Nelson R, Eisenberg D (2006) Recent atomic models of amyloid fibril structure. Curr Opin Struct Biol 16:260–265
19. Nelson R, Eisenberg D (2006) Structural models of amyloid-like fibrils. Adv Protein Chem 73:235–282
20. Eisenberg D, Jucker M (2012) The amyloid state of proteins in human diseases. Cell 148: 1188–1203
21. Laganowsky A, Liu C, Sawaya MR et al (2012) Atomic view of a toxic amyloid small oligomer. Science 335:1228–1231
22. Goldschmidt L, Teng PK, Riek R et al (2010) Identifying the amylome, proteins capable of forming amyloid-like fibrils. Proc Natl Acad Sci U S A 107:3487–3492
23. Ivanova MI, Sievers SA, Guenther EL et al (2014) Aggregation-triggering segments of SOD1 fibril formation support a common pathway for familial and sporadic ALS. Proc Natl Acad Sci U S A 111:197–201
24. Fernandez-Escamilla AM, Rousseau F, Schymkowitz J et al (2004) Prediction of sequence-dependent and mutational effects on the aggregation of peptides and proteins. Nat Biotechnol 22:1302–1306
25. Maurer-Stroh S, Debulpaep M, Kuemmerer N et al (2010) Exploring the sequence determinants of amyloid structure using position-specific scoring matrices. Nat Methods 7: 237–242
26. Tartaglia GG, Pawar AP, Campioni S et al (2008) Prediction of aggregation-prone regions in structured proteins. J Mol Biol 380:425–436

# Chapter 14

# Quenched Hydrogen Exchange NMR of Amyloid Fibrils

Andrei T. Alexandrescu

## Abstract

Amyloid fibrils are associated with a number of human diseases. These aggregatively misfolded intermolecular β-sheet assemblies constitute some of the most challenging targets in structural biology because to their complexity, size, and insolubility. Here, protocols and controls are described for experiments designed to study hydrogen-bonding in amyloid fibrils indirectly, by transferring information about amide proton occupancy in the fibrils to the dimethyl sulfoxide-denatured state. Since the denatured state is amenable to solution NMR spectroscopy, the method can provide residue-level-resolution data on hydrogen exchange for the monomers that make up the fibrils.

**Key words** Hydrogen exchange, NMR, Amyloidogenic proteins, Aggregation, Protein dynamics, Protein stability, Protein folding, Hydrogen-bonding, Protein structure, Secondary structure

## Abbreviations

| | |
|---|---|
| DCA | Dichloroacetic acid |
| DCA-d2 | Deuterated analog of dichloroacetic acid: $Cl_2CDCO_2D$ |
| DCl | Deuterium chloride (deuterated analog of HCl) |
| DMSO | Dimethyl sulfoxide |
| DMSO-d6 | Deuterated analog of dimethyl sulfoxide $(CD_3)_2SO$ |
| $D_2O$ | Deuterium oxide (heavy water) |
| HSQC | Heteronuclear single-quantum coherence |
| IAPP | Islet amyloid polypeptide |
| NMR | Nuclear magnetic resonance |
| NOESY | Nuclear Overhauser enhancement spectroscopy |
| qHX | Quenched hydrogen exchange |
| SDS | Sodium dodecyl sulfate |
| ThT | Thioflavin T |
| TOCSY | Total correlation spectroscopy |

David Eliezer (ed.), *Protein Amyloid Aggregation: Methods and Protocols*, Methods in Molecular Biology, vol. 1345,
DOI 10.1007/978-1-4939-2978-8_14, 

## 1 Introduction

Hydrogen-deuterium isotope exchange experiments provide important information about the solvent accessibility of amide protons in proteins and about the stability of hydrogen-bonded secondary structure [1–3]. NMR offers unparalleled resolution for such studies as the technique can monitor each exchange-labile site [4]. Many interesting biological molecules, however, are not amenable to direct solution NMR spectroscopy. These include transiently formed intermediates with lifetimes shorter than the time required to record an NMR spectrum, species subject to conformational exchange broadening of NMR resonances, and large proteins or complexes that extend beyond the size limit of solution NMR. In these cases it is often possible to transfer amide proton solvent protection data from the NMR-inaccessible state to a conformational state amenable to NMR spectroscopy [5]. Pulse-labeling experiments can be used to trap amide protons protected by nascent secondary structure in short-lived folding intermediates by quickly refolding the protein, and reading out amide proton occupancy from NMR experiments on the native state [6, 7]. Similarly, in "molten globule" intermediates that have substantial secondary structure but a fluctuating tertiary structure leading to NMR line-broadening due to conformational exchange, folding to the NMR-tractable native state can be used to indirectly read out amide proton protection [8].

For large proteins and complexes the denatured rather than the native state can be used to read out amide proton exchange [9]. This is because in the "random coil" limit, segmental motion will be so large that residues will experience rotational diffusion rates approaching those of the free amino acids, rather than a folded protein, where in the rigid limit the overall "global" correlation time affects all residues and increases with molecular size [10]. The short effective correlation times typical of unfolded proteins lead to narrow NMR lines (small R2 transverse relaxation rates), often making unfolded proteins ideally suited for NMR. The lower chemical shift dispersion typical of unfolded proteins can usually be circumvented with the aid of two-dimensional NMR spectroscopy [11].

To relay amide protection data from a folded to an unfolded state requires (1) unfolding the protein to a denatured state suitable for NMR, (2) keeping the amide protons from exchanging through the denatured state for sufficiently long to acquire the NMR data [9]. The solvent dimethyl sulfoxide (DMSO) typically meets these criteria. Although sometimes used as a membrane mimetic, DMSO is a strong denaturant of soluble proteins [12, 13]. Typically amyloid fibrils are extremely stable structures resistant to a variety of harshly denaturing conditions such as heat, urea, SDS and cleavage by proteases [14]. DMSO is one of the few compounds that have been demonstrated to denature a variety of amyloid fibrils to unfolded monomers. Indeed, the solvent has had therapeutical success for some types of amyloidoses, including alleviation of skin amyloid plaques by topical application of DMSO [12]. A second advantage of DMSO is that it is an aprotic solvent with no exchangeable protons. Whereas amide proton exchange in $H_2O$ can occur in seconds to minutes, depending on solution pH, exchange rates can be reduced more than 100-fold in DMSO, allowing amide protons to survive for 30–60 min—a time sufficient to record 2D NMR spectra on the denatured state. As in aqueous solution, hydrogen exchange rates in DMSO are highly pH dependent. In order to control solution pH, it is necessary to use solvent mixtures of 90–95 % DMSO/10–5 % $H_2O$ (v/v) and carefully buffer the solution towards the pH ~5 minimum of hydrogen exchange with an acid such as dichloroacetic acid (DCA). The effects of various solvent and acid combinations on intrinsic exchange rates of unfolded polypeptides have been described in a pioneering paper by Roder and co-workers [15], and the reader is referred to this seminal work. Although the subject of the present review is on using DMSO for quenched hydrogen exchange studies of amyloid fibrils, the approach is more general and has been applied to other systems including protein complexes such as the GroES co-chaperonin [16] and integral membrane proteins in their natural lipid bilayer environment [17].

Amyloids are amongst ~50 types of proteins that undergo aggregative misfolding into fibrillar structures associated with human diseases [18]. Because of their size and complexity, amyloid assemblies are amongst the most challenging systems to study in structural biology. Quenched hydrogen exchange NMR methods to study amyloid fibrils were first described for the *Escherichia coli* protein CspA [9]. Although CspA is not involved in any disease, the relative ease of preparing $^{15}$N-labeled samples of the protein for NMR studies facilitated developing the methodology for quenched hydrogen exchange experiments on model amyloid fibrils formed by this protein. Quenched hydrogen exchange methods have subsequently been applied to a variety of amyloidogenic proteins involved in disease including β-microglobulin [19], prion protein [20], cystatin [21], α-synuclein [22], Aβ [23], and amylin/IAPP [24]. These studies have shed light on the secondary structure of amyloids and have complemented ssNMR data to model the structures of amyloid fibrils in detail [25–28]. It should be possible to further develop the technique to investigate intermediates in the fibril assembly process, follow changes in amyloid secondary structure with solution conditions, and map binding sites for amyloid-directed inhibitors or drugs. In this chapter a general protocol together with controls is presented for quenched hydrogen exchange studies of amyloid fibrils.

## 2 Materials

1. DMSO-h6 (Fisher BioReagents, Fair Lawn, NJ, USA).
2. DMSO-d6 (Cambridge Isotope Laboratories, Andover, MA, USA, CIL).
3. DCA-h2 (Aldrich, St. Louis, MO, USA).
4. DCA-d2 (CDN Isotopes, Point-Claire, Quebec, Canada).
5. Ultrapure ThT (AnaSpec, Fremont, CA, USA).
6. 99.96 % $D_2O$ (CIL).
7. DCl (Aldrich).
8. Low-retention microcentrifuge tubes (Fisher Scientific, Waltham, MA, USA).
9. 0.45 μm cellulose nitrate membrane filter (Millipore Corporation, Billerica, MA, USA).

## 3 Methods

### *3.1 Preparation of Amyloid Fibrils*

1. Prepare 1–10 mM stock solutions of the amyloidogenic protein in a solvent such as 100 % DMSO that will prevent the protein from forming amyloid fibrils (*see* **Note 1**). The stock solutions should be divided into aliquots of the desired volume, or lyophilized, and stored at −80 °C when not in use (*see* **Note 2**).
2. Before adding the amyloidogenic protein, filter all solutions and buffers through a 0.45 μM cellulose nitrate membrane filter and degass them for 20 min using a sonicator (e.g., a Fisher Scientific Model 500 Sonic Dismembrator operating at 75 % amplitude). Dilute the stock solutions to the desired protein concentration (20–250 μM) for the fibrillization reactions using an appropriate pH buffer (e.g., 20 mM sodium phosphate, pH 7.4). If the stock solutions are prepared using a disaggregating solvent such as DMSO, ensure that the final DMSO concentration reaction after dilution (e.g., 1–5 % (v/v)) does not inhibit fibrillization. Include 0.02 % (v/v) of the bacteriostatic compound sodium azide ($NaN_3$) to prevent bacterial growth during the fibrillization reactions. Use low-retention plates or microcentrifuge tubes for the reactions, to avoid sample loss.
3. Incubate the fibrillization reactions at the appropriate temperature (e.g., 37 °C). Depending on the protein and application, the fibrils can be grown under quiescent conditions or using agitation (e.g., orbital shaking at 240–1200 rpm). If agitation is required, this can be done either using the shaking function of a fluorescence plate reader or using an Eppendorf Thermomixer C plate shaker from Fisher (*see* **Note 3**).

4. Incubate the fibrils for a time predetermined from other experiments to form fibrils, or preferably monitor the fibrillization reactions using fluorescence of the amyloid specific dye thioflavin T (ThT) until the reaction is well into the steady-state plateau phase. To follow fibrillization with fluorescence, use 15 μM ultrapure ThT (*see* **Note 4**), and a fluorescence plate reader or fluorometer set near the excitation (450 nm) and emission maxima (482 nm) of amyloid-bound ThT dye.
5. Collect the fibrils by sedimentation in a microcentrifuge for 30 min at 15,000 g (*see* **Note 5**).

### 3.2 Control Experiments to Test the Suitability of Amyloid Proteins for NMR qHX Experiments

1. Demonstrate that the fresh unfibrillized protein is amenable for NMR in DMSO. Prepare a 0.1–1 mM sample of the protein in 95 % DMSO-d6/5 % DCA-h2 (*see* **Note 6**). The apparent pH of the sample should be adjusted with HCl or DCl to between pH 4 and 5 close to the pH-minimum for hydrogen exchange [15]. Record a 2D NMR spectrum (*see* **Note 7**) and check that the fingerprint ($^1$H-$^{15}$N for $^{15}$N-HSQC or HN-Hα for 2D TOCSY) has the expected number of cross peaks predicted from the number of amino acids in the protein (*see* **Note 8**).
2. Demonstrate that amyloid fibrils of the given protein can be dissolved and converted to an NMR-accessible state in DMSO. Prepare and collect fibrils by sedimentation as in Subheading 3.1. Resuspend the fibrils in $H_2O$ and lyophilize the sample. Dissolve the lyophilized samples in 95 % DMSO/5 % DCA-h2 and check that the $^{15}$N-HSQC spectrum is similar to that recorded for the unfibrillized sample in Subheading 3.2, **step 1** (*see* **Note 9**).
3. Obtain backbone amide proton NMR assignments using a 1 mM protein sample in 95 % DMSO-d6/5 % DCA-h2 (*see* **Note 10**).
4. Determine amide proton exchange rates in the solvent (e.g., 95 % DMSO/5 % DCA-d2) to be used for qHX NMR experiments. Rates can be measured by collecting 2D NMR experiments as a function of time in deuterated solvent (use DCA-d2), and fitting the amide proton intensity decay for each amino acid to an exponential function. The rates can be used to set an upper limit on the time to record 2D NMR spectra in the qHX experiments, so that intrinsic amide exchange in DMSO is negligible during the time for NMR data acquisition. In principle, exchange rates in DMSO could also be used to correct for amide proton loss during NMR data acquisition in the qHX experiments, when it is not possible to collect spectra on a time scale faster than exchange.

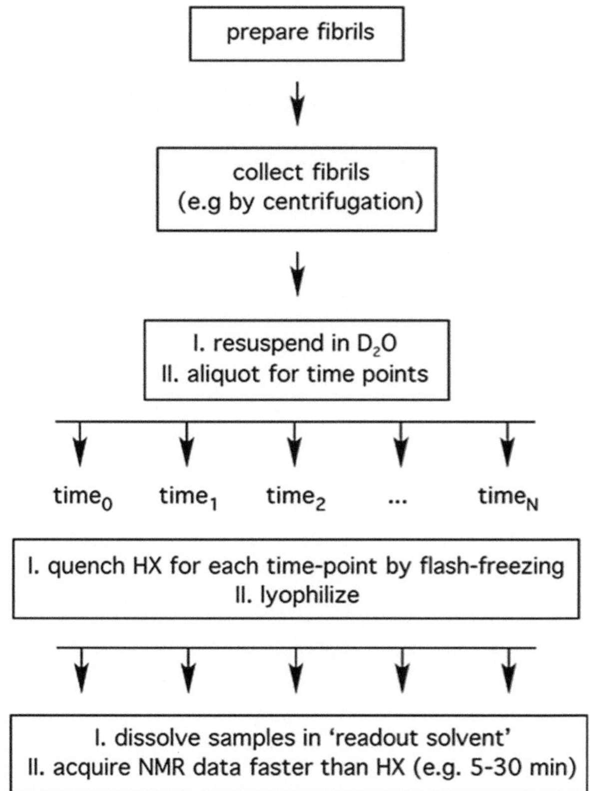


**Fig. 1** Flowchart for qHX NMR experiments on amyloid fibrils

### 3.3 Quenched Hydrogen Exchange Experiments

A scheme for the qHX experiments is shown in Fig. 1 with a more detailed protocol following below.

1. Form fibrils and collect them by sedimentation as in Subheading 3.1.
2. Resuspend the sedimented fibrils in a large excess of $D_2O$ (*see* **Note 11**). Note the time the fibrils are taken up in $D_2O$, as this will be the start of the exchange period.
3. Mix the fibril suspension for 30 s with a Vortex mixer to ensure proper mixing, and withdraw an equal-sized aliquot for each desired exchange time point. Six or more time points should be sampled to characterize the exponential decay of amide proton intensity decay due to deuterium exchange. Flash-freeze the withdrawn aliquot in a dry ice-ethanol bath, lyophilize, and store at −80 °C for subsequent use.
4. Before exchange measurements it is critical to optimize all NMR spectrometer parameters with a standard sample of the protein in 95 % DMSO/5 % DCA-h2. Optimize the shims, pulse widths, spectral windows and optimal number of increments in the indirectly acquired dimension for proper resolution.

5. Dissolve the partially exchanged freeze-dried fibril samples in 95 % DMSO/5 % DCA-d2 (*see* **Note 12**) and collect 2D NMR data in 5–30 min. After collecting the 2D NMR data also collect a 1D 1H-NMR spectrum in case this is needed to normalize amide proton peak intensities (*see* **Note 13**).
6. Process the 2D NMR data and measure amide proton cross-peak intensities as a function of time. Representative qHX data for amyloid fibrils formed from the peptide IAPP are shown in Fig. 2.

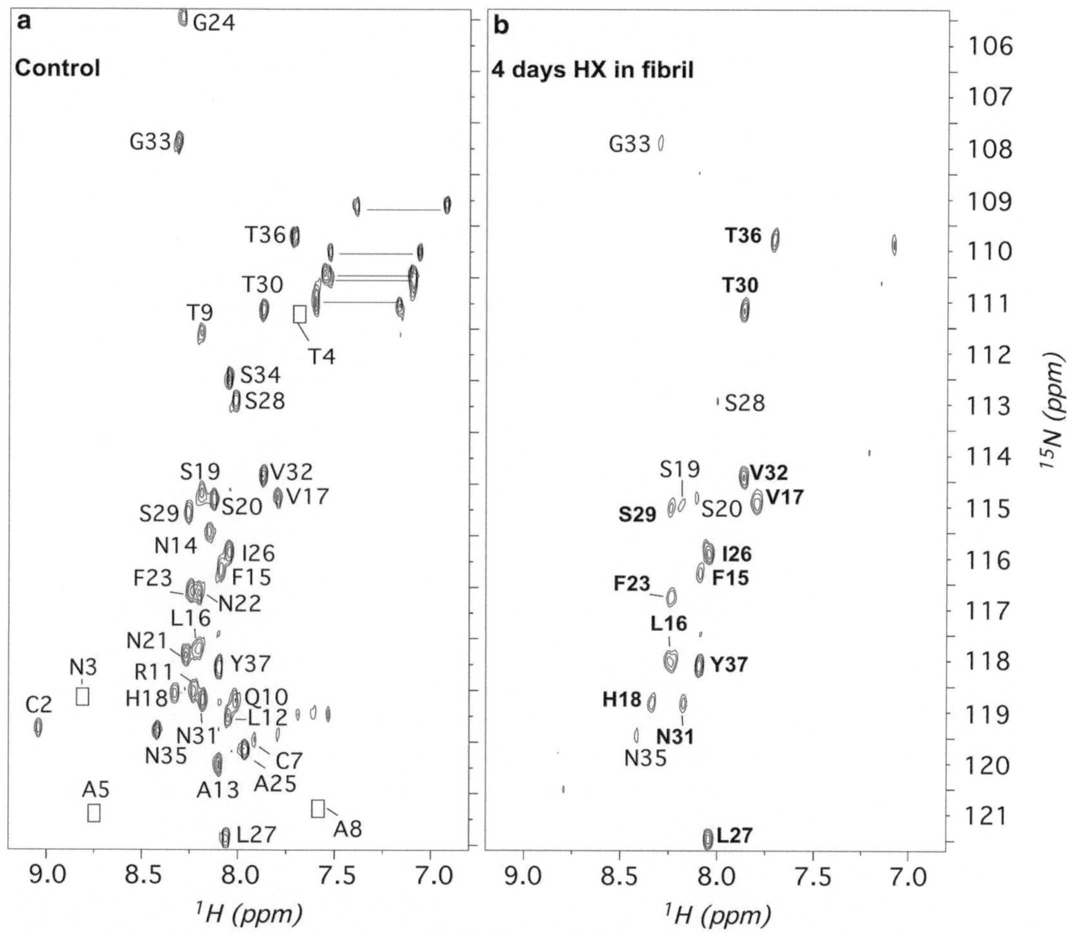


**Fig. 2** Representative qHX NMR data for IAPP amyloid fibrils. (**a**) Control $^1$H-$^{15}$N HSQC spectrum of unfibrillized $^{15}$N-IAPP freshly dissolved in 95 % DMSO-d6/5 % DCA-h2 at 25 °C, pH 3.5. Backbone cross peaks are labeled according to sequence-specific assignments. Residues N3, T4, A5, and A8 are only visible at lower contours than shown. The group of cross peaks connected by horizontal lines between 109 and 111 ppm ($^{15}$N) are unassigned sidechain amide groups from the six Asn and one Gln in amylin. (**b**) Spectrum of a $^{15}$N-amylin after 4 days (99 h) of $D_2O$ exchange in the fibril state, recorded in 95 % DMSO-d6/5 % DCA-d2. The data are reproduced from [24]

7. To obtain exchange rates, fit the amide proton intensity data ($y$-axis) versus $D_2O$ incubation time ($x$-axis) to the exponential decay function: $y = I_0\exp(-xk)$, where the free variables in the nonlinear least squares fit are $I_0$—the initial intensity, and $k$—the exchange rate. Alternatively, the data can be fit to a three-parameter exponential decay to a baseline noise value $y = I_0\exp(-xk) + B$, where $B$ is the baseline intensity in the 2D NMR spectrum.

## 4 Notes

1. DMSO is suggested as a solvent to prepare disaggregated stock solutions of amyloidogenic proteins because DMSO is the most typical solvent used to unfold amyloid fibrils for the qHX experiments, which is the end goal of this protocol. Alternative solvents to prepare disaggregated stock solutions of amyloidogenic proteins include water (if fibrillogenesis is only triggered by a change in solution conditions), the α-helix inducing solvent hexafluoroisopropanol (HIFP), and acetonitrile. A problem with DMSO is that it is a mild oxidant. Specifically, for the Alzheimer's Aβ peptides it promotes oxidation of Met35, which affects fibril formation [29]. Thus for work with Aβ we prepare stock solutions by dissolving the peptide in basic aqueous solutions (0.01 M NaOH).
2. An alternative to storing peptides as frozen stock solutions is to freeze-dry the samples from a disaggregating solvent, if it can be shown that lyophilization does not affect the sample—which is also a requirement for the qHX experiments. This can be tested, for example, by verifying that NMR spectra of the freshly dissolved and lyophilized protein are identical.
3. In our experience continuous shaking in a fluorescence plate reader at orbital shaking speeds >250 rpm, for periods of weeks to months, can damage the shaking mechanism of the instrument. For very long fibrillization reactions it is desirable to use a dedicated shaker (e.g., Eppendorf Thermomixer) and transfer the plates to the fluorescence plate reader for fluorescence measurements.
4. The 15 μM concentration of the ThT dye should not interfere with the properties of the fibrils since the reactions are carried out with a molar excess (20–250 μM) of the protein.
5. The fibrils should form a clear (sometimes gel-like) pellet at the bottom of the microcentrifuge tube. Keep the tube at the same angle as in the microcentrifuge and gently pipette off the supernatant above the pellet without disturbing it. Alternatives to collecting the fibrils by sedimentation are to run the sample through a 100 kDa filter, where the fibrils will be trapped in

the filtrand, or to lyophilize the sample. Filtration or lyophilization may also retain non-fibrillar aggregates, if these are sufficiently populated. This may be desirable for some applications but sedimentation is the best way to collect the sample if the goal is to study only fibrils. Lyophilization will also retain un-aggregated monomers. However, if the amyloidogenic protein is intrinsically unfolded amide protons will not survive hydrogen exchange for more than a few hours in contrast to the aggregated states of the protein.

6. DCA at concentrations between 5 and 10 % (v/v) is used to buffer the pH of samples in DMSO samples [15]. Unlike DMSO, the carboxylic acid proton in DCA is exchange labile. For NMR studies on fully protonated samples (e.g., NMR assignments in DMSO) the protonated version DCA-h2 should be used. For hydrogen-deuterium isotope exchange experiments, the deuterated version of the acid, DCA-d2, must be used since the protons in DCA-h2 can be a source of back-exchange from the solvent back to the protein being studied. We use deuterated DMSO-d6, for qHX experiments for the deuterium solvent lock on the NMR spectrometer. However, deuterated DMSO is not strictly necessary if $^{15}$N-labeled protein is used, since only protons attached to $^{15}$N will be selected in the spectrum and the solvent signals will be suppressed. When working with $^{15}$N-labeled protein it is in principle possible to use mixtures of 90 % DMSO-h6/10 % DMSO-d6 (v/v), where the 10 % deuterated DMSO can serve for the deuterium lock.

7. Ideally, the proteins for qHX NMR experiments should be $^{15}$N-labeled. 2D $^{1}$H-$^{15}$N HSQC experiments on $^{15}$N-labeled proteins offer the advantages of (1) higher sensitivity since the magnetization is transferred through large one-bond $^{1}J_{H\text{-}N}$ couplings, thus requiring less sample, and that (2) the dispersion of $^{15}$N resonances is suitably large even for unfolded proteins [11]. The qHX experiments can be and have been done with unlabeled protein sample using 2D TOCSY $^{1}$H-NMR experiments to measure amide proton occupancy from cross peaks in the $^{3}J_{HN\text{-}H\alpha}$ fingerprint region of the spectrum [30]. For 2D $^{1}$H-NMR experiments the solvent and the buffer components need to be deuterated, so as not to interfere with the spectrum.

8. Although proteins in DMSO are usually unfolded monomers, the structure and oligomerization state of the protein in DMSO does not matter for the qHX experiments, as long as the signals from the amide protons can be detected by NMR and survive exchange for sufficiently long to measure amide proton occupancy. If needed, a monomeric oligomerization state of the protein in DMSO can be verified using NMR pulse-field-gradient diffusion experiments [24].

9. Additional controls can include the following: (1) Check that the fibrils resuspended in $H_2O$ (Subheading 3.2, **step 2**) do not give an NMR spectrum. (2) Lyophilize the supernatant from the sedimentation step in Subheading 3.2, **step 2**, and take it up in 95 % DMSO/5 % DCA-h2. The supernatant component should not give an NMR spectrum if the majority of the monomers have become incorporated into the fibrils. (3) It is useful to check the morphology of the fibrils using electron microscopy, since fibrils with different types of morphologies could have different solvent exchange properties.

10. If $^{13}C/^{15}N$ double-labeled protein is available use NMR experiments such as 3D HNCACB (and HNCO/HN(CA)CO if necessary) to sequentially assign residues by traversing the peptide bond through heteronuclear scalar couplings [31]. If only $^{15}N$-labeled protein is available use 3D NOESY-HSQC and 3D TOCSY-HSQC to obtain assignments. The TOCSY-HSQC experiment provides amino-acid type information about spin systems, the NOESY-HSQC can be used to traverse peptide bonds through sequential $d_{\alpha N}(i,i+1)$ and $d_{NN}(i,i+1)$ distance contacts [31]. If only unlabeled protein is available use 2D TOCSY and NOESY experiments for sequential assignments [31]. Because the protein is unfolded in DMSO and will thus have excellent transverse relaxation properties, and since the goal is to obtain assignments rather than characterize structure, long mixing times of 70 ms and 200–300 ms can be used for the TOCSY and NOESY spectra, respectively.

11. We typically take up the fibrils formed in $H_2O$ in a large >30-fold excess of $D_2O$ to minimize residual $H_2O$ in the sample, that can be a source of deuterium-to-proton back-exchange (for example the sedimented fibrils in a volume of ~40 μl are resuspended in 1.2 ml of $D_2O$. This will give at most 3 % (v/v) of residual $H_2O$ in the sample). An alternative to minimize the amount of $H_2O$ in the sample, is to wash the sedimented fibrils with an excess of $D_2O$ for 1 min, and collect them once again by sedimentation for 30 min in a microcentrifuge operating at 15,000 g. Repeated sedimentation, however, can result in loss of sample.

12. It is necessary to use DCA-d2 rather than DCA-h2 for the qHX NMR measurements to avoid back-exchange of protons from the 5 % DCA acid to the protein.

13. A problem with the qHX NMR measurements is that there can be variability in the concentration of protein in the various time-point aliquots due to sample losses. The differences in protein concentration can cause variability of the amide proton peak intensities used to characterize hydrogen exchange. Similarly, non-ideal spectrometer shims can cause differences in NMR peak intensities. One way to correct for this is to normalize the data according to the intensities of non-exchangeable

protons in 1D $^1$H-NMR spectra of the protein using the same sample (Subheading 3.3, **step 5**). A second elegant way to correct for sample variability and back-exchange is to set up side-by-side incubation reactions during the solvent exchange step (Subheading 3.3, **step 2**) where one fibril sample is incubated in $H_2O$ and the other in $D_2O$ [32]. The protonated sample serves as an internal control to determine proton occupancy after a given exchange incubation time, calculated from the ratio of proton intensity in the $D_2O$ sample divided by the proton intensity in the analogous un-exchanged sample in $H_2O$ [32]. This approach, however, will require twice the amount of protein sample.

## Acknowledgments

This work was supported by Basic Research Award 1-10-BS-04 to A.T.A. from the American Diabetes Association. I thank Rebecca L. Newcomer and Anne R. Kaplan for useful comments on the manuscript.

### References

1. Bai Y, Milne JS, Mayne L, Englander SW (1994) Protein stability parameters measured by hydrogen exchange. Proteins 20(1):4–14
2. Dempsey CE (2001) Hydrogen exchange in peptides and proteins using NMR spectroscopy. Prog Nucl Magn Reson Spectrosc 39:135–170
3. Englander SW, Mayne L, Bai Y, Sosnick TR (1997) Hydrogen exchange: the modern legacy of Linderstrom-Lang. Protein Sci 6(5):1101–1109
4. Bai Y, Feng H, Zhou Z (2007) Population and structure determination of hidden folding intermediates by native-state hydrogen exchange-directed protein engineering and nuclear magnetic resonance. Methods Mol Biol 350:69–81
5. Paterson Y, Englander SW, Roder H (1990) An antibody binding site on cytochrome c defined by hydrogen exchange and two-dimensional NMR. Science 249(4970):755–759
6. Baldwin RL (2008) The search for folding intermediates and the mechanism of protein folding. Annu Rev Biophys 37:1–21
7. Roder H, Wuthrich K (1986) Protein folding kinetics by combined use of rapid mixing techniques and NMR observation of individual amide protons. Proteins 1(1):34–42
8. Alexandrescu AT, Evans PA, Pitkeathly M, Baum J, Dobson CM (1993) Structure and dynamics of the acid-denatured molten globule state of alpha-lactalbumin: a two-dimensional NMR study. Biochemistry 32(7):1707–1718
9. Alexandrescu AT (2001) An NMR-based quenched hydrogen exchange investigation of model amyloid fibrils formed by cold shock protein A. Pac Symp Biocomput 2001:67–78
10. Alexandrescu AT, Jahnke W, Wiltscheck R, Blommers MJ (1996) Accretion of structure in staphylococcal nuclease: an 15N NMR relaxation study. J Mol Biol 260(4):570–587
11. Alexandrescu AT, Abeygunawardana C, Shortle D (1994) Structure and dynamics of a denatured 131-residue fragment of staphylococcal nuclease: a heteronuclear NMR study. Biochemistry 33(5):1063–1072
12. Santos NC, Figueira-Coelho J, Martins-Silva J, Saldanha C (2003) Multidisciplinary utilization of dimethyl sulfoxide: pharmacological, cellular, and molecular aspects. Biochem Pharmacol 65(7):1035–1041
13. Voets IK, Cruz WA, Moitzi C, Lindner P, Areas EP, Schurtenberger P (2010) DMSO-induced denaturation of hen egg white lysozyme. J Phys Chem B 114(36):11875–11883
14. Masters CL, Selkoe DJ (2012) Biochemistry of amyloid beta-protein and amyloid deposits in Alzheimer disease. Cold Spring Harb Perspect Med 2(6):a006262
15. Zhang YZ, Paterson Y, Roder H (1995) Rapid amide proton exchange rates in peptides and

proteins measured by solvent quenching and two-dimensional NMR. Protein Sci 4(4): 804–814

16. Chandak MS, Nakamura T, Makabe K, Takenaka T, Mukaiyama A, Chaudhuri TK, Kato K, Kuwajima K (2013) The H/D-exchange kinetics of the Escherichia coli co-chaperonin GroES studied by 2D NMR and DMSO-quenched exchange methods. J Mol Biol 425(14):2541–2560
17. Czerski L, Vinogradova O, Sanders CR (2000) NMR-Based amide hydrogen-deuterium exchange measurements for complex membrane proteins: development and critical evaluation. J Magn Reson 142(1):111–119
18. Knowles TP, Vendruscolo M, Dobson CM (2014) The amyloid state and its association with protein misfolding diseases. Nat Rev Mol Cell Biol 15(6):384–396
19. Hoshino M, Katou H, Hagihara Y, Hasegawa K, Naiki H, Goto Y (2002) Mapping the core of the beta(2)-microglobulin amyloid fibril by H/D exchange. Nat Struct Biol 9(5):332–336
20. Damo SM, Phillips AH, Young AL, Li S, Woods VL Jr, Wemmer DE (2010) Probing the conformation of a prion protein fibril with hydrogen exchange. J Biol Chem 285(42):32303–32311
21. Morgan GJ, Giannini S, Hounslow AM, Craven CJ, Zerovnik E, Turk V, Waltho JP, Staniforth RA (2008) Exclusion of the native alpha-helix from the amyloid fibrils of a mixed alpha/beta protein. J Mol Biol 375(2):487–498
22. Vilar M, Chou HT, Luhrs T, Maji SK, Riek-Loher D, Verel R, Manning G, Stahlberg H, Riek R (2008) The fold of alpha-synuclein fibrils. Proc Natl Acad Sci U S A 105(25):8637–8642
23. Luhrs T, Ritter C, Adrian M, Riek-Loher D, Bohrmann B, Dobeli H, Schubert D, Riek R (2005) 3D structure of Alzheimer's amyloid-beta(1-42) fibrils. Proc Natl Acad Sci U S A 102(48):17342–17347
24. Alexandrescu AT (2013) Amide proton solvent protection in amylin fibrils probed by quenched hydrogen exchange NMR. PLoS One 8(2): e56467
25. Hoshino M, Katou H, Yamaguchi K, Goto Y (2007) Dimethylsulfoxide-quenched hydrogen/deuterium exchange method to study amyloid fibril structure. Biochim Biophys Acta 1768(8):1886–1899
26. Maji SK, Wang L, Greenwald J, Riek R (2009) Structure-activity relationship of amyloid fibrils. FEBS Lett 583(16):2610–2617
27. Kheterpal I, Wetzel R (2006) Hydrogen/deuterium exchange mass spectrometry – a window into amyloid structure. Acc Chem Res 39(9):584–593
28. Lee YH, Goto Y (2012) Kinetic intermediates of amyloid fibrillation studied by hydrogen exchange methods with nuclear magnetic resonance. Biochim Biophys Acta 1824(12): 1307–1323
29. Hou L, Shao H, Zhang Y, Li H, Menon NK, Neuhaus EB, Brewer JM, Byeon IJ, Ray DG, Vitek MP, Iwashita T, Makula RA, Przybyla AB, Zagorski MG (2004) Solution NMR studies of the A beta(1-40) and A beta(1-42) peptides establish that the Met35 oxidation state affects the mechanism of amyloid formation. J Am Chem Soc 126(7):1992–2005
30. Whittemore NA, Mishra R, Kheterpal I, Williams AD, Wetzel R, Serpersu EH (2005) Hydrogen-deuterium (H/D) exchange mapping of Abeta 1-40 amyloid fibril secondary structure using nuclear magnetic resonance spectroscopy. Biochemistry 44(11):4434–4441
31. Cavanagh J, Fairbrother WJ, Palmer AG III, Rance M, Skelton NJ (2006) Protein NMR spectroscopy principles and practice, 2nd edn. Elsevier Inc., Amsterdam
32. Olofsson A, Lindhagen-Persson M, Sauer-Eriksson AE, Ohman A (2007) Amide solvent protection analysis demonstrates that amyloid-beta(1-40) and amyloid-beta(1-42) form different fibrillar structures under identical conditions. Biochem J 404(1):63–70

# Part V

## Computational Approaches

# Chapter 15

# Studying the Early Stages of Protein Aggregation Using Replica Exchange Molecular Dynamics Simulations

Joan-Emma Shea and Zachary A. Levine

## Abstract

The simulation of protein aggregation poses several computational challenges due to the disparate time and lengths scales that are involved. This chapter focuses on the use of atomistically detailed simulations to probe the initial steps of aggregation, with an emphasis on the Tau peptide as a model system, run under a replica exchange molecular dynamics protocol.

**Key words** Protein folding, Protein aggregation, Intrinsically disordered proteins, Molecular dynamics simulations, Replica exchange molecular dynamics, Amyloid fibrils, Alzheimer's disease, Tau protein

## 1 Introduction

The energy landscape of proteins is characterized by deep wells separated by high barriers. Protein configurations that are sampled using conventional, constant temperature molecular dynamics (MD) simulations tend to get trapped in deep energy wells; therefore it becomes necessary to resort to enhanced sampling techniques in order to fully explore a protein's conformational space. Moreover, the energy landscapes that emerge during protein aggregation are further complicated by the fact that multiple proteins can assemble over a large range of length scales (up to hundreds of nanometers) and time scales (sometimes exceeding hours).

Protein aggregation can be studied using a variety of computational techniques, from coarse-grained simulations [1] to atomistically detailed models [2] augmented with enhanced sampling methodologies [3, 4]. Coarse-grained simulations enable a study of aggregation from monomers to fibrils, but their lack of detailed sequence information makes the comparison to experiments on a particular system problematic. Atomistic simulations on the other hand, where both protein and solvent are described in detail, enable a (nearly) one-to-one comparison with experiment,

David Eliezer (ed.), *Protein Amyloid Aggregation: Methods and Protocols*, Methods in Molecular Biology, vol. 1345, DOI 10.1007/978-1-4939-2978-8_15, 

but these simulations can be prohibitively costly. As a result, atomistic simulations using enhanced sampling approaches have been limited to the study of the early stages of aggregation [3].

In this chapter we focus on the use of enhanced sampling methods, in particular replica exchange molecular dynamics simulations, to study the folding behavior and potential oligomerization of an intrinsically disordered Tau peptide that is implicated in the onset of Alzheimer's disease.

## 2 Materials

Replica exchange molecular dynamics (REMD) is a computational technique that enables chemical systems to escape from low-lying energetic traps, all while preserving a canonical thermodynamic ensemble (*see* **Note 1**). This method (which is outlined below) was first applied in 1999 to proteins by Sugita and coworkers [5]. Several variations of this method exist, and one can choose to bias a number of independent variables such as the temperature or the Hamiltonian [6] in order to better traverse an energy landscape. We will focus here on temperature-based replica exchange molecular dynamics simulations [3] (*see* **Notes 2** and **3**). The method is presented in a point-by-point manner below, but we will summarize it here briefly. A schematic of the process is shown in Fig. 1 for a system involving two Tau peptides. Simulations are launched in parallel, each at different temperature, where neighboring replicas may then be swapped at regular intervals based on a Metropolis criterion (or swapping probability $\Delta$) between replicas $i$ and $j$. Here we define

$$\Delta = \left(\beta_i - \beta_j\right)\left(V_i - V_j\right)$$

where $\beta_i = {}^{1}\!/\!_{\left(k_{\mathrm{B}}T\right)}$ and $V_i$ is the potential energy of replica $i$.

A swap is performed either when $\Delta \leq 0$, or when $\Delta > 0$ with probability $P = \exp(-\Delta)$. This ensures that the entire potential energy landscape is traversed, rather than only the steepest energy wells (*see* **Note 4**).

In the next section we will review the major computational steps that are required to carry out REMD simulations using the GROMACS 5 software package [7] (*see* **Notes 5** and **6**) and Avogadro 1.1 [8]. Installation instructions for these packages can be found at http://www.gromacs.org/Documentation/Installation_Instructions and http://avogadro.cc/wiki/Main_Page, respectively. For simplicity we will focus on using the Linux/Bash terminal environment to investigate the folding behavior of the Tau[273–284] peptide monomer ($_{273}$GKVQIINKKLDL$_{284}$)

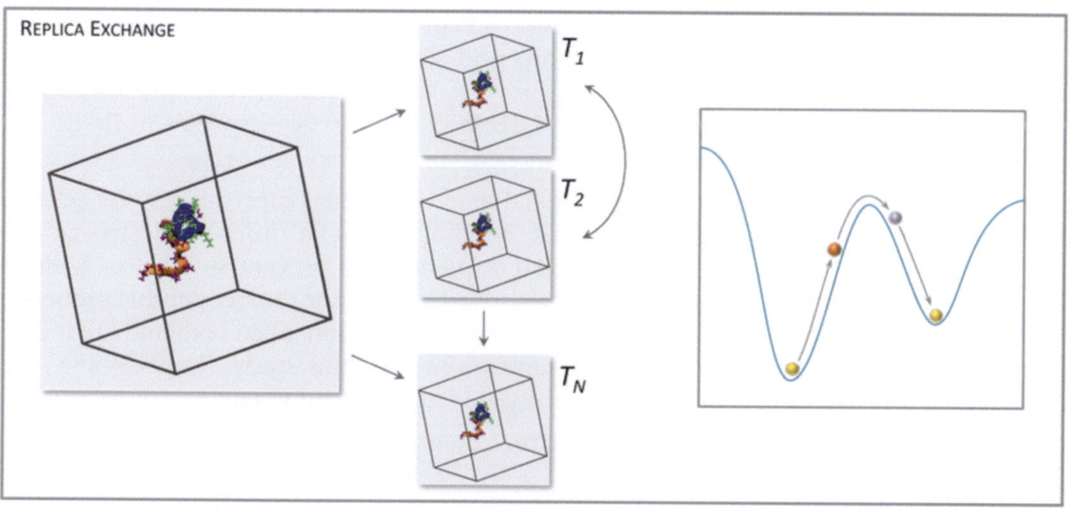


**Fig. 1** A general overview of REMD which contains multiple system replicas at varying temperature (*left*), followed by the traversal of a potential energy landscape (*right*)

[cite] [2, 9]; however larger simulations containing dimers and higher-order oligomers can further elucidate information about the aggregation propensities of various proteins in a variety of complex biochemical environments. This particular example however will focus on the folding of a peptide monomer in the presence of bulk water at neutral pH. Additionally, while a personal workstation will be required to set up and analyze these simulations, the use of a distributed computer cluster is strongly encouraged since they are capable of significantly speeding up the computation time required to carry out these studies.

## 3 Methods

### 3.1 Creating a Molecular Box

1. First, a geometric representation of the molecule of interest must be constructed. The preferred format for molecular representations is a PDB (or "Protein Data Bank") file, which contains the three-dimensional atomic positions of all atoms (in Angstroms) for a given system. Existing PDB structures can be obtained from the Protein Data Bank (http://www.rcsb.org); however novel peptides must be synthesized in silico. To create the Tau[273–284] peptide from scratch, the following steps must be performed.
2. Open the interactive Avogadro software package and create arbitrary unit-cell dimensions, which can be found under the "Crystallography" menu.

3. Select Build > Insert > Peptide from the main menu, and select the desired amino acid sequence you wish to simulate, from the N-terminus to the C-terminus. For Tau[273–284] the sequence is "GKVQIINKKLDL", or "Gly-Lys-Val-Gln-Ile-Ile-Asn-Lys-Lys-Leu-Asp-Leu".
4. Various N-terminal and C-terminal caps can also be selected from the peptide building menu in order to electrostatically cap an amino acid sequence and prevent simulation artifacts. This is particularly advantageous for shorter peptide sequences where abruptly terminated backbones can contaminate molecular dynamics simulations. For this study Tau[273–284] was amidated and acetylated at its N and C termini, respectively.
5. Export the structure by going to File > Save As, and select the PDB format.
6. Once a PDB file, or molecular geometry is obtained, a corresponding molecular force field must be generated which identifies the bonds, partial charges, masses, and overall topology of the molecule. For the most common amino acids, the GROMACS tool "pdb2gmx" can be utilized to generate a customized per-molecule force field if standard amino acid names are written to the PDB file. While we will cover basic usage of the pdb2gmx tool below, additional documentation can be found at http://www.gromacs.org/Documentation/Gromacs_Utilities/pdb2gmx. For more complicated molecules containing nonstandard residues, a force field must be constructed ab initio using quantum chemistry optimization (e.g., density functional theory).
7. For simple molecules, pdb2gmx can be invoked from the terminal (where GROMACS 5.X is installed) by typing "gmx pdb2gmx –h". Note that in earlier versions of GROMACS (version 4.6.X and below), pdb2gmx and other associated MD tools could be called by typing, e.g., "pdb2gmx"; however GROMACS 5.X requires the "gmx" prefix before each command. Additionally if GROMACS was installed with parallel support (which utilizes the Message Passing Interface, or "MPI"), then the "gmx_mpi" command might be required instead which contains the "_mpi" suffix. The "-h" flag brings up a list of all available options that can be used in conjunction with pdb2gmx.
8. To process the PDB file that was just created as an input into pdb2gmx, one can type "gmx pdb2gmx –f FILE.pdb –inter –ter –renum", where FILE.pdb is the input PDB file, and the other flags tell GROMACS to be "interactive", "ask about the molecular termini", and "renumber each residue sequentially" (*see* **Note 7**).

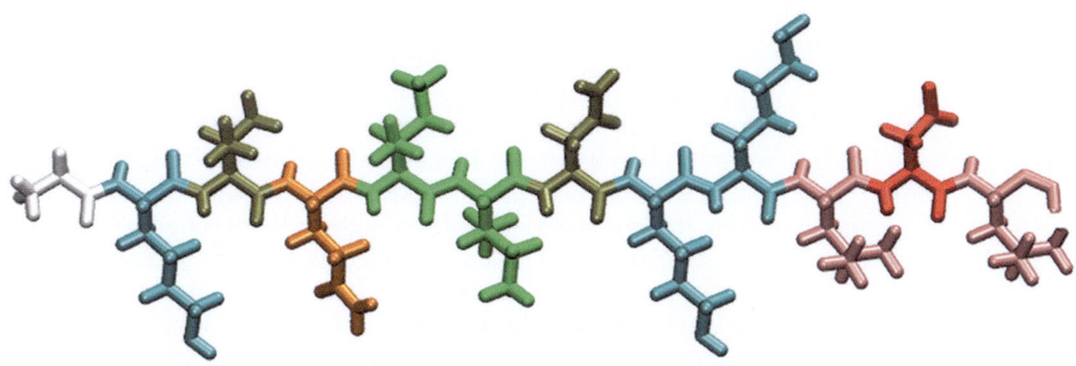

**Fig. 2** A molecular rendering of an unequilibrated Tau[273–284] peptide in VMD

9. GROMACS will then ask a series of questions including the general molecular force field you would like to utilize. There is no inherently correct force field to use; however careful consideration should be taken as to the strengths and weaknesses of each respective force field. Some force fields are incompatible with other force fields, while some are not consistent with various experiments [10]. For the purpose of this example, the "AMBER03" force field [11] will be used henceforth for the Tau[273–284] peptide, and the "TIP3P" force field [12] will be used for water molecules (which we will add later); however the literature should always be extensively searched for the most appropriate force field in each new situation. Additional details about the use of pdb2gmx can be found in Subheading 4 below.
10. If pdb2gmx is successful, it will output two standard files—topol.top and conf.gro. Gro files are nearly identical to pdb files, other than the fact that gro files are written in nanometers and exclude any superfluous (non-geometric) pdb information. Gro files are also the standard coordinate files used in GROMACS, so we will work with them often. The file topol.top is the resulting topology file, which contains a customized force field for our input peptide based on the general AMBER03 model.
11. External viewers such as VMD (Visual Molecular Dynamics) [13] can be useful for graphically visualizing pdb or gro files. As a final check of our molecular geometry, it is worthwhile to inspect the resulting structures visually to ensure that the proper molecular sequence was constructed successfully (Fig. 2).

### 3.2 Solvation, Energy Minimization, and Equilibration

1. Now that our peptide has been created and parameterized under a standard molecular force field, it must be solvated in water, charge-neutralized, energy-minimized, and equilibrated. To begin this process, the GROMACS command "gmx editconf -f conf.gro -o conf_centered.gro -c -resnr 1 -box X Y Z" can be used to set the unit-cell size and check for any

formatting errors that might be incompatible with the GROMACS format. In the above command, conf.gro is the gro file that was produced in Subheading 3.1, and conf_centered.gro is the re-centered output file. The dimensions *X Y Z* should be chosen (in nm) such that (a) they're large enough that the fully extended peptide can only span, at most, half the box length, and (b) small enough so that computation is not wasted by the inclusion of unnecessary water molecules. One must also ensure that enough water can exist in the box to fully hydrate the peptide; therefore an adequate box size must be selected which is both computationally efficient and large enough to avoid dehydration or periodic boundary effects. The ends of the box will be simulated with periodic boundary conditions, so atoms which lie near the edge will "see" the opposite side of the box as if it is continuous. Therefore if a peptide fully extends to 4 nm, the desired box size should be somewhere around $8 \times 8 \times 8$ nm$^3$.

2. The peptide must then be solvated by running "gmx solvate -cp conf_centered.gro -cs -p -o conf_hydrated.gro". This will not only solvate the system, which can be viewed by plotting the output file (conf_hydrated.gro) in VMD (Fig. 3), but will also append the topol.top topology file so that the TIP3P water model we specified earlier will apply to all of the new water molecules that were just added. The end of topol.top should now list the total number of water molecules expected in the unit-cell, e.g.:

```
==============
==> topol.top <==
==============
[...]
[ molecules ]
; Compound        #mols
Protein_chain_A    1
SOL           32973
[...]
```

where "32973" reflects the current number of water molecules present in the solvated system (though this number will likely be different in each solvated box). Note that if you're using an older version of GROMACS, the "genbox" tool should be substituted for the "gmx solvate" tool.

3. Next, a parameter input (.mdp) file must be generated to fine-tune the remaining properties of the simulation. A standard template for this file can be found online at http://manual.gromacs.org/online/mdp.html, and additional descriptions of each field can be found at http://manual.gromacs.org/online/mdp_opt.html. Typical mdp options include:

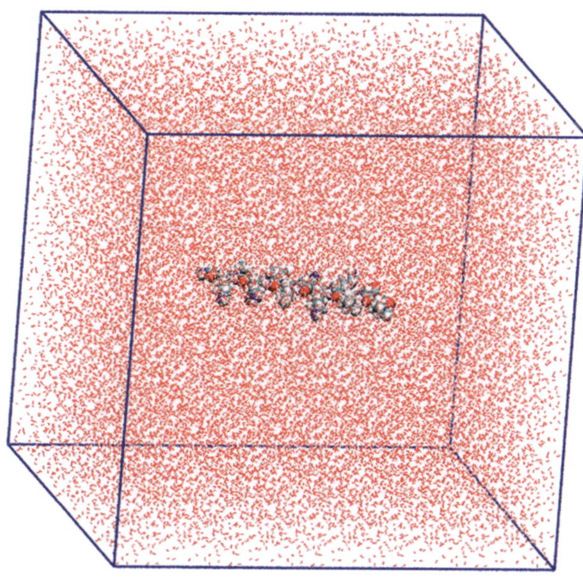

**Fig. 3** A solvated Tau[273–284] peptide in a 10 × 10× 10 nm³ simulation box

```
integrator              = steep       ;steepest-descent energy
minimization
tinit                   = 0           ;initial time = 0
dt                      = 0.002       ;MD time step = .002 ps
(2 fs)
nsteps                  = 15000       ;simulate for 15000
time steps (30 ps)
emtol                   = 100         ;energy minimization
tolerance (in Newtons)
emstep                  = 0.01        ;energy minimization step
size
niter                   = 20          ;number of energy mini-
mization iterations
nstxout                 = 5000        ;number of steps for
outputting highres x,y,z coords
nstvout                 = 5000        ;number of steps for
outputting Vx,Vy,Vz coords
nstfout                 = 5000        ;number of steps for
outputting Fx,Fy,Fz coords
nstlog                  = 500         ;number of steps for
outputting log information
nstenergy               = 1000        ;number of steps for
outputting energy information
nstxtcout               = 5000        ;number of steps for
outputting lowres x,y,z coords
xtc-precision           = 1000        ;numerical precision for
the lowres trajectory
nstlist                 = 10          ;number of steps to update
atomic neighbor list
```

```
ns_type                 = grid
pbc                     = xyz            ;periodic boundary
condition type
rlist                   = 1              ;nearest neighbor
radius (nm)
coulombtype             = pme            ;long-range electro-
static algorithm is PME
rcoulomb-switch         = 0
rcoulomb                = 1              ;short-range electro-
static cutoff radius (nm)
vdwtype                 = cut-off        ;vdw radius trunca-
tion method
rvdw                    = 1              ;vdw radius (nm)
fourierspacing          = 0.12           ;long-range PME
spacing in Fourier space
pme_order               = 4
ewald_rtol              = 1e-5           ;ewald tolerance
ewald_geometry          = 3d
tcoupl                  = nose-hoover    ;thermostat type
tc-grps                 = Protein SOL    ;thermostat groups
tau_t                   = 1.0 1.0        ;thermostat time
constants (ps)
ref_t                   = 300 300        ;thermostat tempera-
tures (K)
Pcoupl                  = Berendsen      ;barostat type
tau-p                   = 1              ;barostat time con-
stant (ps)
compressibility         = 4.5E-5         ;isobaric compress-
ibility (inverse bar)
ref-p                   = 1              ;reference pressure
(bar)
gen_vel                 = yes            ;randomized velocity
generation
gen_temp                = 300            ;initial temperature of
the system
gen_seed                = 173529         ;random seed for
velocity generation
constraints             = all-bonds      ;selective bond
constraints
constraint-algorithm    = Lincs          ;bond constraint
algorithm
```

Note that we set the “integrator” option to “steep”, indicating that we wish to perform steepest descent energy minimization. Other options might include conjugate-gradient (cg) energy minimization, or production level molecular dynamics (MD) simulations. Save these parameters in a file called “grompp.mdp”.

4. We currently have all of the files we need to begin compiling the existing human-readable files to machine-readable binary. In GROMACS 5.X, this can be done by running "gmx grompp -c conf_hydrated.gro -p topol.top -f grompp.mdp". Essentially we are collecting the information found in the molecular geometry file (conf_hydrated.gro), the force field topology file (topol.top), and the parameter input file (grompp.mdp) in order to produce a single output binary file (topol.tpr). This binary file contains all of the information we just generated; however grompp also acts a preprocessor that notifies us of any outstanding warnings (hence the "pp" in "grompp"). Therefore while the above command should successfully generate a binary input file, there may also be a warning which states "System has non-zero total charge: 2.0000" (or some other integer). If this is the case, as it is for Tau[273–284], then counter ions must be introduced to neutralize the system. This must be done because the long-range electrostatics algorithm in GROMACS utilizes the Particle Mesh Ewald (PME) methodology [14] which expects periodic unit-cells to be charge neutral.
5. In order to add counter ions to the solvated unit-cell, the tool "genion" must be used. To do this, the command "gmx genion -s topol.tpr -o conf_hydrated_with_ions.gro -p topol.top -nn 2 –nname CL" can be used where conf_hydrated_with_ions.gro is the output (net-neutral) gro file, –nn represents the number of negative ions to add (two here since our peptide has a net charge of plus two), and –nname is the name of the negative counter ion added (chloride). If positive counter ions were required instead, then –nn and –nname could be replaced by –np and –pname. GROMACS will then ask which atom group it should substitute the counter ions into; therefore the "SOL" or solvent group should be selected. conf_solvated_with_ions.gro is shown in VMD (Fig. 4) without water for clarity.
6. Now that our peptide is solvated and charge-neutralized, we can run grompp again with the updated gro file. As before the command "gmx grompp -c conf_hydrated_with_ions.gro -p topol.top -f grompp.mdp" generates the output binary file topol.tpr. This time the warning about non-zero total charge should be absent.
7. With topol.tpr generated, GROMACS is now ready to run the parameters specified in grompp.mdp. Since steepest descent energy minimization was specified earlier, it can be initialized by running the "mdrun" command, which is the primary workhorse of GROMACS. "mdrun" is invoked by using the command "gmx mdrun -v" in GROMACS 5.X, or "mdrun -v" in GROMACS 4.6.X and below. The –v flag stands for verbose and offers additional information about the running simulation.

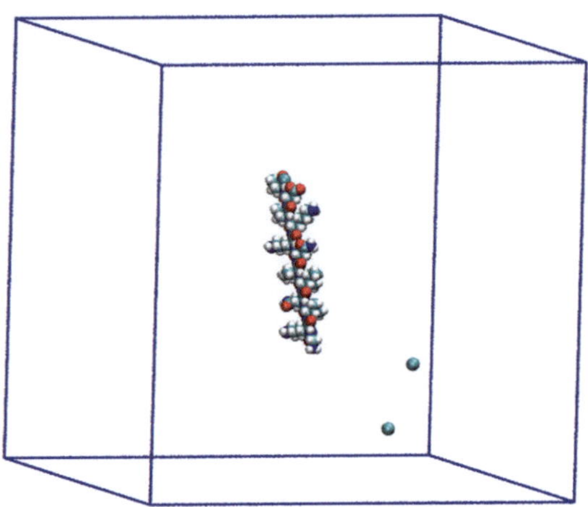

**Fig. 4** A neutral simulation box containing Tau[273–284] (+2e) and two CL counter ions (−2e). Water is not shown for clarity

8. As steepest descent energy minimization is performed, GROMACS will either (a) minimize the system until the potential energy converges, or the maximum force (Fmax) reaches the "emtol" tolerance specified in the mdp file, or (b) until the total number of integration time steps "dt" is carried out in the absence of total energy minimization. Depending on the complexity of the system, the number of energy minimization steps will vary, though a few thousand steps should be sufficient for the current example (e.g., nsteps = 7000).
9. When energy minimization completes, GROMACS will produce an output file called confout.gro. It is wise to rename this file to something along the lines of "energy_minimized.gro" as each new simulation will output its own unique confout.gro file. To avoid constantly overwriting this same file name, either one can utilize multiple simulation directories while managing a multitude of confout.gro files or one can instead implement an efficient and consistent naming system within a single directory. In this example, confout.gro will be renamed to energy_minimized.gro for simplicity.
10. Now that energy minimization is completed, the integrator field in the grompp.mdp parameter file must be switched over to "md", or molecular dynamics. While energy minimization will adjust only a handful of energetically unfavorable atoms, molecular dynamics will integrate Newton's equations of motion for every atom in the simulation box. Therefore running all-atom MD will take quite a bit longer than energy minimization. *While energy minimization can be efficiently performed on a personal computer, MD should ideally be performed on a supercomputing cluster with MPI support. For*

*efficiency we will assume that for the remainder of this example MD will be performed on a distributed computing cluster*. Please be conscious of your cluster's rules and regulations in order to optimize MD simulations over a distributed environment. Performance times can also vary significantly based on the computing environment that is in effect. At present, running REMD in GROMACS for a few hundred nanoseconds on a large supercomputing cluster can take anywhere from 2 days to 2 months, so it is a good idea to setup an efficient computing environment early on.

11. Due to the addition of our counter ions earlier, the two temperature groups we specified previously in grompp.mdp "tc-grps = Protein SOL" must now be updated to include the additional ions. Therefore change this line in grompp.mdp to "tc-grps = Protein Water_and_ions". The water_and_ions group is a default group in GROMACS that incorporates water and any newly added counter ions. Additionally change the "nsteps" field to a value between 5,000,000 and 50,000,000 (without commas). "nsteps" is measured in units of integration time steps (dt) which is currently set to 2 fs. Therefore 5,000,000 dt equals 10 ns while 50,000,000 equals 100 ns.
12. Again we have to preprocess our files by running "gmx grompp -c energy_minimized.gro -p topol.top -f grompp.mdp"; however there may also be warning about utilizing the "Berendsen" barostat [15]. While the Berendsen weak-coupling barostat has been shown to be thermodynamically inaccurate over long timescales [16], it is quite useful for equilibrating new dynamical systems that are typically more unstable than fully equilibrated systems. Therefore because this is an equilibration run before REMD is performed, this warning can be ignored by adding the "-maxwarn 1" flag to the grompp command above. This allows GROMACS to ignore a single warning, though this should NEVER be used to circumvent important warnings.
13. Finally, molecular dynamics can be run by invoking "mdrun" in the normal way, i.e., "gmx mdrun -v". If GROMACS was installed with MPI-support, mdrun can be run across multiple nodes with "mpirun –np N gmx_mpi mdrun –v" where mpirun is the built-in MPI executor, *N* is the desired number of cores, and gmx_mpi represents the GROMACS binary containing the _mpi suffix which indicates that it supports distributed computation.
14. The peptide will then be equilibrated when a number of measurements appear to converge. For instance "gmx energy" can be called after the simulation has completed, to analyze the energies from the energy file "ener.edr" and check the convergence of the system's potential energy, box dimensions,

pressure, and other metrics that should appear constant after equilibration has occurred. Protein secondary structures may also be inspected for convergence from the molecular trajectory files (where “traj.trr” contains double-precision trajectory data and “traj.xtc” contains single-precision data). VMD can visualize both the initial gro coordinates (e.g., energy_minimized.gro) and subsequent trajectories to help verify when equilibration has occurred, though energy convergence is often the most useful indicator of peptide equilibration. This can be done by typing “vmd energy_minimized.gro –xtc traj.xtc”. If equilibration does not appear to occur during the given simulation time, then the simulation must be continually extended until equilibration is apparent (see http://www.gromacs.org/Documentation/How-tos/Extending_Simulations). The final confout.gro file should be retained for use in the next step below.

### 3.3 Replica Exchange Molecular Dynamics in a Canonical (NVT) Ensemble

1. Now that we have a charge-neutralized, energy-minimized, and equilibrated peptide structure that is fully hydrated, we can begin to construct a canonical ensemble to perform REMD on. An overview of REMD can be found in Fig. 5. Step 1 of Fig. 5 was mostly covered in Subheading 3.2. In this section, **steps 2** and **3** will be covered which span the MD production runs. We will conclude then with Subheading 3.4 which will go over REMD analyses that are typically carried out during **step 4** at room temperature. Additional information on REMD can be found at http://www.gromacs.org/Documentation/Tutorials/GROMACS_USA_Workshop_and_Conference_2013/An_introduction_to_replica_exchange_simulations%3A_Mark_Abraham,_Session_1B and http://www.gromacs.org/Documentation/How-tos/REMD.
2. The next step is to create a set of replicate systems that are similar to the one created from Subheading 3.2; however we will want to run each replica at a unique temperature. The range of optimal temperatures often varies, since we must heat our peptide high enough to observe energetically unfavorable protein behaviors. On the lower end of the spectrum, room temperature replicas must also be considered for physiological analysis; therefore the current example will consider temperatures ranging from 290 to 500 K. The idea is that we will want to simulate in parallel a handful of replica systems, and compare the energetics of adjacent systems (in temperature-space). We will then swap an adjacent peptide conformation if the replica’s energy can be further decreased by the swap. This encourages the active propagation of energetically favorable conformations across the replica spectrum, while also steering simulations away from unfavorable states. However the active Metropolis-Hastings algorithm that is used [17] ensures that

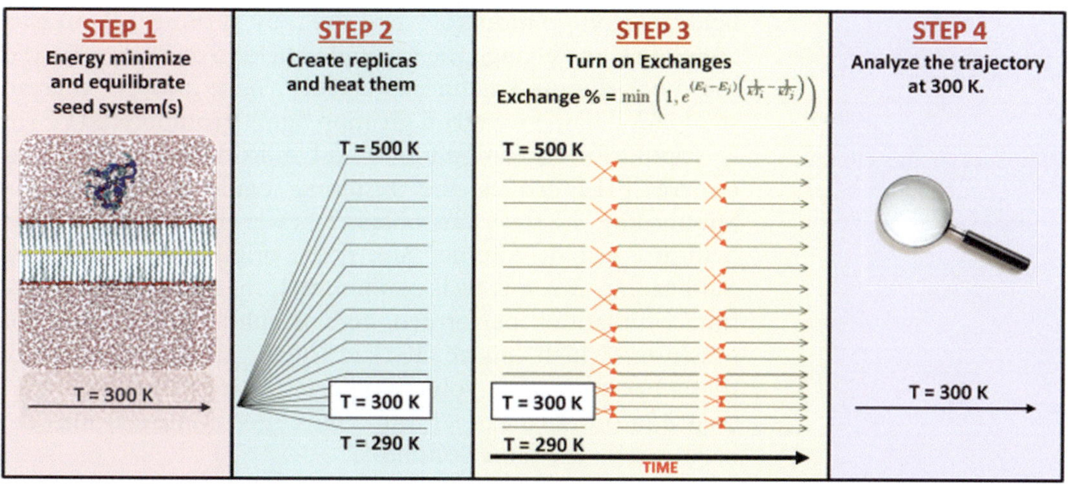


**Fig. 5** The overall framework of the replica exchange molecular dynamics (REMD) methodology that utilizes enhanced Metropolis sampling

there is also some finite probability of sampling unfavorable conformations, thereby providing opportunities to climb out of energy minima in order to sample new energy states that might normally be difficult to access entropically. This allows a much better survey to be taken of the entire protein landscape.

3. To perform these simulations a number of "seed" conformation must be extracted from the original equilibrated system that are more-or-less dissimilar from one another. This way the canonical ensemble we wish to populate can be constructed from a multitude of different protein conformations, rather than utilizing only a single geometric structure. To extract these seed geometries, the tool "trjconv" can be used to extract individual gro files from whole trajectories. For example the Tau[273–284] conformation at 5 ns can be extracted from the high-resolution traj.trr trajectory by running "gmx trjconv –f traj.trr -o 5ns.gro -s topol.tpr -dump 5000 -pbc mol". Here the –dump flag dumps out the frame closest to the 5000 ps time step, and –pbc mol outputs the frame (5ns.gro) with periodic boundary conditions removed across molecules, thus making molecules whole across the unit-cell boundary.

4. Unfortunately though the unit-cell volumes for each of the starting seed geometries must precisely match one another since we cannot easily compare two systems with differing volumes (unless we want to perform NPT enthalpy exchanges); therefore we will need to extend the original NPT equilibration simulation under a constant volume, or under a canonical NVT ensemble. To do this we will need to change the "Pcouple"

field in grompp.mdp to “no” (thereby turning off pressure coupling), and change the “tcouple” field (from Berendsen) to the more thermodynamically correct “nose-hoover” thermostat [18]. Then if we rerun grompp “gmx grompp -c FILE.gro -p topol.top -f grompp.mdp” and mdrun “gmx mdrun -v” (where FILE.gro is the resulting confout.gro file from Subheading 3.2), we can obtain a newer molecular trajectory held at constant volume. Also if the original NPT system in Subheading 3.2 was well equilibrated, then the constant volume simulations performed here should exhibit reasonable pressures (which can be checked using “gmx energy”). After these newer constant volume trajectories are generated, repeat **step 3** in order to extract four to five seed conformations that have exactly the same volumes.

5. After four or five unique “seed” geometries are extracted from Tau[273–284], they can be used to create our REMD canonical ensemble. To do this we will need to know how many replicas we want to create, and at what temperatures each replica should be held at (since the energies at higher temperatures follow an exponential Boltzmann distribution). There is a very useful tool at http://folding.bmc.uu.se/remd/ [19] that can quickly assist in this process. For this example we can enter in all of our previous information into this tool, and set the ideal transition probabilities between different states at 0.25 (or 25 %). Additionally we will be utilizing rigid waters and rigid protein constraints, and we will also need to set the “Simulation type” field on the website to “NPT”, despite the fact that we will be simulating under an NVT ensemble. When all of the fields are properly filled in and submitted, a number of replicas and replicate temperatures should be generated from which we will construct our canonical ensemble.
6. We will ideally only need about 40–60 replicas for REMD to be successful; however the online tool can often generate up to three times as many replicas and replicate temperatures. Therefore many of the higher temperature replicas must be ignored, as the temperature differentials between them are significantly smaller than the differentials observed at lower temperatures (due to the exponential increase in energy at higher temperatures). However this is often a difficult task to carry out since larger temperature gaps between replicas will reduce the probabilities of adjacent conformation swapping. Conversely, significant temperature proximity between adjacent replicas will result in unnecessarily high probability transitions. The only way to go about the replica elimination process is to run a number of parallel simulations over a subset of about 50–60 of the total replicas, and then observe the transition probabilities between them. After the transition probabilities

between replicas are checked, iterative adjustments in temperature can be made to reweigh the resulting transition probabilities between neighbors so that they approach the ideal value of about 0.25.

7. To carry out this procedure efficiently, it is helpful to use an external programming language (e.g., Perl, Python, or Bash) and write a loop which will (a) randomly copy one of the multiple seed conformations, (b) name the new replica start0.gro, (c) copy the existing NVT grompp.mdp file and rename it to grompp0.mdp, (d) change both the "ref_t" and "gen_temp" values in grompp0.mdp from 300 (Kelvin) to the lowest temperature specified in the replicate temperature list (usually around 290 K), (e) set "nsteps" in grompp0.mdp to 5,000,000 (10 ns), and (f) run grompp on the first replica in order to output a binary tpr file with a similar "0" suffix, e.g., "gmx grompp -c start0.gro -p topol.top -f grompp0.mdp –o topol0.tpr". Then iterate the loop variable and change the 0 suffix to a 1, repeating the process for the second lowest temperature. Repeat this process for the higher temperatures as well. As the temperatures become larger, more of the replicas can be skipped such that we end up with only about 40–60 total replicas across a range of 290 K to about 500 K. There are no objectively right or wrong temperatures to select from; however be sure to include a room temperature (300 K) replica into your final replica subset.
8. At this time there should be a set of topolN.tpr files, where *N* is an integer that ranges from 0 to 60 (or less). Each one of these replicas should correspond to a complimentary gromppN.mdp file that specifies a unique temperature. A quick way to check this in a bash shell is to use the "grep" command in the following way "grep ref-t 'ls grompp*.mdp | sort -V'". Note that while the single quotation marks should be ignored, the apostrophe around the ls command should be part of the input statement. If this is done successfully, an ascending list of all the replica temperatures should be displayed as follows:

```
grompp0.mdp:ref-t      = 290
grompp1.mdp:ref-t      = 292.5
grompp2.mdp:ref-t      = 295
grompp3.mdp:ref-t      = 297.5
grompp4.mdp:ref-t      = 300
grompp5.mdp:ref-t      = 302.5
[…]
```

9. At this point when all of the separate tpr files are generated, GROMACS can run the replicas at the same time by using the "–multi N" flag after calling mdrun. For instance if all of the topol[0…60].tpr files are present and in ascending order

without missing a file number (e.g., 0…60 must be continuous even if the temperatures they represent are not continuous from the original temperature set), then the command "gmx mdrun –v –multi 61" will run all 61 simulations in parallel. If this task is not run in parallel on multiple compute nodes, then these jobs will likely crash a single node if they are all started at the same time.

10. After MD is finished, it is time to check the resulting transition probabilities. Various log files will be generated named mdN.log, where *N* is the replica number. However at the end of these files contains statistical information about the replica transition probabilities under the "Replica exchange statistics" section. For instance the end of md0.log might contain something like this:

```
    Replica exchange statistics
    Repl  5666 attempts, 2833 odd, 2833 even
    Repl  average probabilities:
    Repl    0   1   2   3   4   5   6   7   8   9  10  11  12
13  14  15  16  17  18  19  20  21  22  23  24  25  26
27  28  29  30  31  32  33  34  35  36  37  38  39  40
41  42  43  44  45  46  47  48  49  50  51  52  53  54
55  56  57  58  59  60  61
    Repl     .25 .26 .26 .22 .23 .22 .24 .24 .25 .25 .27
.26 .28 .28 .28 .28 .20 .22 .23 .23 .23 .23 .23 .24 .26
.26 .27 .26 .28 .29 .29 .22 .23 .23 .24 .24 .25 .27 .26
.27 .28 .28 .22 .24 .23 .24 .25 .26 .25 .28 .28 .27 .28
.20 .18 .21 .21 .22 .23 .24 .24
```

11. The above information represents 62 MD replicas all held at different temperatures, and the associated transition probabilities between each adjacent system. Note that in this ensemble most of the transition probabilities are reasonably close to 0.25 (i.e., between 0.2 and 0.3). If this is the case, then the canonical ensemble is completed. However more likely than not, some of the transition probabilities will be too high or too low (i.e., less than 0.2 or greater than 0.3).

12. Consider for instance if there were only five replicas numbered from 0 to 4. If the transition probability from 0 ⇔ 1 was 0.25, then the temperature differential between these systems would not need to change. However if the transition probability from system 1 ⇔ 2 was 0.4, then they would require a larger temperature differential between them to reduce their transition probability. This can be accomplished by increasing the temperature of system 2 away from system 1; however then the temperatures of systems 3 and 4 must also be increased by the same amount so as to not perturb the other transition probabilities (e.g., 2 ⇔ 3 and 3 ⇔ 4). One could also lower

the temperatures of systems 1 and 2 if there are significantly less replicas to modify. Both of these techniques are equally valid, though we must ensure (a) that we maintain the original room temperature (300 K) replica, and (b) that we change temperatures by different amounts based on whether they are high or low (since larger changes at higher temperatures have different energetic outcomes compared to similar changes at lower temperatures). To easily satisfy condition (a) it is wise to move temperatures away from the 300 K replica, and to satisfy (b) multiple temperature changes need to be considered depending on how close they are to the room temperature replica. At lower temperatures near 300 K, changes in temperature can be as small as 0.2 K or as big as 1 K. Larger shifts in temperature are typically too forceful at small temperatures, which can sometimes lead transition probabilities to veer past their target of 0.25, sending them far off into the other direction. In contrast, at higher temperatures far from 300 K, temperature changes as high as 3–5 K can be used without significant effects on adjacent transition probabilities.

13. For each of the *N* replicas, it is helpful to use a loop to (a) modify all of the gromppN.mdp files so that "gen-vel = no" instead of "gen-vel = yes" (which ensures that we no longer generate random velocity across each trial as we extend the simulations longer and longer), (b) change "nstlog = 1500", "nstenergy = 1500", and "nstcalcenergy = 1500" (which we will explain below), (c) change "gen-temp" and "ref_t" to the new temperatures which will shift the transition probabilities back to 0.25, and (d) use the output gro files (confoutN.gro) from each of the previous *N* replicas as an input to grompp in order to make a new set of updated topolN.tpr files that reflect the updated gromppN.mdp parameter files (e.g., "gmx grompp -c confoutN.gro -p topol.top -f gromppN.mdp –o topolN.tpr"). At this point you should now have a new set of topolN.tpr files that will continue running for another 10 ns without initial velocity generation.
14. Repeat **step 9** (by running "gmx mdrun –v –multi 61"); however this time also include the "-replex 1500" flag as well which will turn on replica exchange sampling. Now exchanges will be attempted every 1500 dt (or 1500 × 2 fs = 3 ps). This is why we changed the energy log settings above to refresh after each 3 ps interval (or after 1500 time steps), so that we can accurately determine if an exchange is energetically favorable after 3 ps has elapsed.
15. **Steps 10–14** must then be iteratively repeated, where the transition probabilities determined from **step 14** will be used to determine new temperatures in **step 10**. This will subsequently create an ensemble where each adjacent replica is able to

transition to its nearest neighbor's conformation with roughly a 25 % probability, all of which can be accomplished using the 50 replicas we setup that span a temperature range of 290–500 K.

16. When all of the replicas report an adjacent transition probability between 0.2 and 0.3, then a REMD production can begin (which can be initialized with "gmx mdrun –v –multi 61 –replex 1500" using the final ascending topolN.tpr files). Production runs are similar to the simulations performed in **steps 10–14**; however they should typically last for a few hundred nanoseconds in order to collect adequate statistics. Therefore you should change "nsteps" in your grompp.mdp file to something like 150,000,000 (300 ns). Then rerun grompp and mdrun to carry out the full simulation. Ideally it is advantageous to collect about 300 ns or more across each replica, all while ignoring the first 100 ns so that the ensemble has adequate time to mix together. The final 200 ns can then be used for our subsequent analysis.
17. One might wonder how long a simulation needs to run to obtain adequate statistics? There are in fact no objective criteria for evaluating the convergence of protein folding behavior. Some researchers prefer to use "gmx energy" to observe if the room temperature replica exhibits continually decreasing energies; however there is no guarantee that a peptide is not stuck in some local energy well. The GROMACS Perl script "demux.pl md0.log" can also be used (which can be called from anywhere) which demultiplexes the replicas and writes out the location of a given conformation (Fig. 6) over time. This can be useful to observe the extent of mixing, though it is not always clear when convergence has occurred. Typically if there are multiple expected peptide conformations, e.g., states A and B (which could represent a folded and extended peptide state), then one can plot the transition probability from state A ⇔ state B and see if that transition rate (between macrostates) converges over time. However this assumes that all of the outcomes are already known, and that the probability of transitioning between these unknown states converges at some constant value, though it is always possible that a number of hidden states are not accessed on the timescales of these simulations, and that longer timescales (on the order of microseconds) are required.

### 3.4 Protein Folding Analysis and Peptide Clustering

1. When REMD has been carried out for a few hundred nanoseconds, it is time to analyze the data in order to observe which energy-minimized structures emerged as a result of the surrounding environment. One of the most common clustering methods involves the use of the Daura algorithm [20], which

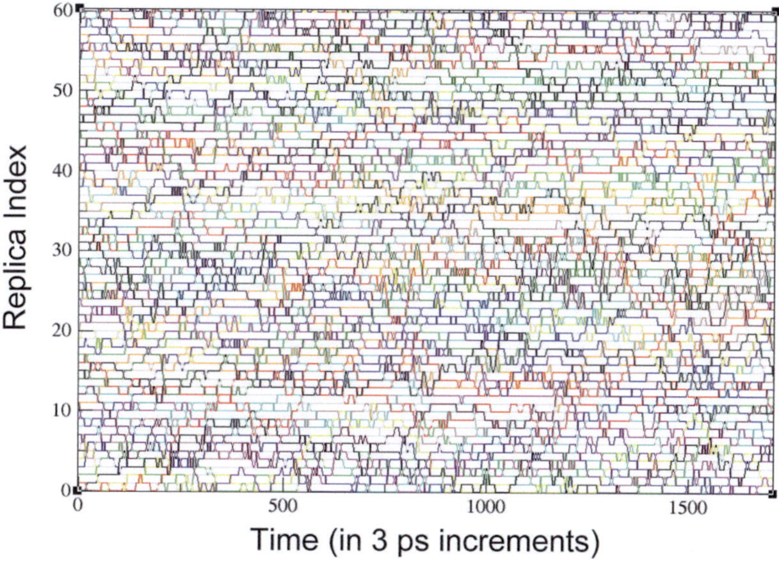


**Fig. 6** A demultiplexed set of 60 REMD replicas that has propagated various peptide conformations across multiple adjacent replicas. This graphic was generated using the GROMACS tool "demux.pl"

compares protein backbones (excluding terminal amino acids) and groups them together based on their root mean square values. Before this can be done however, the REMD trajectory files (trajN.trr) must be processed by centering the protein(s) of interest and removing periodic cutoffs near the borders of the unit-cell. As indicated in Fig. 5, *we will focus the remainder of our analysis on the room temperature replica at 300 K*, which is likely to correspond to one of the first few replicas. *Therefore from now on, all references to the trajectory file will refer specifically to the room-temperature trajectory (e.g., traj4.trr), and all references to the binary topol.tpr file will refer to the room-temperature binary file (e.g., topol4.tpr).* To center the trajectory file, use the "trjconv" GROMACS tool by typing "gmx trjconv -f trajN.trr -s topolN.tpr -pbc whole -center -o traj_centered.trr" (where *N* corresponds to the room temperature replica number). This will output a trajectory "traj_centered.trr" in which the Tau(273–284) peptide is centered, thus simplifying clustering about a similar geometric origin. GROMACS will ask which atoms should be written out during the re-centering procedure, and if only protein-related studies are desired, then it is efficient to only output protein atoms (excluding subsequent analysis on water). However if there is a desire to analyze both water and protein structures, then all atoms may be written out, though this will take up considerably more disk space.

2. Now that a centered trajectory has been produced, we must also create an atomic backbone selection for which to consider when clustering different peptide conformations together. For instance we will want to exclude a majority of the amino acid side chains, and also ignore the backbone caps that we placed earlier in Subheading 3.1. To do this we can use the "make_ndx" GROMACS tool, which can create customized index (ndx) files for later analysis. This will be useful for a number of later GROMACS analyses, so it is best to create this file now. To invoke make_ndx in GROMACS 5.X, type "gmx make_ndx –f topolN.tpr" where *N* is the room temperature replica. This will initialize the "make_ndx" interface, which uses logical operators to select or deselect specific subsets of atoms. This interface can be subsequently exited by typing "q" and "enter" into the interface, thereby producing a default index.ndx file.
3. Open the index.ndx file in a text editor, and notice that it is simply a list of atom numbers under various group names:

```
[ System ]
1 2 3 4 5 6 7 8 9 10 11 12 13 14 15
16 17 18 19 20 21 22 23 24 25 26 27 28 29 30
31 32 33 34 35 36 37 38 39 40 41 42 43 44 45
46 47 48 49 50 51 52 53 54 55 56 57 58 59 60
61 62 63 64 65 66 67 68 69 70 71 72 73 74 75
76 77 78 79 80 81 82 83 84 85 86 87 88 89 90
91 92 93 94 95 96 97 98 99 100 101 102 103 104
105
106 107 108 109 110 111 112 113 114
[…]
```

This file can be modified manually by adding in new group names, and placing in corresponding atom numbers (as identified by the gro files) under these group names to add atoms into the group. Therefore we will add a new group at the bottom of the index file called "[CN_backbone_atoms]" and then we will manually add below this all carbon and nitrogen backbone atoms in the peptide (excluding atoms from the endcaps). Therefore there should be $2X$ atom numbers pasted into this new group, where $X$ is the number of amino acids that were simulated. Once this is done, go ahead and save the file. The bottom of the file should then look something like this:

```
[CN_backbone_atoms]
17 34 36 55 57 74 76 91 93 115 117 134 136 156
158
168 170 175 177 182 184 196 198 207 209 218 220 225
227 249
251 263 265 277 279 291
```

though the specific atom numbers will vary based on your unique system. *See* **Note 8** for additional ways of adding these atoms to the index file using the "make_ndx" GROMACS tool.

4. Now we can apply the Daura clustering algorithm to the backbone group we just created. This can be accomplished using the "cluster" command in GROMACS by running the command "gmx cluster -f traj_centered.trr -s topolN.tpr -n index.ndx -method gromos -wcl 20 -cutoff 0.14 -sz". This analyzes the newly created traj_centered.trr room temperature trajectory file, the room-temperature topolN.tpr file, and the recently created index group index.ndx, and produces the top 20 peptide conformations based on an RMSD cutoff of 0.14 nm. The cutoff threshold can also be increased to 0.2 nm if there are a number of structurally similar peptide conformations, though the magnitude of this cutoff is often up to the discretion of the researcher. Be sure to select the "CN_backbone_atoms" group for the RMSD comparisons when you are prompted, then select all of the peptide atoms when prompted on which atoms to write out.
5. The primary outputs from "gmx cluster" will be (a) clust-size.xvg, which provides a histogram of the number of individual cluster frames (which can be used to generate normalized cluster probabilities), (b) cluster.log, which provides a list of all of the trajectory time steps used to construct each cluster, (c) clusters.pdb, which is a pdb file containing a summary of all of the dominant peptide conformations (where frame 0 corresponds to cluster 1, frame 1 corresponds to cluster 2, etc.), and (d) clusters.pdb00N.pdb, which represents a collection of the *N* most dominant peptide conformations. Therefore while the first frame of clusters.pdb corresponds to the most dominant peptide conformation's average structure, clusters.pdb001.pdb will contain all of the peptide conformations that went into constructing that first cluster. Taken together, these files can be utilized to extract the most dominant protein morphologies (Fig. 7) found in a given environment.
6. There exist multiple analysis tools within the GROMACS framework that are capable of further investigating proteins in REMD simulations. This includes extracting rate measurements of forward and backward protein transitions (gmx kinetics), radii of gyration (gmx gyrate), proximities of various amino acid chains (gmx distance or gmx traj), hydrogen bond affinities (gmx hbond), and many others which are documented at http://www.gromacs.org/Documentation/Gromacs_Utilities, though we will not cover these analyses here.

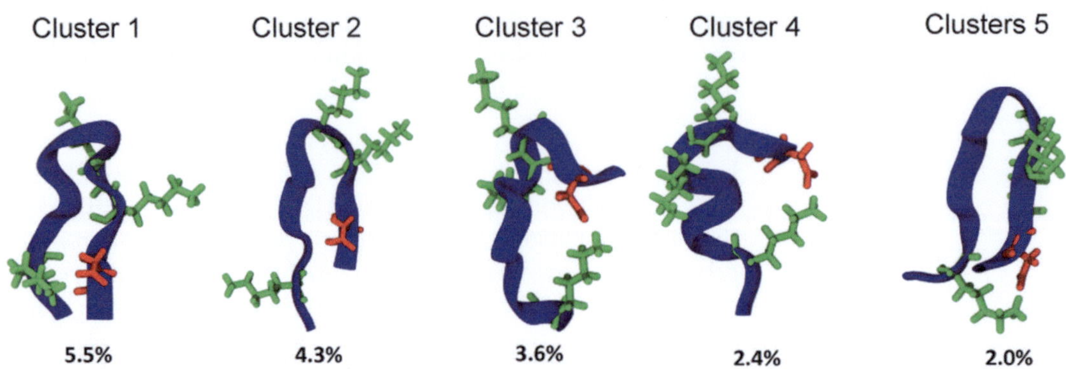


**Fig. 7** An example of some of the most dominant Tau[273–284] clusters which emerge over the final 200 ns of REMD. Percentages correspond to the amount of total time spent in each conformation. For intrinsically disordered peptides (like Tau[273–284]), these percentages are often low since the peptide does not normally adopt a natively folded structure

This chapter was meant to provide the reader with a general overview of how protein folding can be investigated through the use of REMD simulations, though there is much more that can be simulated and analyzed. We have covered here only the most basic steps, though there are multiple ways to construct the many systems developed in this chapter. For more information on REMD and Tau[273–284], including the different analyses that can be performed on replicate protein systems, *see* ref. 2 for additional details. Simulations containing dimers or larger oligomers can also be performed in the same way as we have documented above, thereby providing information on the aggregation propensity of proteins in the presence of various environments.

As computation becomes more available to the research community, and as the computational efficiency of large computing clusters becomes increasingly optimized, simulations will continue to provide more detailed analysis on the dynamics and inner workings of biological systems, thus providing an indispensable tool that can bridge experimental observations in the lab to existing theoretical models.

## 4 Notes

1. Replica exchange molecular dynamics cannot be used to extract the kinetics of molecules.
2. The replica exchange protocol can be applied to coarse-grained systems [21], as well as atomistically detailed simulations.
3. The replica exchange protocol can be applied to both molecular dynamics simulations [5] and Monte Carlo simulations [22].

4. Additional enhanced sampling methods include Metadynamics and Umbrella Sampling [23].

5. While GROMACS 4.6.X (and earlier versions) can also be used to carry out this example, careful consideration must be taken when invoking GROMACS since version 5.0.X introduced a number of substantial changes. A complete list of these changes can be found at http://www.gromacs.org/About_Gromacs/Release_Notes/Versions_5.0.x. Note that some of the flags used in this example might vary between older and newer versions of GROMACS. Please be aware of these changes if you plan to use an older version of this software.

6. Other MD packages can also be used to perform REMD such as CHARMM [24], NAMD [25], and AMBER [26]. The use of GROMACS in this example is only meant to construct a self-consistent list of commands under a single molecular dynamics package; however the authors make no claim that one molecular dynamics platform is better suited for protein studies compared to its competitors.

7. The tool pdb2gmx uses a lookup table in order to match existing atom and/or residue names found in GROMACS to the names encountered in a PDB file. When a force field is selected in pdb2gmx which uses different nomenclatures from the input PDB file, errors can result, e.g., "Fatal error: Atom HN2 in residue GLY 1 was not found in rtp entry NGLY". To fix this, the PDB file must contain standardized atom names that exactly match the existing force field names. For example the above error message might occur during the following atom naming mismatch:

```
=================
==> FILE.pdb <==
=================
[...]
ATOM   1 N   GLY A 1   -16.776 -0.383  0.000 1.00 0.00   N1+
ATOM   2 CA  GLY A 1   -15.329 -0.383  0.000 1.00 0.00   C
ATOM   3 C   GLY A 1   -14.808 -0.383  1.433 1.00 0.00   C
ATOM   4 O   GLY A 1   -15.537 -0.383  2.422 1.00 0.00   O
ATOM   5 HA1 GLY A 1   -14.977 -1.279 -0.522 1.00 0.00   H
ATOM   6 HA2 GLY A 1   -14.976  0.510 -0.524 1.00 0.00   H
ATOM   7 H   GLY A 1   -17.248 -0.386  0.900 1.00 0.00   H
ATOM   8 HN  GLY A 1   -17.115  0.470 -0.435 1.00 0.00   H
ATOM   9 HN2 GLY A 1   -17.115 -1.236 -0.435 1.00 0.00   H
[...]
```

```
=======================
==> aminoacids.rtp <==
=======================
[...]
[ NGLY ]
 [ atoms ]
  N     N3    -0.600766  1
  H1    H      0.450255  2
  H2    H      0.450255  3
  H3    H      0.450255  4
  CA    CT     0.126891  5
  HA1   HP     0.036849  6
  HA2   HP     0.036849  7
  C  C         0.648768  8
  O  O        -0.599357  9
[...]
```

Here aminoacids.rtp (located in the default GROMACS topology directory—$GMX/share/top/FORCEFIELD.ff/) contains the default force field atom names. This error can be fixed by changing the names of atoms "H, HN, and HN2" in FILE.pdb to "H1, H2, and H3" as observed in the default FORCEFIELD file.

8. The "make_ndx" tool can be utilized to quickly select a subset of atoms in order to streamline **step 3** in Subheading 3.4. Additional documentation for "make_ndx" can be found at http://www.gromacs.org/Documentation/Gromacs_Utilities/make_ndx. If for instance we wanted to select only C and N atoms for residues 2–9 (assuming that residues 1 and 10 are the peptide end caps), we would start make_ndx by typing "gmx make_ndx –f topolN.tpr" (where *N* is the room temperature replica), and then we would use an atom selection criteria such as "a C | a N". This would create a group of atoms that is either a "C" (carbon) type or a "N" (nitrogen) type. If we press enter, the list of groups will be updated with our new "C and N" group displayed at the bottom. This group will also have a corresponding group number (such as a "6") that will be useful for creating more complex atom selections. For instance we can create yet another group from the argument "6 & r 2–9", which says that we will start with the atoms in group 6 (our "C and N" atom group), and only select atoms from that which are part of residues 2–9. Therefore the resulting group is now constructed from C and N atoms that are a part of the uncapped peptide backbone structure, thus accomplishing the same outcome as in **step 3** from Subheading 3.4.

## References


1. Morriss-Andrews A, Brown FLH, Shea J-E (2014) A coarse-grained model for peptide aggregation on a membrane surface. J Phys Chem B 118(28):8420–8432. doi:10.1021/jp502871m
2. Larini L, Gessel MM, LaPointe NE, Do TD, Bowers MT, Feinstein SC, Shea JE (2013) Initiation of assembly of tau(273–284) and its Delta K280 mutant: an experimental and computational study. Phys Chem Chem Phys 15(23):8916–8928. doi:10.1039/C3cp00063j
3. Kim J, Straub JE, Keyes T (2012) Replica exchange statistical temperature molecular dynamics algorithm. J Phys Chem B 116(29):8646–8653. doi:10.1021/Jp300366j
4. Kim J, Straub JE, Keyes T (2006) Statistical-temperature Monte Carlo and molecular dynamics algorithms. Phys Rev Lett 97(5):050601. doi:10.1103/Physrevlett.97.050601
5. Sugita Y, Okamoto Y (1999) Replica-exchange molecular dynamics method for protein folding. Chem Phys Lett 314(1–2):141–151. doi:10.1016/S0009-2614(99)01123-9
6. Laghaei R, Mousseau N, Wei GH (2011) Structure and thermodynamics of amylin dimer studied by Hamiltonian-temperature replica exchange molecular dynamics simulations. J Phys Chem B 115(12):3146–3154. doi:10.1021/Jp108870q
7. Hess B (2009) GROMACS 4: algorithms for highly efficient, load-balanced, and scalable molecular simulation. Abstracts of Papers of the American Chemical Society 237
8. Hanwell MD, Curtis DE, Lonie DC, Vandermeersch T, Zurek E, Hutchison GR (2012) Avogadro: an advanced semantic chemical editor, visualization, and analysis platform. J Cheminform 4:17. doi:10.1186/1758-2946-4-17
9. von Bergen M, Barghorn S, Li L, Marx A, Biernat J, Mandelkow EM, Mandelkow E (2001) Mutations of tau protein in frontotemporal dementia promote aggregation of paired helical filaments by enhancing local beta-structure. J Biol Chem 276(51):48165–48174. doi:10.1074/Jbc.M105196200
10. Guvench O, MacKerell A Jr (2008) Comparison of protein force fields for molecular dynamics simulations. In: Kukol A (ed) Molecular modeling of proteins, vol 443, Methods molecular biology™. Humana Press, Totowa, NJ, pp 63–88. doi:10.1007/978-1-59745-177-2_4
11. Beauchamp KA, Lin Y-S, Das R, Pande VS (2012) Are protein force fields getting better? A systematic benchmark on 524 diverse NMR measurements. J Chem Theory Comput 8(4):1409–1414. doi:10.1021/ct2007814
12. Jorgensen WL, Chandrasekhar J, Madura JD, Impey RW, Klein ML (1983) Comparison of simple potential functions for simulating liquid water. J Chem Phys 79(2):926–935. doi:10.1063/1.445869
13. Humphrey W, Dalke A, Schulten K (1996) VMD: visual molecular dynamics. J Mol Graph 14(1):33–38
14. Essmann U, Perera L, Berkowitz ML, Darden T, Lee H, Pedersen LG (1995) A smooth particle mesh Ewald method. J Chem Phys 103 (19):8577–8593
15. Berendsen HJC, Postma JPM, Vangunsteren WF, Dinola A, Haak JR (1984) Molecular-dynamics with coupling to an external bath. J Chem Phys 81(8):3684–3690
16. Bussi G, Donadio D, Parrinello M (2007) Canonical sampling through velocity rescaling. J Chem Phys 126(1):014101. doi:10.1063/1.2408420
17. Chib S, Greenberg E (1995) Understanding the Metropolis-Hastings algorithm. Am Stat 49(4):327–335. doi:10.2307/2684568
18. Hoover WG (1985) Canonical dynamics: equilibrium phase-space distributions. Phys Rev A 31(3):1695–1697
19. Patriksson A, van der Spoel D (2008) A temperature predictor for parallel tempering simulations. Phys Chem Chem Phys 10(15):2073–2077. doi:10.1039/B716554d
20. Daura X, Suter R, van Gunsteren WF (1999) Validation of molecular simulation by comparison with experiment: rotational reorientation of tryptophan in water. J Chem Phys 110(6):3049–3055. doi:10.1063/1.477900
21. Morriss-Andrews A, Bellesia G, Shea JE (2011) Effects of surface interactions on peptide aggregate morphology. J Chem Phys 135(8):085102. doi:10.1063/1.3624929
22. Swendsen RH, Wang JS (1986) Replica Monte-Carlo simulation of spin-glasses. Phys Rev Lett 57(21):2607–2609. doi:10.1103/Physrevlett.57.2607
23. Torrie GM, Valleau JP (1977) Nonphysical sampling distributions in Monte Carlo free-energy estimation: umbrella sampling. J Comput Phys 23(2):187–199, http://dx.doi.org/10.1016/0021-9991(77)90121-8
24. Brooks BR, Brooks CL, Mackerell AD, Nilsson L, Petrella RJ, Roux B, Won Y, Archontis G, Bartels C, Boresch S, Caflisch A, Caves L, Cui Q, Dinner AR, Feig M, Fischer S, Gao J, Hodoscek M, Im W, Kuczera K,

Lazaridis T, Ma J, Ovchinnikov V, Paci E, Pastor RW, Post CB, Pu JZ, Schaefer M, Tidor B, Venable RM, Woodcock HL, Wu X, Yang W, York DM, Karplus M (2009) CHARMM: the biomolecular simulation program. J Comput Chem 30(10):1545–1614. doi:10.1002/Jcc.21287

25. Nelson MT, Humphrey W, Gursoy A, Dalke A, Kale LV, Skeel RD, Schulten K (1996) NAMD: a parallel, object oriented molecular dynamics program. Int J Supercomput Appl High Perform Comput 10(4):251–268

26. Pearlman DA, Case DA, Caldwell JW, Ross WS, Cheatham TE, Debolt S, Ferguson D, Seibel G, Kollman P (1995) Amber, a package of computer-programs for applying molecular mechanics, normal-mode analysis, molecular-dynamics and free-energy calculations to simulate the structural and energetic properties of molecules. Comput Phys Commun 91(1–3):1–41

# Chapter 16

# Computational Methods for Structural and Functional Studies of Alzheimer's Amyloid Ion Channels

**Hyunbum Jang, Fernando Teran Arce, Joon Lee, Alan L. Gillman, Srinivasan Ramachandran, Bruce L. Kagan, Ratnesh Lal, and Ruth Nussinov**

## Abstract

Aggregation can be studied by a range of methods, experimental and computational. Aggregates form in solution, across solid surfaces, and on and in the membrane, where they may assemble into unregulated leaking ion channels. Experimental probes of ion channel conformations and dynamics are challenging. Atomistic molecular dynamics (MD) simulations are capable of providing insight into structural details of amyloid ion channels in the membrane at a resolution not achievable experimentally. Since data suggest that late stage Alzheimer's disease involves formation of toxic ion channels, MD simulations have been used aiming to gain insight into the channel shapes, morphologies, pore dimensions, conformational heterogeneity, and activity. These can be exploited for drug discovery. Here we describe computational methods to model amyloid ion channels containing the β-sheet motif at atomic scale and to calculate toxic pore activity in the membrane.

**Key words** Amyloid channel, β-Sheet channel, Lipid bilayer, Molecular dynamics simulations, CHARMM, NAMD

## 1 Introduction

Alzheimer's disease (AD) is characterized by the presence of extracellular plaques, intracellular neurofibrillary tangles, and the loss of synapses and neurons in the brain of AD patients [1]. As a subclass of fatal protein deposition diseases [2–8], termed amyloidosis, AD is caused by misfolded, water insoluble aggregates of amyloid-β (Aβ) peptides [5]. During their self-assembly into mature fibrils, Aβ peptides explore various organizations including small oligomers (globular and fibril-like) and protofibrils (straight, bent, and annular) [9, 10]. Although early studies pointed to fibrillar deposits of Aβ peptides in the extracellular plaques as directly associated with the cause of the disease [11], the current amyloid cascade hypothesis in AD points to small Aβ oligomers as

David Eliezer (ed.), *Protein Amyloid Aggregation: Methods and Protocols*, Methods in Molecular Biology, vol. 1345, DOI 10.1007/978-1-4939-2978-8_16, 

the main toxic species [12–16]. However, the mechanism of the amyloid toxicity is still not entirely understood.

The interaction of Aβ with the cell membrane is a fundamental chemical feature in the mechanism of AD pathogenesis [17–19]. Upon binding to the cell membrane, Aβ undergoes conformational changes to insoluble β-sheet-rich aggregates ranging from small oligomers to fibrils [20–22]. The oligomeric Aβ aggregates are responsible for disrupting cellular function, inducing cytotoxicity [23] through ion channel formation [24]. The evidence for the presence of amyloid ion channel was first reported two decades ago, by exploiting planar lipid bilayer (PLB) measurements [25–28]. The experiments discovered that Aβ induced unregulated ionic flux across model membranes through the formation of non-gated ion channels. Subsequently, atomic force microscopy (AFM) provided the images of amyloid channels formed by Aβ peptides [12, 29] and by other disease-related amyloid species [12], suggesting that channel formation is a general feature for amyloids. The AFM images revealed that the amyloid channels exhibited various shapes from rectangular with four subunits to octahedral with eight. The heterogeneity in the Aβ channel conformations was further confirmed by recent extensive molecular dynamics (MD) simulations [30–42]. These showed that Aβ channels consisted of β-sheet-rich subunits with morphologies and dimensions in good agreement with the imaged AFM channels [12, 29]. The simulations of other amyloids and β-hairpin peptides showed that the subunit-assembly morphology is a common feature for the membrane embedded β-sheet channels [43–45].

To form an ion channel, small oligomers of Aβ insert into the membrane and assemble into common β-sheet-rich structural motifs. Recent studies indicated that small fibril-like Aβ oligomers [46] with a solvent exposed hydrophobic face [47] and parallel β-sheet structures [48] could induce neurotoxicity, providing an Aβ oligomer morphology with potential relevance to AD. These membrane-inserted small oligomers can easily align to form the toxic amyloid ion channels. While experimental tools are limited in defining the channel structure in the membrane environment, computational studies can provide their three-dimensional, atomic-level conformation. Here, we detail the computational methods of how to model β-sheet channels and to calculate pore activity in the membrane.

## 2 Materials: Recruiting Monomer Conformations

### 2.1 U-Shaped Peptides with the β-Strand-Turn-β-Strand Motif

Amyloids tend to aggregate to form a highly ordered fibrillar structure. In this organization, peptides fold into the U-shaped *β-strand-turn-β-strand* motif, which associates into stacked β-sheets with intermolecular hydrogen bonds (H-bonds). Recent computational and NMR studies defined several amyloid peptides with such U-shaped motif [49–53]. Using their reported atomic coordinates, these peptides were recruited in computational studies for the atomistic modeling of amyloid channels in aqueous and lipid environments [30–43, 46, 54–57].

1. $A\beta_{16-35}$ peptide: The U-shaped Aβ peptide was first introduced by a computational model using molecular dynamics (MD) simulations [49]. The $A\beta_{16-35}$ peptide contains an intramolecular salt bridge between residues Asp23 and Lys28 near a turn at Val24-Asn27 (Fig. 1a).

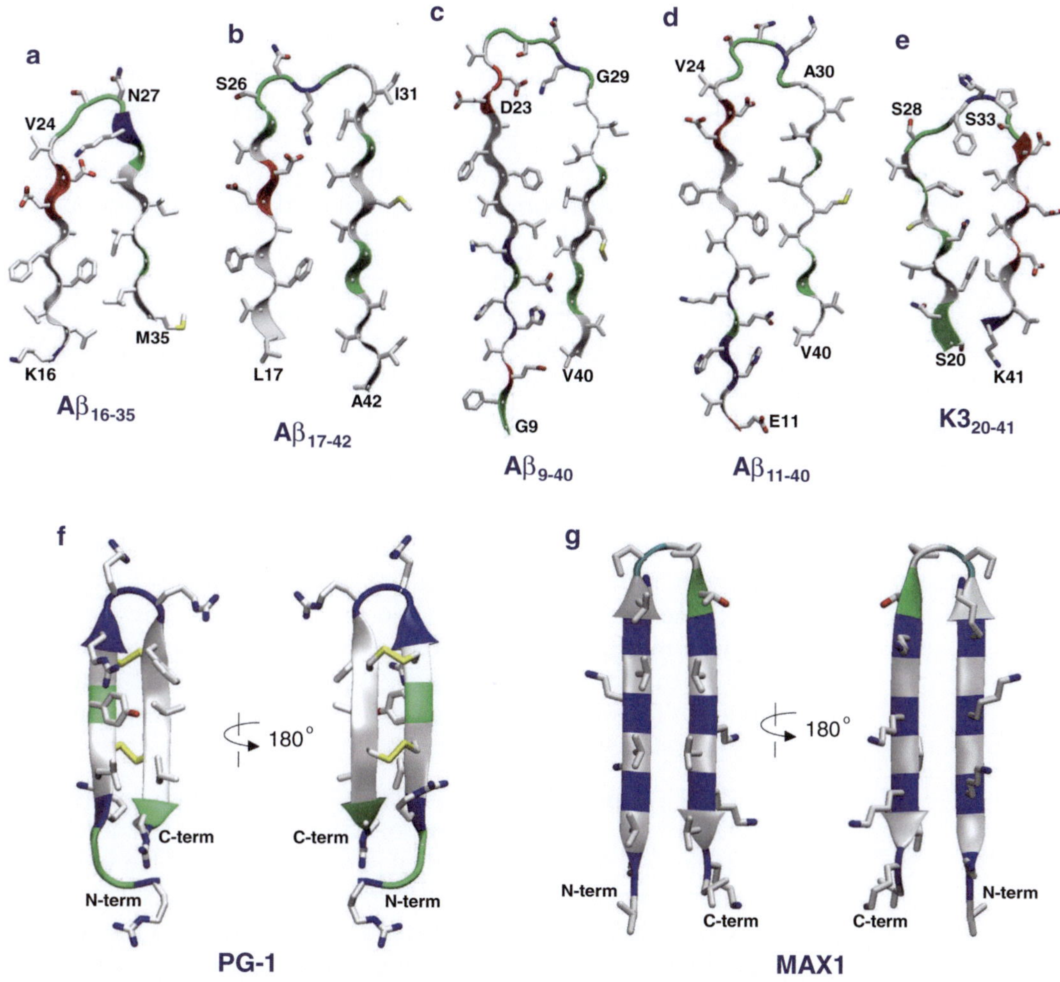


**Fig. 1** Monomer conformations recruited for the molecular dynamics (MD) simulations of β-sheet channels in the lipid bilayers. The U-shaped amyloid peptides with the *β-strand-turn-β-strand* motif for (**a**) the computational $A\beta_{16-35}$, (**b**) the NMR-derived $A\beta_{17-42}$, (**c**) the ssNMR $A\beta_{9-40}$, (**d**) the ssNMR $A\beta_{11-40}$, and (**e**) the ssNMR $K3_{20-41}$ peptides. The β-hairpin motif for the synthetic (**f**) protegrin-1 (PG-1) and (**g**) MAX1 peptides. In the peptide ribbon, hydrophobic, polar/Gly, positively charged, and negatively charged residues are colored *white*, *green*, *blue*, and *red*, respectively. *Yellow sticks* in PG-1 denote the disulfide bonds

2. A$\beta_{17-42}$ peptide: A combination of hydrogen/deuterium-exchange nuclear magnetic resonance (NMR) data, side-chain packing constraints from pairwise mutagenesis, solid-state NMR (ssNMR), and electron microscopy (EM) defined A$\beta_{1-42}$ fibril (pdb code: 2BEG) [50]. The A$\beta_{1-42}$ peptide provided the coordinates for residues 17–42, while the N-terminal coordinates (residues 1–16) were missing due to disorder. The A$\beta_{17-42}$ peptide has a turn at Ser26-Ile31 and the salt bridge of Asp23/Lys28 (Fig. 1b).
3. A$\beta_{9-40}$ peptide: Studies using ssNMR defined small A$\beta_{1-40}$ protofibrils (pdb codes: 2LMN and 2LMO) [51]. The N-terminal coordinates (residues 1–8) were missing due to disorder. The A$\beta_{9-40}$ peptide has a turn at Asp23-Gly29 and the same salt bridge of Asp23/Lys28 (Fig. 1c).
4. A$\beta_{11-40}$ peptide: Comprehensive ssNMR techniques defined A$\beta_{1-40}$ fibrils [52]. The N-terminal coordinates (residues 1–10) were missing due to disorder. The A$\beta_{11-40}$ peptide has a turn at Val24-Ala30, but has a shifted inter β-strand contacts within the U-shaped motif. Unlike the previous NMR models (**item 3** above), the peptide did not contain the salt bridge of Asp23/Lys28 (Fig. 1d).
5. The U-shaped motif is a general feature of amyloid organization. Other amyloids, such as $\beta_2$-microglobulin fragment (K3 peptide, pdb code: 2E8D) [53] (Fig. 1e) and the second WW domain of CA150 (pdb code: 2NNT) [58], also exhibit the U-shaped motif with the *β-strand-turn-β-strand* motif.

### 2.2 β-Hairpin Peptides

Monomeric or dimeric amyloids tend to form a β-hairpin, an aggregate intermediate that facilitates membrane insertion [46]. Conversion to the U-shaped structure in the oligomerization process with β-hairpin monomers or small oligomers followed by membrane insertion takes place in the membrane. The membrane insertion mechanism of amyloid β-hairpins is similar to that of the cytolytic cationic β-hairpins, such as protegrin-1 (PG-1) and MAX peptides. These β-hairpins are also capable of forming ion channels. [44, 45, 59].

1. PG-1 peptide: A small cationic β-hairpin peptide consisting of 18 amino acids is capable of forming β-sheet channels [44, 45]. PG-1 is an antimicrobial peptide (AMP) with a great antibiotic potency [60]. NMR determined the PG-1 β-hairpin structure in solution with the data confirming to the presence of two antiparallel β-strands linked by a β-turn and stabilized by two disulfide bonds [61–64] (Fig. 1f).
2. MAX peptides: Synthetic amphiphilic cationic peptides, MAX1 and MAX35, can form β-barrels inducing membrane leakage [59]. The MAX peptides consisting of 20 amino acids and contain alternative hydrophobic (Val or Ile) and hydrophilic (Lys) residues connected by a reciprocal turn, -V$^{D}$PPT-, where $^{D}$P denotes the D-amino acid proline (Fig. 1g).

### 2.3 Lipids

To simulate amyloid channels in a membrane environment, a unit cell containing two layers of lipids is constructed. In the middle of the unit cell, simple van der Waals (vdW) spheres representing lipid headgroups are placed in two parallel planes (or membrane surfaces) separated by expected bilayer thickness [65, 66]. Dynamics are performed on the spheres with constraints on their respective planes and with the embedded channel held rigid, resulting in vdW spheres that are randomly distributed onto the planes and well packed around the channel. The lipid molecules are randomly selected from the library of pre-equilibrated states and replaced with pseudo-vdW spheres at the positions of the lipid headgroup constituting the lipid bilayer topology. Simulations employ both zwitterionic and anionic lipid bilayers with various lipids in the liquid phase. Each lipid used in the simulations exhibits different phase transitions yielding different physical properties for the cross-sectional area per lipid, $A_{cross}$, and headgroup distance, $D_{HH}$. Thus, with a proper choice for the number of lipid molecules, the optimal value of the lateral cell dimensions can be determined. The following list shows the lipid molecules used in the amyloid channel simulations.

1. DOPC: 1,2-dioleoyl-*sn-glycero*-3-phosphocholine, zwitterionic, $A_{cross}=72.4$ Å$^2$ and $D_{HH}=36.7$ Å at 30 °C [67].
2. DOPS: 1,2-dioleoyl-*sn-glycero*-3-phosphoserine, anionic, $A_{cross}=65.3$ Å$^2$ and $D_{HH}=38.4$ Å at 30 °C [68].
3. POPC: 1-palmitoyl-2-oleyl-*sn-glycero*-3-phosphatidylcholine, zwitterionic, $A_{cross}=68.3$ Å$^2$ and $D_{HH}=37.0$ Å at 30 °C [67].
4. POPE: 1-palmitoyl-2-oleoyl-*sn-glycero*-3-phosphoethanolamine, zwitterionic, $A_{cross}=56.0$ Å$^2$ and $D_{HH}=41.3$ Å at 30 °C [69].
5. POPS: 1-palmitoyl-2-oleoyl-*sn-glycero*-3-phosphoserine, anionic, $A_{cross}=55.0$ Å$^2$ at 27 °C [70].
6. POPG: 1-palmitoyl-2-oleyl-*sn-glycero*-3-phosphatidylglycerol, anionic, $A_{cross}=62.8$ Å$^2$ and $D_{HH}=36.0$ Å at 37 °C [71].

The zwitterionic lipid bilayer is constituted with DOPC lipids. Various mixed lipid bilayers with combination of each zwitterionic and anionic lipid molecule, DOPS:POPE (1:2 mole ratio), POPC: POPS (3:1 mole ratio), and POPC:POPG (4:1 mole ratio) are used for representing the anionic bilayer system. For the mixed lipid bilayers, averaged values of $A_{cross}$ and $D_{HH}$ are taken based on a mole ratio.

## 3 Methods

Atomistic MD simulations with explicit atom representations for protein, lipid, water, and ion are performed using the CHARMM [72] program with the NAMD [73] parallel computing code on a

Biowulf cluster at the National Institute of Health, Bethesda, MD (http://biowulf.nih.gov). Updated CHARMM [72] all-atom additive force field for lipids (C36) [74] and the modified TIP3P water model [75] are used to construct the set of starting points and to relax the systems to a production-ready stage. The bilayer system containing an Aβ channel/barrel, lipids, salts, and water normally has 200,000 atoms depending on the size of Aβ channel/barrel. In the pre-equilibrium stages, a series of minimizations is performed for the initial configurations to remove overlaps of the alkane chains in the lipids and to gradually relax the solvents around the harmonically restrained peptides. The initial configurations are gradually relaxed through dynamic cycles with electrostatic cutoffs (12 Å). The harmonic restraints are gradually diminished with the full Ewald electrostatics calculation and constant temperature (Nosé–Hoover) thermostat/barostat at 303 K. For $t < 30$ ns, our simulation employ the NPAT (constant number of atoms, pressure, surface area, and temperature) ensemble with a constant normal pressure applied in the direction perpendicular to the membrane. After $t = 30$ ns, the simulations employ the NPT ensemble. Production runs are generally performed up to 100 ns, and averages are taken after 30 ns, discarding initial transients.

### 3.1 Constructing Amyloid Channels

The initial channel models are constructed by using the U-shaped *β-strand-turn-β-strand* motifs and β-hairpins. The U-shaped peptide or the β-hairpin is subject to a multifold rotational symmetry operation with respect to the pore axis, creating the annular channel conformation. Depending on the direction of the rotation, the U-shaped peptide generates two different channel topologies: CNpNC (where C and N represent C- and N-terminal β-strands respectively, and p denotes a central pore) and NCpCN channels (Fig. 2a). In Aβ channels, the CNpNC channel preserves a central pore, while the NCpCN channel collapses the pore due to the hydrophobic mismatch of the charged N-terminal strands with the lipid bilayer hydrophobic core [30, 31] (*see* **Note 1**). However, in contrast to Aβ channels, K3 channels preserve the pore with NCpCN topology, while CNpNC K3 channel collapses the solvated pore due to the hydrophobic mismatch [43] (Fig. 2b). The U-shaped peptides yield a double-layered annular β-sheet [32, 35]. The designed channels have a perfectly annular shape with the pore-lining inner strands forming a β-sheet through intermolecular backbone hydrogen bonds (H-bonds), but the outer strands do not form a β-sheet due to the larger curvature at the channel periphery. In contrast, β-hairpins generate a single layered annular β-sheet [44, 45, 59] (Fig. 2c, d). Backbone H-bond formation is monitored during the simulations (*see* **Note 2**).

### 3.2 Generating Full-Length $Aβ_{1–42}$ Peptides

The NMR-derived U-shaped Aβ peptides only provide N-terminally truncated coordinates due to conformational disorder [50, 51]. To create full-length Aβ peptides, the $Aβ_{1–16}$ coordinates in the absence

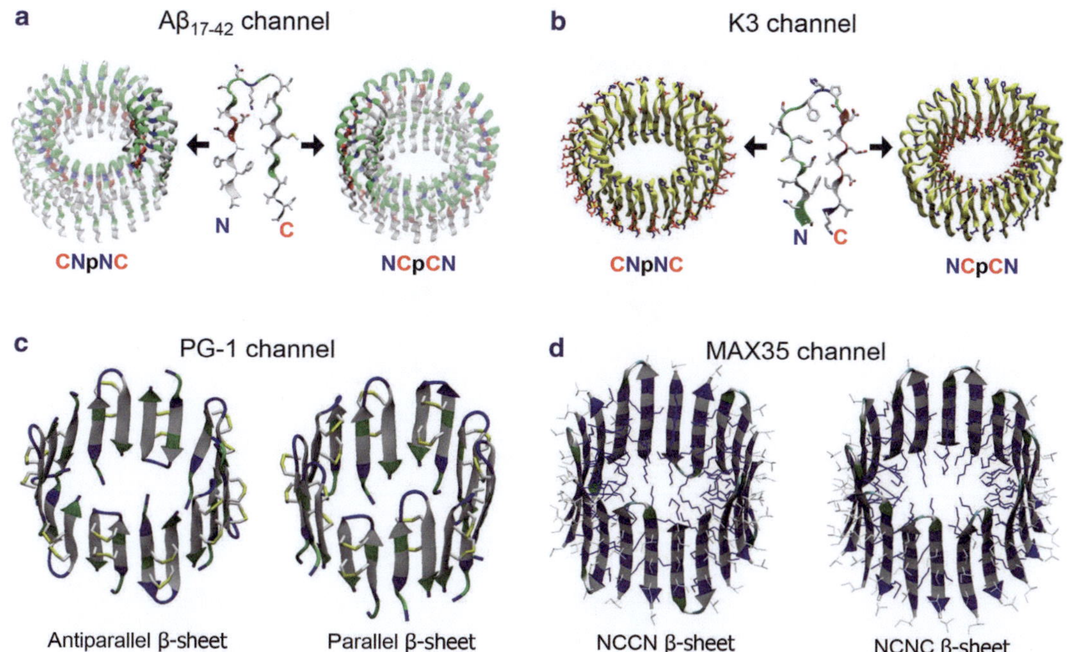


**Fig. 2** Computational models of β-sheet channel with the U-shaped and β-hairpin motifs. (**a**) Building annular channel structures in the membrane using (**a**) the NMR-based $A\beta_{17-42}$ and (**b**) the ssNMR $K3_{20-41}$ peptides. In the double-layered β-sheet channels, two different directions of peptide addition along the curvature yield the CNpNC (*left*) and NCpCN (*right*) channels (here, C: C-terminal, N: N-terminal, p: pore). Unlike the U-shaped peptides, β-hairpins generate a single layered annular β-sheet for (**c**) protegrin-1 (PG-1) and (**d**) MAX 35 channels. The PG-1 channels contain the antiparallel (turn-next-to-tail, *left*) and parallel (turn-next-to-turn, *right*) β-sheet arrangements in an NCCN packing mode. In the MAX channels, the β-hairpin arrangements give rise to two potential β-sheet motifs; turn-next-to-tail β-hairpins in NCCN packing mode (*left*) and turn-next-to-turn β-hairpins in NCNC packing mode (*right*). In both cases, the MAX β-hairpins form antiparallel β-sheets, positioning the positively charged Lys side chains into the central pore

of $Zn^{2+}$ (pdb code: 1ZE7) [76] are used for the missing N-terminal portions of the peptides. For each combination of the N-terminal structure with the NMR U-shaped motifs of $A\beta_{17-42}$ and $A\beta_{9-40}$, two $A\beta_{1-42}$ conformers were generated (Fig. 3). Conformer 1 has a turn at Ser26-Ile31, and conformer 2 at Asp23-Gly29. In the latter conformer, two C-terminal residues, Ile41 and Ala42, were added to create $A\beta_{1-42}$. Both $A\beta_{1-42}$ conformers retained the U-shaped β-strand-turn-β-strand motif and can be divided into four domains: the extramembranous N-terminal fragment (residues 1–16 and 1–8 for conformer 1 and 2, respectively), solvated pore-lining β-strand (residues 17–25 and 9–22 for conformer 1 and 2, respectively), turn (residues 26–31 and 23–29 for conformer 1 and 2, respectively), and lipid-interacting C-terminal β-strand (residues 32–42 and 30–42 for conformer 1 and 2, respectively).

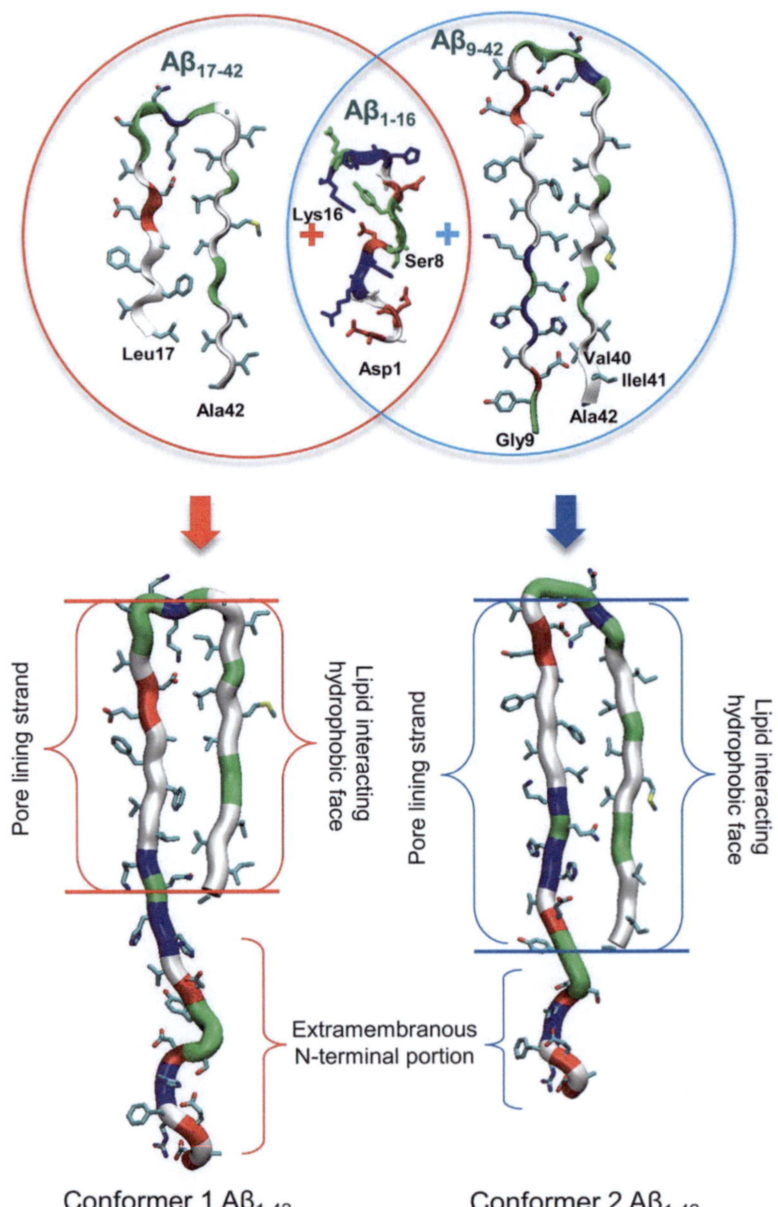


**Fig. 3** Schematic diagram for the constructions of full-length $A\beta_{1-42}$ peptides. The U-shaped Aβ monomers, $A\beta_{17-42}$ and $A\beta_{9-42}$, recover the missing N-terminal portions through the covalent connection with the solution structure of $A\beta_{1-16}$ (pdb code: 1ZE7), generating two $A\beta_{1-42}$ conformers (conformer 1 and 2) with different turns (from Jang et al. [8], reprinted with permission)

### 3.3 β-Barrel Topology of Aβ Channels

Amyloid channels can have conventional annular β-sheet channel and β-barrel topologies. To construct the channel structure with the conventional β-strands arrangement, monomers (U-shaped peptides or β-hairpins) were inserted without inclination with

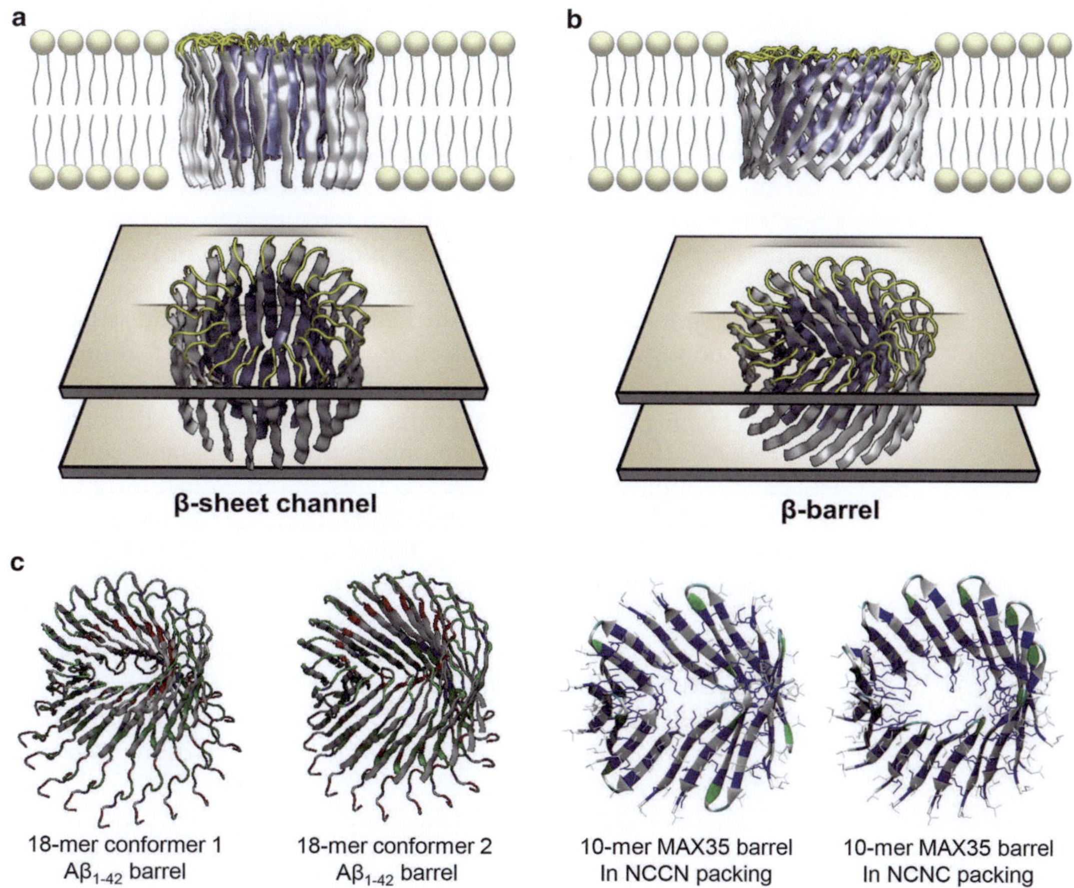


**Fig. 4** Constructing the conventional β-sheet channel and β-barrel in the membrane. (**a**) The conventional β-sheet channel has the β-strands that orient parallel to the membrane normal, (**b**) while the β-strands that orient obliquely to the membrane normal generate β-barrel structure (from Jang et al. [8], reprinted with permission). Both form a barrel-stave pore. (**c**) Examples are shown for the 18-mer conformer 1 and 2 $A\beta_{1-42}$ barrels, and 10-mer MAX35 barrels in NCCN and NCNC packing modes

respect to the membrane normal, generating the annular channel topology (Figs. 2 and 4a). To construct the β-barrel structure, the monomers were inclined ~37° with respect to the pore axis, creating the β-barrel topology [34] (Fig. 4b). The β-barrel morphology mimics naturally occurring β-barrels observed in transmembrane proteins that are found frequently in the outer membranes of bacteria, mitochondria, and chloroplasts. The β-barrel motif is a large β-sheet composed of an even number of β-strands. Some known structures of β-barrel membrane proteins have β-strands ranging in number from 8 to 22 [77, 78] (*see* **Note 3**). Examples are shown here for the U-shaped amyloid β-barrels and β-hairpin barrels (Fig. 4c). In the simulations, the initial annular conformation is gradually lost during the relaxation of the lipid bilayer. No peptide dissociation from the barrels is observed at these time scales.

The relaxed barrel conformation with localized β-sheet optimization leads to subunit formations. Heterogeneity in barrel conformations can be evaluated by several criteria (*see* **Note 4**).

### 3.4 D-Enantiomer Aβ$_{1–42}$ Peptides

The standard CHARMM force field [72] is primarily designed for L-amino acids ("left-handed" isomers). To simulate D-amino acids ("right-handed" isomers), a protein force field for asymmetric isomers is required. D-amino acid is a mirror image of L-amino acid, indicating that they are identical, except for their backbone chirality. In the simulations, the L- and D-amino acids share the same backbone bonds and angles, indicating that the standard L-amino acids parameters can be used for the D-amino acids. However, the parameters include the dihedral angle cross term map (CMAP) correction [79], which was created for only L-amino acids, and cannot be directly applied to D-amino acids. Thus, in the simulation, a mirror-image CMAP term for D-amino acids reflecting the phi-psi CMAP matrix should be used [80]. Current version of CHARMM36 force field supports D-amino acids simulations.

### 3.5 Aβ Mutants

Naturally occurring point mutations in the amyloid precursor protein (APP) clustered around the central region of the Aβ residues are related to familial forms of AD [81]. However, designed synthetic point substitutions significantly alter the channel activity, suppressing Aβ toxicity.

1. F19P and F20C point substitutions: Two phenylalanine residues, Phe19 and Phe20, were replaced with Pro19 and Cys20, respectively (Fig. 5a). The F19P substitution in both truncated Aβ$_{17–42}$ and full-length Aβ$_{1–42}$ channels/barrels prevents pore activity and hence cellular toxicity, while the F20C substitution preserves the solvated pore with channel activity comparable to the wild type [33, 37, 39].
2. Unlike point substitution, Osaka mutant (ΔE22) eliminates residue Glu22 from the pore-lining β-strand [40]. As a result, pore-lining residues 10–21 for both conformers flip their side chains, while the other domains remain intact (Fig. 5b). The ΔE22 barrels show the membrane embedded β-sheet channel topology, indistinguishable from the wild-type Aβ$_{1–42}$ barrels.

### 3.6 Pyroglutamate (pE) Modified Aβ Peptides

Pyroglutamate-modified Aβ (Aβ$_{pE3–42}$) peptide is particularly associated with cytotoxicity in AD [82, 83]. The peptide is posttranslationally generated by cleavage of the first two N-terminal amino acids (Asp1 and Ala2) of Aβ$_{1–42}$, leaving an exposed Glu3 residue. Intramolecular dehydration catalyzed by the glutaminyl cyclase (QC) enzyme generates a lactam ring in Glu3, converting to the pyroglutamate (pE) residue [82, 83]. To simulate pyroglutamate (pE), the first two residues, Asp1 and Ala2, from each conformer 1

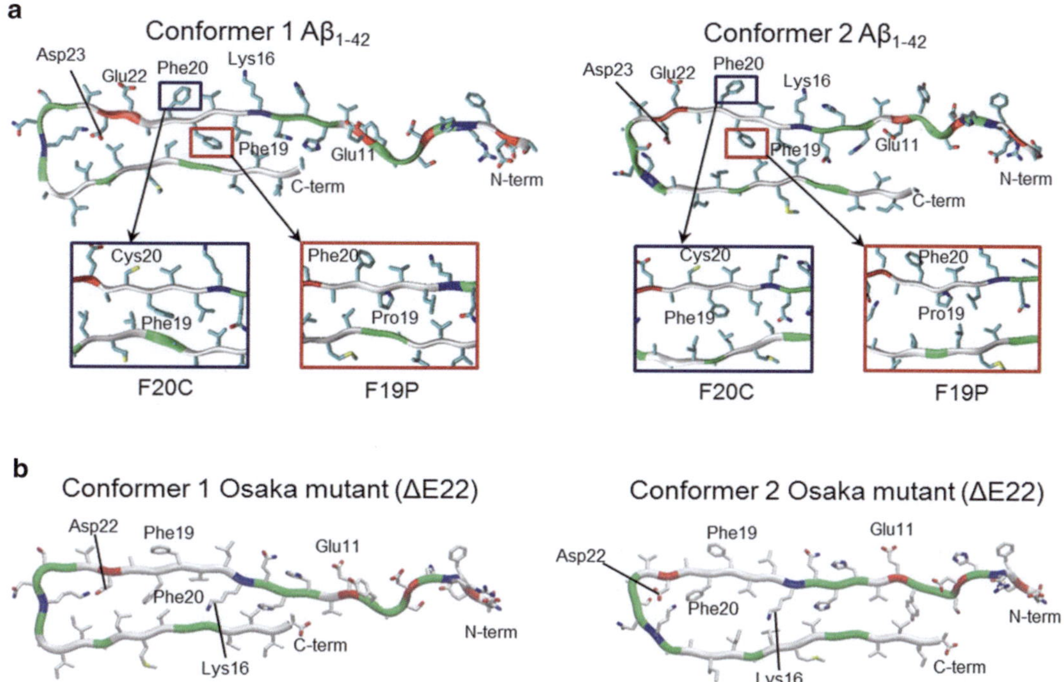


**Fig. 5** Aβ mutants conformations. (**a**) Monomer conformations of the $A\beta_{1-42}$ wild type (*top panels*) and F19P and F20C mutants (highlighted in *rectangular insets*) with two different conformers, conformer 1 (*left*) with turn at Ser26-Ile31 and conformer 2 (*right*) with turn at Asp23-Gly29. (**b**) Monomer conformations of the Osaka mutant (ΔE22) with two different conformers, conformer 1 (*left*) and conformer 2 (*right*). Several important residues in the pore-lining strand are marked. In the peptide ribbon, hydrophobic, polar/Gly, positively charged, and negatively charged residues are colored *white*, *green*, *blue*, and *red*, respectively

and 2 of $A\beta_{1-42}$ peptide are removed. The Glu3 residue is then converted into pE3 by generating the lactam ring (Fig. 6). The pE molecular topology was generated by the Avogadro software [84], since the pE residue is not included in the standard CHARMM [72] force field protocol. The Gaussian09 program [85] can be used to calculate parameters including partial charges, bond lengths, angles, and torsional angles for the atoms in the pE residue. The calculated parameters can be directly adopted in the CHARMM [72] program.

### 3.7 Calculating Aβ Pore Activity

Amyloid channels/barrels preserve a large pore, ~1–2 nm, wide enough for conducting ions and water. In addition to counter ions to the system for neutralization, the bilayer systems contain $Mg^{2+}$, $K^{+}$, $Ca^{2+}$, and $Zn^{2+}$ at the same concentration of 25 mM to satisfy a total cation concentration near 100 mM, as well as $Cl^{-}$, which mimics the physiological salt concentration. Cations can be trapped by the negatively charged amino acids in the solvated pore.

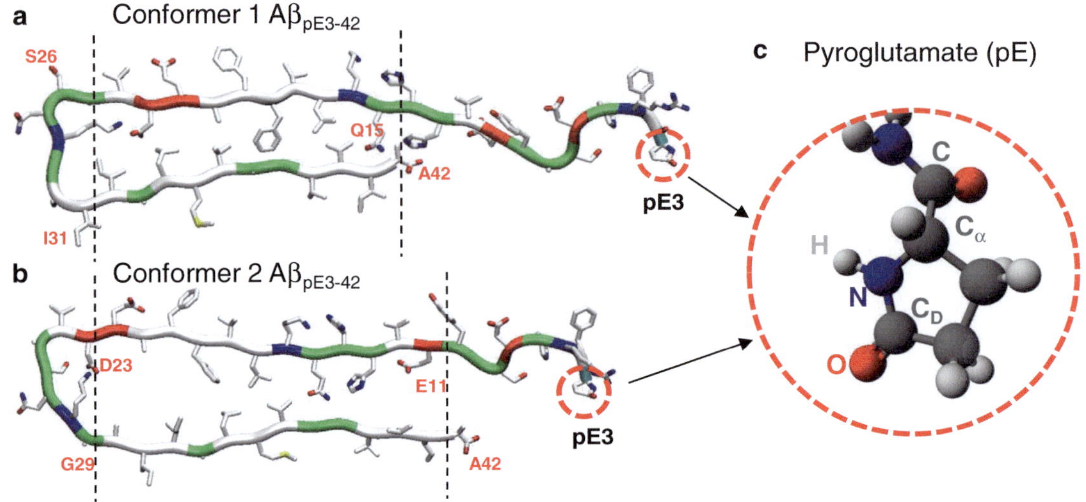


**Fig. 6** Conformations of pyroglutamate-modified Aβ ($A\beta_{pE}$) peptide. Monomer conformations of $A\beta_{pE3-42}$ with two different conformers, (**a**) conformer 1 with turn at Ser26-Ile31 and (**b**) conformer 2 with turn at Asp23-Gly29. (**c**) Molecular topology of the pyroglutamate at residue 3 (pE3). From Gilman et al. [42], reprinted with permission

The motions of the cations in the pore, which reflect the electrostatic interaction, can delineate the electrophysiological currents across the membrane. Several quantities calculated from the simulations can provide detailed information of the mechanism for ion permeation across the bilayer.

1. Probability distribution functions, $P$, for ions representing ionic permeation as a function of the distance along the pore center axis can be calculated over the simulations. Peaks in the distribution curve represent the cationic binding sites [30, 34, 40, 45, 59].
2. In order to see the ions' behavior in the pore, the potential of mean force (PMF), $\Delta G_{PMF}$, representing the relative free energy profile for each ion across the bilayer can be calculated [32–35, 44] (*see* **Note 5**).
3. To observe ion fluctuation across the pore, the change in total charge, $\Delta Q$ ($C$/ns), in the pore as a function of the simulation time can be calculated. In the calculations, different pore lengths with different cutoffs along the pore axis are used [36, 37, 40, 42].
4. To correlate the charge fluctuations with experimental ion conductance, the maximum conductance, $g_{max}$, representing the ion transport can be calculated from the equilibrium simulations [36, 42] (*see* **Note 6**).

## 4 Notes

1. In the simulations, pore structures can be examined by HOLE program [86]. The program allows us to visualize and analyze the pore or cavity in biomolecules such as ion channels.
2. The time-dependent fraction of intermolecular (or intramolecular) backbone H-bonds, $Q_{\text{H-bond}}(t)$, for the β-sheet channels in the lipid bilayer can be calculated by

$$Q_{\text{H-bond}}(t) = \frac{N_{\text{H-bond}}}{N_{\text{H-bond}}^{\max}}$$

where $N_{\text{H-bond}}$ is the number of intermolecular H-bonds at each time frame, and $N_{\text{H-bond}}^{\max}$ is the maximum possible number of the backbone H-bonds as monitored in the initial configuration.
3. Aβ channels/barrels were modeled with 12–36 Aβ peptides. Different numbers of Aβ monomers produced channels/barrels with different outer and pore dimensions. Preferred sizes of Aβ channels/barrels were found to be in the range of 16–24 Aβ peptides, i.e., 16–24 β-strands lining the pores. The smaller one (12-mer) collapsed and the larger one (36-mer) was not supported by the bilayer [32, 35]. This range was also found to hold for other toxic β-sheet channels; K3 channels with 24 β-strands [43], 18- and 24-mer human islet amyloid polypeptide (hIAPP) channels [87], PG-1 channels with 16–20 β-strands [44, 45], and MAX channels with 20 β-strands [59].
4. The β-sheet secondary structure was determined by the STRIDE program [88]. "Straightness" of the strand was calculated by β-strand order parameter, $S_\beta = \frac{1}{N_V}\sum_{k=1}^{N_V}\left(\frac{3\cos^2\theta_\alpha - 1}{2}\right)$ where $\theta_\alpha$ is the angle between the positional vectors connecting two $C_\alpha$ atoms, and $N_v$ is the total number of vector pairs. The averaged β-strand *B*-factor (or Temperature factor) was calculated from the root-mean-squared (RMS) fluctuations [89] relative to the starting point during the simulations with a simple correlation of $B = 8\pi^2 \frac{\langle \text{RMSF} \rangle^2}{3}$ where ‹ › denotes averaging over simulation time. Percent of β-sheet content based on the intermolecular backbone H-bonds between β-strands was calculated.
5. The PMF is calculated by using the equation of $\Delta G_{\text{PMF}}(\vec{r}) = -k_B T \ln(\rho(\vec{r}) / \rho_{\text{bulk}})$. Here, $k_B$ is the Boltzmann constant, $T$ is the simulation temperature, $\rho(\vec{r})$ is the equilibrium number density of ions, and $\rho_{\text{bulk}}$ is the averaged ion density in the bulk region. Accurate equilibrium PMF relevant to ion permeation should be obtained from free energy calculations

with the umbrella sampling method [90]. Nevertheless, given the simulation trajectory without additional multiple equilibrium runs for sampling, ion-density-based PMF calculations are useful to obtain an estimate of the relative free energy changes for ions, providing an outline for pore ion permeation [91, 92].

6. For the equilibrium all-atom MD simulations in the absence of membrane potentials, the maximum conductance, $g_{\max}$, [93] representing the ion transport can be described as

$$g_{\max} = \frac{q_e^2}{k_B T L^2} \langle D(z) e^{G_{\text{PMF}}(z)/k_B T} \rangle^{-1} \langle e^{-G_{\text{PMF}}(z)/k_B T} \rangle^{-1}$$

where $q_e^2$ is the elementary charge, $k_B$ denotes the Boltzmann's constant, $T$ is the simulation temperature, and $L$ represents the pore length. In the bracket, $D(z)$ and $\Delta G_{\text{PMF}}(z)$ denote the one-dimensional diffusion coefficient and the one-dimensional potential of mean force for ions, respectively. The bracket averages over the pore length $L$.

## Acknowledgements

This project has been funded in whole or in part with Federal funds from the Frederick National Laboratory for Cancer Research, National Institutes of Health, under contract HHSN261200800001E. This research was supported [in part] by the Intramural Research Program of NIH, Frederick National Lab, Center for Cancer Research. The content of this publication does not necessarily reflect the views or policies of the Department of Health and Human Services, nor does mention of trade names, commercial products or organizations imply endorsement by the US Government.

### References

1. Monsonego A, Zota V, Karni A, Krieger JI, Bar-Or A, Bitan G, Budson AE, Sperling R, Selkoe DJ, Weiner HL (2003) Increased T cell reactivity to amyloid beta protein in older humans and patients with Alzheimer disease. J Clin Invest 112:415–422
2. Cohen FE, Kelly JW (2003) Therapeutic approaches to protein-misfolding diseases. Nature 426:905–909
3. Temussi PA, Masino L, Pastore A (2003) From Alzheimer to Huntington: why is a structural understanding so difficult? EMBO J 22:355–361
4. Dobson CM (2003) Protein folding and misfolding. Nature 426:884–890
5. Chiti F, Dobson CM (2006) Protein misfolding, functional amyloid, and human disease. Annu Rev Biochem 75:333–366
6. DeToma AS, Salamekh S, Ramamoorthy A, Lim MH (2012) Misfolded proteins in Alzheimer's disease and type II diabetes. Chem Soc Rev 41:608–621
7. Eisenberg D, Jucker M (2012) The amyloid state of proteins in human diseases. Cell 148: 1188–1203

8. Jang H, Arce FT, Ramachandran S, Kagan BL, Lal R, Nussinov R (2014) Disordered amyloidogenic peptides may insert into the membrane and assemble into common cyclic structural motifs. Chem Soc Rev 43:6750–6764
9. Marshall KE, Serpell LC (2010) Fibres, crystals and polymorphism: the structural promiscuity of amyloidogenic peptides. Soft Matter 6:2110–2114
10. Miller Y, Ma B, Nussinov R (2010) Polymorphism in Alzheimer Abeta amyloid organization reflects conformational selection in a rugged energy landscape. Chem Rev 110: 4820–4838
11. Sipe JD, Cohen AS (2000) Review: history of the amyloid fibril. J Struct Biol 130:88–98
12. Quist A, Doudevski I, Lin H, Azimova R, Ng D, Frangione B, Kagan B, Ghiso J, Lal R (2005) Amyloid ion channels: a common structural link for protein-misfolding disease. Proc Natl Acad Sci U S A 102:10427–10432
13. Glabe CG (2008) Structural classification of toxic amyloid oligomers. J Biol Chem 283: 29639–29643
14. Bernstein SL, Dupuis NF, Lazo ND, Wyttenbach T, Condron MM, Bitan G, Teplow DB, Shea JE, Ruotolo BT, Robinson CV, Bowers MT (2009) Amyloid-beta protein oligomerization and the importance of tetramers and dodecamers in the aetiology of Alzheimer's disease. Nat Chem 1:326–331
15. Butterfield SM, Lashuel HA (2010) Amyloidogenic protein-membrane interactions: mechanistic insight from model systems. Angew Chem Int Ed Engl 49:5628–5654
16. Matsumura S, Shinoda K, Yamada M, Yokojima S, Inoue M, Ohnishi T, Shimada T, Kikuchi K, Masui D, Hashimoto S, Sato M, Ito A, Akioka M, Takagi S, Nakamura Y, Nemoto K, Hasegawa Y, Takamoto H, Inoue H, Nakamura S, Nabeshima Y, Teplow DB, Kinjo M, Hoshi M (2011) Two distinct amyloid beta-protein (Abeta) assembly pathways leading to oligomers and fibrils identified by combined fluorescence correlation spectroscopy, morphology, and toxicity analyses. J Biol Chem 286:11555–11562
17. Relini A, Cavalleri O, Rolandi R, Gliozzi A (2009) The two-fold aspect of the interplay of amyloidogenic proteins with lipid membranes. Chem Phys Lipids 158:1–9
18. Garner B (2010) Lipids and Alzheimer's disease. Biochim Biophys Acta 1801:747–749
19. Williams TL, Serpell LC (2011) Membrane and surface interactions of Alzheimer's A beta peptide—insights into the mechanism of cytotoxicity. FEBS J 278:3905–3917
20. Verdier Y, Zarandi M, Penke B (2004) Amyloid beta-peptide interactions with neuronal and glial cell plasma membrane: binding sites and implications for Alzheimer's disease. J Pept Sci 10:229–248
21. Shirwany NA, Payette D, Xie J, Guo Q (2007) The amyloid beta ion channel hypothesis of Alzheimer's disease. Neuropsychiatr Dis Treat 3:597–612
22. Buchete NV (2012) Unlocking the atomic-level details of amyloid fibril growth through advanced biomolecular simulations. Biophys J 103:1411–1413
23. Walsh DM, Klyubin I, Fadeeva JV, Cullen WK, Anwyl R, Wolfe MS, Rowan MJ, Selkoe DJ (2002) Naturally secreted oligomers of amyloid beta protein potently inhibit hippocampal long-term potentiation in vivo. Nature 416:535–539
24. Kagan BL (2012) Membrane pores in the pathogenesis of neurodegenerative disease. Prog Mol Biol Transl Sci 107:295–325
25. Arispe N, Pollard HB, Rojas E (1993) Giant multilevel cation channels formed by Alzheimer disease amyloid beta-protein [A beta P-(1–40)] in bilayer membranes. Proc Natl Acad Sci U S A 90:10573–10577
26. Arispe N, Rojas E, Pollard HB (1993) Alzheimer disease amyloid β protein forms calcium channels in bilayer membranes: blockade by tromethamine and aluminum. Proc Natl Acad Sci U S A 90:567–571
27. Arispe N, Pollard HB, Rojas E (1994) The ability of amyloid beta-protein [a-Beta-P(1–40)] to form $Ca^{2+}$ channels provides a mechanism for neuronal death in Alzheimer's disease. Ann N Y Acad Sci 747:256–266
28. Arispe N, Pollard HB, Rojas E (1996) $Zn^{2+}$ interaction with Alzheimer amyloid β protein calcium channels. Proc Natl Acad Sci U S A 93:1710–1715
29. Lin H, Bhatia R, Lal R (2001) Amyloid β protein forms ion channels: implications for Alzheimer's disease pathophysiology. FASEB J 15:2433–2444
30. Jang H, Zheng J, Nussinov R (2007) Models of β-amyloid ion channels in the membrane suggest that channel formation in the bilayer is a dynamic process. Biophys J 93:1938–1949
31. Jang H, Zheng J, Lal R, Nussinov R (2008) New structures help the modeling of toxic amyloidβ ion channels. Trends Biochem Sci 33:91–100
32. Jang H, Arce FT, Capone R, Ramachandran S, Lal R, Nussinov R (2009) Misfolded amyloid ion channels present mobile β-sheet subunits in contrast to conventional ion channels. Biophys J 97:3029–3037

33. Jang H, Arce FT, Ramachandran S, Capone R, Azimova R, Kagan BL, Nussinov R, Lal R (2010) Truncated β-amyloid peptide channels provide an alternative mechanism for Alzheimer's disease and down syndrome. Proc Natl Acad Sci U S A 107:6538–6543
34. Jang H, Arce FT, Ramachandran S, Capone R, Lal R, Nussinov R (2010) β-Barrel topology of Alzheimer's β-amyloid ion channels. J Mol Biol 404:917–934
35. Jang H, Teran Arce F, Ramachandran S, Capone R, Lal R, Nussinov R (2010) Structural convergence among diverse, toxic β-sheet ion channels. J Phys Chem B 114:9445–9451
36. Capone R, Jang H, Kotler SA, Connelly L, Teran Arce F, Ramachandran S, Kagan BL, Nussinov R, Lal R (2012) All-D-enantiomer of β-amyloid peptide forms ion channels in lipid bilayers. J Chem Theory Comput 8:1143–1152
37. Capone R, Jang H, Kotler SA, Kagan BL, Nussinov R, Lal R (2012) Probing structural features of Alzheimer's amyloid-β pores in bilayers using site-specific amino acid substitutions. Biochemistry 51:776–785
38. Connelly L, Jang H, Arce FT, Capone R, Kotler SA, Ramachandran S, Kagan BL, Nussinov R, Lal R (2012) Atomic force microscopy and MD simulations reveal pore-like structures of all-D-enantiomer of Alzheimer's β-amyloid peptide: relevance to the ion channel mechanism of AD pathology. J Phys Chem B 116:1728–1735
39. Connelly L, Jang H, Arce FT, Ramachandran S, Kagan BL, Nussinov R, Lal R (2012) Effects of point substitutions on the structure of toxic Alzheimer's β-amyloid channels: atomic force microscopy and molecular dynamics simulations. Biochemistry 51:3031–3038
40. Jang H, Arce FT, Ramachandran S, Kagan BL, Lal R, Nussinov R (2013) Familial Alzheimer's disease Osaka mutant (ΔE22) β-barrels suggest an explanation for the different Aβ1-40/42 preferred conformational states observed by experiment. J Phys Chem B 117:11518–11529
41. Jang H, Connelly L, Arce FT, Ramachandran S, Lal R, Kagan BL, Nussinov R (2013) Alzheimer's disease: which type of amyloid-preventing drug agents to employ? Phys Chem Chem Phys 15:8868–8877
42. Gillman AL, Jang H, Lee J, Ramachandran S, Kagan BL, Nussinov R, Teran Arce F (2014) Activity and architecture of pyroglutamate-modified amyloid-β (AβpE3-42) pores. J Phys Chem B 118:7335–7344
43. Mustata M, Capone R, Jang H, Arce FT, Ramachandran S, Lal R, Nussinov R (2009) K3 fragment of amyloidogenic β2-microglobulin forms ion channels: implication for dialysis related amyloidosis. J Am Chem Soc 131: 14938–14945
44. Jang H, Ma B, Lal R, Nussinov R (2008) Models of toxic β-sheet channels of protegrin-1 suggest a common subunit organization motif shared with toxic Alzheimer β-amyloid ion channels. Biophys J 95:4631–4642
45. Capone R, Mustata M, Jang H, Arce FT, Nussinov R, Lal R (2010) Antimicrobial protegrin-1 forms ion channels: molecular dynamic simulation, atomic force microscopy, and electrical conductance studies. Biophys J 98:2644–2652
46. Jang H, Connelly L, Arce FT, Ramachandran S, Kagan BL, Lal R, Nussinov R (2013) Mechanisms for the insertion of toxic, fibril-like β-amyloid oligomers into the membrane. J Chem Theory Comput 9:822–833
47. Ladiwala AR, Litt J, Kane RS, Aucoin DS, Smith SO, Ranjan S, Davis J, Vannostrand WE, Tessier PM (2012) Conformational differences between two amyloid beta oligomers of similar size and dissimilar toxicity. J Biol Chem 287: 24765–24773
48. Chimon S, Shaibat MA, Jones CR, Calero DC, Aizezi B, Ishii Y (2007) Evidence of fibril-like beta-sheet structures in a neurotoxic amyloid intermediate of Alzheimer's beta-amyloid. Nat Struct Mol Biol 14:1157–1164
49. Ma B, Nussinov R (2002) Stabilities and conformations of Alzheimer's β-amyloid peptide oligomers (Aβ16–22, Aβ16–35, and Aβ10–35): sequence effects. Proc Natl Acad Sci U S A 99:14126–14131
50. Lührs T, Ritter C, Adrian M, Riek-Loher D, Bohrmann B, Doeli H, Schubert D, Riek R (2005) 3D structure of Alzheimer's amyloid-β(1–42) fibrils. Proc Natl Acad Sci U S A 102: 17342–17347
51. Petkova AT, Yau WM, Tycko R (2006) Experimental constraints on quaternary structure in Alzheimer's β-amyloid fibrils. Biochemistry 45:498–512
52. Bertini I, Gonnelli L, Luchinat C, Mao J, Nesi A (2011) A new structural model of Aβ40 fibrils. J Am Chem Soc 133:16013–16022
53. Iwata K, Fujiwara T, Matsuki Y, Akutsu H, Takahashi S, Naiki H, Goto Y (2006) 3D structure of amyloid protofilaments of β2-microglobulin fragment probed by solid-state NMR. Proc Natl Acad Sci U S A 103: 18119–18124
54. Zheng J, Jang H, Ma B, Nussinov R (2008) Annular structures as intermediates in fibril formation of Alzheimer Aβ17–42. J Phys Chem B 112:6856–6865
55. Zheng J, Jang H, Nussinov R (2008) β2-microglobulin amyloid fragment organization and

morphology and its comparison to Aβ suggests that amyloid aggregation pathways are sequence specific. Biochemistry 47:2497–2509

56. Arce FT, Jang H, Ramachandran S, Landon PB, Nussinov R, Lal R (2011) Polymorphism of amyloid β peptide in different environments: implications for membrane insertion and pore formation. Soft Matter 7:5267–5273
57. Lee J, Gillman AL, Jang H, Ramachandran S, Kagan BL, Nussinov R, Teran Arce F (2014) Role of the fast kinetics of pyroglutamate-modified amyloid-β oligomers in membrane binding and membrane permeability. Biochemistry 53:4704–4714
58. Ferguson N, Becker J, Tidow H, Tremmel S, Sharpe TD, Krause G, Flinders J, Petrovich M, Berriman J, Oschkinat H, Fersht AR (2006) General structural motifs of amyloid protofilaments. Proc Natl Acad Sci U S A 103: 16248–16253
59. Gupta K, Jang H, Harlen K, Puri A, Nussinov R, Schneider JP, Blumenthal R (2013) Mechanism of membrane permeation induced by synthetic β-hairpin peptides. Biophys J 105: 2093–2103
60. Miyasaki KT, Lehrer RI (1998) Beta-sheet antibiotic peptides as potential dental therapeutics. Int J Antimicrob Agents 9:269–280
61. Roumestand C, Louis V, Aumelas A, Grassy G, Calas B, Chavanieu A (1998) Oligomerization of protegrin-1 in the presence of DPC micelles. A proton high-resolution NMR study. FEBS Lett 421:263–267
62. Fahrner RL, Dieckmann T, Harwig SS, Lehrer RI, Eisenberg D, Feigon J (1996) Solution structure of protegrin-1, a broad-spectrum antimicrobial peptide from porcine leukocytes. Chem Biol 3:543–550
63. Jang H, Ma B, Woolf TB, Nussinov R (2006) Interaction of protegrin-1 with lipid bilayers: membrane thinning effect. Biophys J 91: 2848–2859
64. Jang H, Ma B, Nussinov R (2007) Conformational study of the protegrin-1 (PG-1) dimer interaction with lipid bilayers and its effect. BMC Struct Biol 7:21
65. Woolf TB, Roux B (1994) Molecular dynamics simulation of the gramicidin channel in a phospholipid bilayer. Proc Natl Acad Sci U S A 91:11631–11635
66. Woolf TB, Roux B (1996) Structure, energetics, and dynamics of lipid-protein interactions: a molecular dynamics study of the gramicidin A channel in a DMPC bilayer. Proteins 24: 92–114
67. Kucerka N, Tristram-Nagle S, Nagle JF (2005) Structure of fully hydrated fluid phase lipid bilayers with monounsaturated chains. J Membr Biol 208:193–202
68. Petrache HI, Tristram-Nagle S, Gawrisch K, Harries D, Parsegian VA, Nagle JF (2004) Structure and fluctuations of charged phosphatidylserine bilayers in the absence of salt. Biophys J 86:1574–1586
69. Rand RP, Parsegian VA (1989) Hydration forces between phospholipid-bilayers. Biochim Biophys Acta 988:351–376
70. Mukhopadhyay P, Monticelli L, Tieleman DP (2004) Molecular dynamics simulation of a palmitoyl-oleoyl phosphatidylserine bilayer with $Na^+$ counterions and NaCl. Biophys J 86: 1601–1609
71. Rog T, Murzyn K, Pasenkiewicz-Gierula M (2003) Molecular dynamics simulations of charged and neutral lipid bilayers: treatment of electrostatic interactions. Acta Biochim Pol 50:789–798
72. Brooks BR, Bruccoleri RE, Olafson BD, States DJ, Swaminathan S, Karplus M (1983) CHARMM—a program for macromolecular energy, minimization, and dynamics calculations. J Comp Chem 4:187–217
73. Phillips JC, Braun R, Wang W, Gumbart J, Tajkhorshid E, Villa E, Chipot C, Skeel RD, Kale L, Schulten K (2005) Scalable molecular dynamics with NAMD. J Comp Chem 26: 1781–1802
74. Klauda JB, Venable RM, Freites JA, O'Connor JW, Tobias DJ, Mondragon-Ramirez C, Vorobyov I, MacKerell AD, Pastor RW (2010) Update of the CHARMM all-atom additive force field for lipids: validation on six lipid types. J Phys Chem B 114:7830–7843
75. Durell SR, Brooks BR, Bennaim A (1994) Solvent-induced forces between two hydrophilic groups. J Phys Chem 98:2198–2202
76. Zirah S, Kozin SA, Mazur AK, Blond A, Cheminant M, Segalas-Milazzo I, Debey P, Rebuffat S (2006) Structural changes of region 1–16 of the Alzheimer disease amyloid β-peptide upon zinc binding and in vitro aging. J Biol Chem 281:2151–2161
77. Schulz GE (2002) The structure of bacterial outer membrane proteins. Biochim Biophys Acta 1565:308–317
78. Sansom MS, Kerr ID (1995) Transbilayer pores formed by beta-barrels: molecular modeling of pore structures and properties. Biophys J 69:1334–1343
79. Mackerell AD, Feig M, Brooks CL (2004) Extending the treatment of backbone energetics in protein force fields: limitations of gas-phase quantum mechanics in reproducing protein conformational distributions in

molecular dynamics simulations. J Comput Chem 25:1400–1415

80. Connelly L, Jang H, Teran Arce F, Capone R, Kotler SA, Ramachandran S, Kagan BL, Nussinov R, Lal R (2012) Atomic force microscopy and MD simulations reveal pore-like structures of all-D-enantiomer of Alzheimer's beta-amyloid peptide relevance to the ion channel mechanism of AD pathology. J Phys Chem B. doi:10.1021/jp2108126
81. de Groot NS, Aviles FX, Vendrell J, Ventura S (2006) Mutagenesis of the central hydrophobic cluster in Abeta42 Alzheimer's peptide. Side-chain properties correlate with aggregation propensities. FEBS J 273:658–668
82. Jawhar S, Wirths O, Bayer TA (2011) Pyroglutamate amyloid-β (Aβ): a hatchet man in Alzheimer disease. J Biol Chem 286: 38825–38832
83. Saido TC, Iwatsubo T, Mann DMA, Shimada H, Ihara Y, Kawashima S (1995) Dominant and differential deposition of distinct β-amyloid peptide species, a-beta(N3(Pe)), in senile plaques. Neuron 14:457–466
84. Hanwell MD, Curtis DE, Lonie DC, Vandermeersch T, Zurek E, Hutchison GR (2012) Avogadro: an advanced semantic chemical editor, visualization, and analysis platform. J Cheminform 4:17
85. Frisch MJ, Trucks GW, Schlegel HB, Scuseria GE, Robb MA, Cheeseman JR, Scalmani G, Barone V, Mennucci B, Petersson GA, Nakatsuji H, Caricato M, Li X, Hratchian HP, Izmaylov AF, Bloino J, Zheng G, Sonnenberg JL, Hada M, Ehara M, Toyota K, Fukuda R, Hasegawa J, Ishida M, Nakajima T, Honda Y, Kitao O, Nakai H, Vreven T, Montgomery JA, Peralta JE, Ogliaro F, Bearpark M, Heyd JJ, Brothers E, Kudin KN, Staroverov VN, Kobayashi R, Normand J, Raghavachari K, Rendell A, Burant JC, Iyengar SS, Tomasi J, Cossi M, Rega N, Millam JM, Klene M, Knox JE, Cross JB, Bakken V, Adamo C, Jaramillo J, Gomperts R, Stratmann RE, Yazyev O, Austin AJ, Cammi R, Pomelli C, Ochterski JW, Martin RL, Morokuma K, Zakrzewski VG, Voth GA, Salvador P, Dannenberg JJ, Dapprich S, Daniels AD, Farkas Ö, Foresman JB, Ortiz JV, Cioslowski J, Fox DJ (2009) Gaussian 09, Revision B.01. Wallingford CT
86. Smart OS, Goodfellow JM, Wallace BA (1993) The pore dimensions of gramicidin-A. Biophys J 65:2455–2460
87. Zhao J, Luo Y, Jang H, Yu X, Wei G, Nussinov R, Zheng J (2012) Probing ion channel activity of human islet amyloid polypeptide (amylin). Biochim Biophys Acta 1818:3121–3130
88. Frishman D, Argos P (1995) Knowledge-based protein secondary structure assignment. Proteins 23:566–579
89. Wriggers W, Mehler E, Pitici F, Weinstein H, Schulten K (1998) Structure and dynamics of calmodulin in solution. Biophys J 74: 1622–1639
90. Allen TW, Andersen OS, Roux B (2006) Molecular dynamics—potential of mean force calculations as a tool for understanding ion permeation and selectivity in narrow channels. Biophys Chem 124:251–267
91. de Groot BL, Grubmuller H (2001) Water permeation across biological membranes: mechanism and dynamics of aquaporin-1 and GlpF. Science 294:2353–2357
92. Leontiadou H, Mark AE, Marrink SJ (2007) Ion transport across transmembrane pores. Biophys J 92:4209–4215
93. Allen TW, Andersen OS, Roux B (2004) Energetics of ion conduction through the gramicidin channel. Proc Natl Acad Sci U S A 101:117–122

# Chapter 17

# Analyzing Ensembles of Amyloid Proteins Using Bayesian Statistics

**Thomas Gurry, Charles K. Fisher, Molly Schmidt, and Collin M. Stultz**

## Abstract

Intrinsically disordered proteins (IDPs) are notoriously difficult to study experimentally because they rapidly interconvert between many dissimilar conformations during their biological lifetime, and therefore cannot be described by a single structure. The importance of studying these systems, however, is underscored by the fact that they form toxic aggregates that play a role in the pathogenesis of many disorders. The first step towards a comprehensive understanding of the aggregation mechanism of these proteins involves a description of their thermally accessible states under physiologic conditions. The resulting conformational ensembles correspond to coarse-grained descriptions of their energy landscapes, where the number of structures in the ensemble is related to the resolution in which one views the free energy surface. Here, we provide step-by-step instructions on how to use experimental data to construct a conformational ensemble for an IDP using a Variational Bayesian Weighting (VBW) algorithm. We further discuss how to leverage this Bayesian approach to identify statistically significant ensemble-wide observations that can form the basis of further experimental studies.

**Key words** Conformational ensemble, Intrinsically disordered proteins, Bayesian weighting, Variational Bayesian Method, Degeneracy

## 1 Introduction

Intrinsically disordered proteins (IDPs) sample a large number of structurally dissimilar conformations under physiologic conditions. While understanding the thermodynamics of these systems is interesting from the standpoint of protein folding in general, many of these proteins are particularly prone to aggregation into amyloids and are intimately related to the pathology of a number of human diseases [1]. Studies that improve our understanding of the thermally accessible states of these heteropolymers will likely provide insight into fundamental mechanisms underlying important human diseases.

A description of an IDP ensemble consists of a set of distinct, representative conformations, $S = \{s_i\}_{i=1}^{n}$, and their associ-

David Eliezer (ed.), *Protein Amyloid Aggregation: Methods and Protocols*, Methods in Molecular Biology, vol. 1345, DOI 10.1007/978-1-4939-2978-8_17, 

ated relative stabilities, $\vec{w} = (w_1, \ldots, w_n)$, where $wi$ is the probability that the protein adopts structure $si$. The number of structures, $n$, in the ensemble is related to the resolution in which one wishes to view that IDP's energy landscape. While it is unrealistic to expect to enumerate all of the states that an IDP can adopt in solution, relatively low resolution descriptions of their energy landscapes have proven useful in practice [2–4]. The final ensemble, $\{S, \vec{w}\}$, is usually constructed in a manner such that it agrees with existing experimental data—a notion we make more precise below.

In this work we demonstrate how to use a procedure called Variational Bayesian Weighting (VBW) to construct conformational ensembles using experimental data. In principle, experimental data that represent ensemble averages over the many different conformations of the IDP can be incorporated in the method. Indeed, most experiments on IDPs correspond to ensemble averages because the experimental time scale exceeds the timescale of IDP conformational transitions by several orders of magnitude. Experimental observables such as SAXS profiles, Chemical Shifts, Residual Dipolar Couplings (RDCs), and scalar J-couplings are typically employed, but in principle, many more experimental observables could be included.

The process of constructing a conformational ensemble from experimental measurements, $\vec{m}$, using VBW involves two parts: (a) generating a representative set of structures, $S$, which captures the dominant conformations that the IDP adopts in solution; (b) computing their associated weights, $\vec{w}$, using available experimental data (Fig. 1).

A major issue in the field of modeling IDPs is that the problem of constructing an IDP ensemble is inherently degenerate; i.e., one can construct many different ensembles that equally well agree with any given set of experimental constraints. This degeneracy arises from the fact that the number of experimental observables that are typically used to construct the model pales in comparison to the number of degrees of freedom in the system. Degeneracy is further exacerbated by errors in the experimental measurements themselves, and in the prediction of experimental observables. These considerations are highlighted by the fact that it is possible to construct ensembles that agree with experiment, which are clearly incorrect [5]. As a result, it is desirable to provide metrics that quantify one's uncertainty in the constructed ensemble. VBW provides a formalism that enables us to construct metrics that quantify our uncertainty in the ensemble.

The VBW algorithm returns a probability distribution for the weights, $\vec{w}$, rather than point estimates. This probability distribution provides quantitative estimates of the extent to which structures in $S$ can be weighted differently while retaining agreement

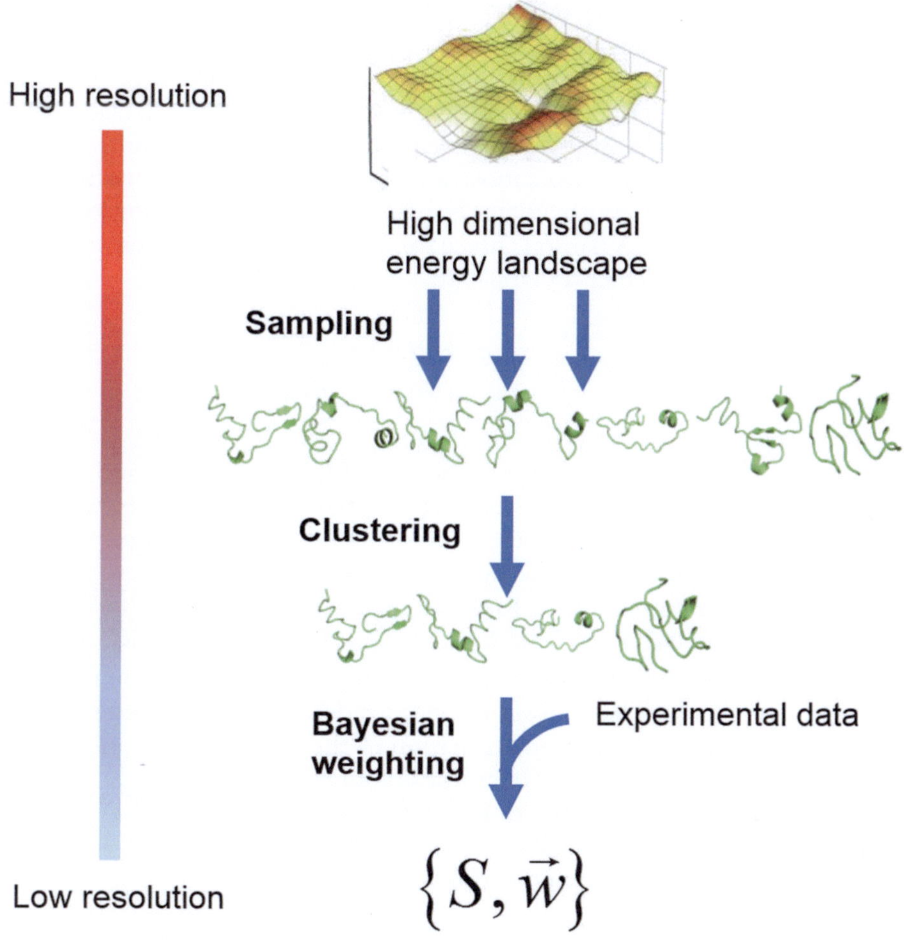


**Fig. 1** Schematic of the Bayesian Weighting procedure for constructing a conformational ensemble

with the experimental data, $\vec{m}$. This is known as the posterior distribution, $f_{\vec{W}|\vec{M},S}(\vec{w}\mid\vec{m},S)$, which is computed using Bayes' rule:

$$f_{\vec{W}|\vec{M},S}(\vec{w}\mid\vec{m},S)=\frac{f_{\vec{M}|\vec{W},S}(\vec{m}\mid\vec{w},S)\,f_{\vec{W}|S}(\vec{w}\mid S)}{f_{\vec{M}|S}(\vec{m}\mid S)} \tag{1}$$

where the term $f_{\vec{W}|S}(\vec{w}\mid S)$ is the prior distribution and $f_{\vec{M}|\vec{W},S}(\vec{m}\mid\vec{w},S)$ is the likelihood function for the experimental data, $\vec{m}$. Complete descriptions of these individual terms can be found in the original publication of the method [5]. Ensemble averages can be calculated using this form of the posterior distribution, and this is precisely how the original Bayesian Weighting method was formulated. However, calculating ensemble averages using the probability distribution cannot be done analytically. Monte Carlo simulations using this method, in general, converge very slowly, making the full Bayesian Weighting method impractical for many applications.

Here we describe how to use a variational approximation to the method that allows ensemble averages to be calculated analytically, thereby significantly increasing the computational efficiency [6]. The idea is to choose a simpler probability density function (PDF) which approximates the full posterior distribution calculated from Eq. 1. A natural choice for the approximate posterior distribution is a Dirichlet distribution:

$$g(\vec{w} \mid \vec{\alpha}, S) = \frac{\Gamma(\alpha_0)}{\sum_{i=1}^{n} \Gamma(\alpha_i)} \prod_{i=1}^{N} w_i^{\alpha_i - 1} \tag{2}$$

where $\{\alpha_i\}_{i=1}^{n}$ are the Dirichlet parameters and $\alpha_0 = \sum \alpha_i$. If $g(\vec{w} \mid \vec{\alpha}, S)$ is chosen appropriately, then each $\alpha_i$ will be proportional to the weight of structure $i$. The function $g(\vec{w} \mid \vec{\alpha}, S)$ or equivalently the appropriate Dirichlet parameters are chosen by minimizing the Kullback–Leibler distance (i.e., the KL divergence) between $g(\vec{w} \mid \vec{\alpha}, S)$ and $f_{\vec{W}|\vec{M},S}(\vec{w} \mid \vec{m}, S)$. (The KL divergence is a metric that quantifies how different two probability distributions are.) The final set of optimized Dirichlet parameters, $\vec{\alpha}' = \{\alpha_i'\}_{i=1}^{N}$, provides an approximation to the true posterior and allows one to easily calculate quantities of interest.

The Bayes estimate for the weights, $\vec{w}^B = \{w_i^B\}$, can then be computed as the expected value of the vector of weights over the approximate posterior distribution:

$$\vec{w}^B = \int d\vec{w} g(\vec{w} \mid \vec{\alpha}', S) \vec{w}. \tag{3}$$

Individual Bayes estimates for the structure weights can easily be calculated from the Dirichlet PDF according to:

$$w_i^B = \frac{\alpha_i'}{\alpha_0'}, \tag{4}$$

where $\alpha_0' = \sum_i \alpha_i'$. The procedure also returns an uncertainty parameter $0 \le \sigma_{\vec{w}^B} \le 1$, called the posterior expected divergence, which corresponds to the average distance from the Bayes weights over the entire space of weights:

$$\sigma_{\vec{w}_B} = \sqrt{\int d\vec{w} \Omega^2(\vec{w}^B, \vec{w}) g\left(\vec{w} \mid \hat{\alpha}', S\right)} \tag{5}$$

where $\Omega^2(\vec{w}_B, \vec{w})$ is the Jensen-Shannon divergence, a metric which quantifies the distance between vectors $\vec{w}_B$ and $\vec{w}$ [5]. Empirical studies suggest that when the uncertainty parameter is close to zero, it is likely that the resulting ensemble is correct.

Conversely, when it is close to one, there is a high level of uncertainty in the resulting model [5]. A strength of the method is that even in the case of high uncertainty, we can still compute observables from the ensemble and quantify our uncertainty with the calculated data by including confidence intervals.

## 2 Materials

The VBW algorithm computes a posterior distribution for the weights $g(\vec{w}\,|\,\vec{\alpha},S)$ using a set of experimental measurements $\vec{m}$ and structures $S$. Thus, the minimal inputs required to run the Variational Bayesian Weighting algorithm are an experimental data file that contains $\vec{m}$, and a set of representative structures $S$ that has previously been generated and clustered down to a heterogeneous set. The VBW package can be downloaded at http://www.rle.mit.edu/cbg/data.htm.

### 2.1 Experimental Data File

The experimental data file is a tab-delimited text file summarizing $\vec{m}$ that the user must create in the following format:

| # | Type | HA | H | N | CA | CB | C | RDC | J |
|---|---|---|---|---|---|---|---|---|---|
| 1 | ASP | 4.13 | 8.60 | NA | 52.5 | 40.65 | 176.49 | NA | NA |
| 2 | ALA | 4.50 | 9.39 | 123.6 | 52.81 | 19.32 | NA | NA | NA |
| 3 | GLU | 3.86 | 8.96 | 120.7 | 56.39 | 30.44 | NA | 1.46 | 6.43 |
| 4 | PHE | 5.40 | 8.71 | 122.1 | 57.53 | 39.44 | NA | 0.67 | 6.61 |
| … | … | … | … | … | … | … | … | … | … |
| 40 | VAL | 4.34 | 9.22 | 128.6 | 63.79 | 33.27 | NA | 3.45 | 8.87 |
| | | | | | | | | | |
| RG: | 12.1 | | | | | | | | |

Shown above is a truncated experimental data file constructed for the 40 residue Amyloid-β40 peptide. For each residue number (column 1), there is a column for each experimental measurement type. In this example, the measurement types (columns 3–10) are: chemical shifts for Hα (HA), backbone amide hydrogen (H), backbone amide nitrogen (N), Cα (CA), Cβ (CB), and backbone carbonyl carbon (C) nuclei (in ppm), as well as residual dipolar couplings (RDC) and J-couplings (J) (in Hz). Such published measurements were acquired from the publically available Biological Magnetic Resonance Databank (BMRB) [7]. In addition, radius of gyration measurements obtained from techniques such as SAXS can be specified at the end of the file on the line starting with "RG:". In cases where a particular measurement is not available or has no meaning (e.g., Cβ chemical shifts for Glycine), replace the measurement by "NA." The procedure can be extended to include new, experimental data of a quantitative nature, provided they can be predicted from a PDB file for a structure (*see* **Note 1**).

### 2.2 Obtaining a Set of Representative Structures

The user has several options for obtaining a representative set of conformations (in PDB format) to input to VBW. The steps involved in generating *S* may be separated into two separate tasks. The first involves sampling a sufficiently large set of distinct conformations, while the second involves clustering these structures into a smaller subset of structures that preserves the heterogeneity obtained in the first step. For the first task, one can use any method that generates a set of distinct conformations. A quick and easy method is a statistical coil model, such as the one made publically available by the University of Chicago (http://unfolded.uchicago.edu/index.html), which generates structures by sampling an empirical probability distribution for backbone dihedral angles [8], or the Flexible Meccano program (http://www.ibs.fr/science-213/scientific-output/software/flexible-meccano?lang=en), which can also include certain types of user-specified restraints [9]. An alternative approach involves sampling from a molecular mechanics potential energy function with a Monte Carlo approach or with molecular dynamics. To improve the sampling efficiency, one should employ one of the many available enhanced sampling methods such as Replica Exchange Molecular Dynamics (REMD).

An automated pipeline for generating a set of structures is available through our lab upon request. While this method is not part of the VBW package, it is complementary to it and will eventually be available via a web server. In this section, we only briefly outline some of its functionalities. More details of the various options available to the investigator are described in the accompanying documentation. The process of generating structures only requires an amino acid sequence as input (option "-i"):

```
python pipeline.py --setup -i DAEFRHDSGYEVHHQKLV
FFAEDVGSNKGAIIGLMVGGVV
```

The above command will create a model of the protein having the amino acid sequence above and then sample different conformations using the CHARMM force field [10]. To ensure that a variety of different structures are sampled, one can apply biasing potentials that enable the protein to sample conformations with varying degrees of structure. Indeed, although IDPs are, on average, lacking in secondary structure, the ensemble may contain conformations that have residual structure [11]. Sampling conformations that have residual structure is accomplished by applying biasing potentials similar to the one described in a previous study [12].

In this manner the system is restrained to adopt different contents of helix and strand, thus obtaining a set of conformations that may include regions of secondary structure that would otherwise be sampled infrequently in unrestrained simulations. These restrained simulations can be run simply by invoking:

```
python pipeline.py --simulations
```

Once a set of heterogeneous conformations has been obtained, it is desirable to reduce the set of structures in order to improve the computational efficiency of the weighting procedure, and to reduce the extent of the degeneracy problem, which increases with the number of degrees of freedom. However, modeling an IDP, which by definition contains a large amount of structural diversity, inherently requires a sizable set of structures, so there is a balance to be struck between these two competing interests. Clustering or pruning can be performed using any method available to the investigator, including hierarchical clustering using pairwise RMSDs, or some other measure of dissimilarity. The previously mentioned automated pipeline can also be used to perform the clustering step after the sampling is performed, by invoking:

```
python pipeline.py --cluster
```

This will by default compute backbone RMSDs between all pairs of structures in the initial structural library, and perform hierarchical clustering until each combination of fraction of helix and strand contains ten structures or less, resulting in a pruned structural library. Note that while generating a set of structures that spans a range of secondary structure contents is important when modeling an IDP system, the unfolded conformations with low secondary structure content can be particularly diverse and require additional sampling.

### 2.3 Required Packages

In order to run VBW, one requires Python version 2.6 or later, along with the NumPy and SciPy packages [13]. In addition, the following GNU libraries must be installed on the user's system: MPFR, GMP, MPC, GCC, and GSL.

## 3 Methods

### 3.1 Computing the Weights

First, unzip the VBW_Release_1.0 folder where you would like to use it. Assuming the user has obtained a set of PDBs and created an experimental data file in the correct format, the VBW procedure can be run, from start to finish, in a BASH shell with a single command:

```
python RunVBW.py -i path1/pdb_folder -d path2/
experimental_data_file.txt   -o   path3/output_
directory
```

The RunVBW.py script accepts several options, including:

| Option flag | Description |
|---|---|
| -i | Path to directory containing the set of representative structures in PDB format |
| -o | Path to desired output directory |
| -d | Path to experimental data file |
| -f | Force program to run if the output directory already exists |
| -m | Alignment medium used to measure Residual Dipolar Couplings experimentally (bic/pf1, where "bic" corresponds to bicelles and "pf1" corresponds to Pf1 bacteriophage; default: bic) |
| -c | Alignment medium concentration (default: 0.05M) |
| -k | Karplus constants for J-couplings (defaults: A=9.5 Hz/degree$^2$, B=-1.4 Hz/degree, C=0.3 Hz) |
| -s | Flag for performing a backward elimination procedure which removes nonessential structures with very low weights (default: no structure selection; for more details about this procedure, see the description in the text) |

Only the first three options must be specified for the program to run. The remaining options have default values that are adopted by the program if the user does not specify them. Internally, RunVBW.py proceeds by calling the CreateVBWFiles.py, which creates all the necessary files from the user-defined inputs. It will calculate predicted experimental data from the set of PDBs using the CalcMeas.py script, which calls the relevant programs, including ShiftX for chemical shifts [14] and PALES for residual dipolar couplings [15]. At present, there is no option to use ShiftX2 [16] because our initial work suggested that the use of SHIFTX2 did not improve our results. Future iterations of the code, however, will allow the user to specify what algorithm they wish to use to compute various experimental observables from structural data.

Once the CalcMeas.py script is finished, the master program (RunVBW.py) will call the VBW_parallel executable, which computes the posterior distribution in Eq. 2. If the flag "-s" is passed as an option to RunVBW.py, in addition to computing the posterior distribution, the VBW_parallel will use Bayesian variable selection techniques to perform an additional backward elimination procedure on the set of structures, to ensure that the ensemble excludes *nonessential* structures. Formally speaking, the set of nonessential structures is the largest set of structures such that the probability that each structure's weight in $I$ is below a cutoff $c$ exceeds a chosen confidence level $\theta$; i.e., using a confidence level of $\theta$, we can say that the weight of every structure in $I$ is below $c$. The cutoff ($c$) and confidence level ($\theta$) are set by default to 0.005

and 0.05, respectively. This procedure is repeated iteratively until convergence, where nonessential structures are eliminated when identified. When the set $I$ is empty (there are no nonessential structures), the posterior distribution and pruned set of structures are returned.

### 3.2 Analyzing the Ensemble

Once Run_VBW.py has completed, the user-specified output directory will contain three subdirectories (called "data," "pruned," and "structures"). In addition, it will contain a dated, compressed directory containing the data specifying the ensemble. This compressed directory contains all the information required to analyze the resulting ensemble and can be moved to a separate location for analysis if required. It contains the subset of structures obtained from the backwards elimination procedure in PDB format, which specify our final set of conformations $S$, and a predicted experimental data file for each of these conformations. In addition, it contains two summary files of interest: "ensemble_summary.txt" and "ensemble_fit_statistics.dat". "ensemble_summary.txt" contains a table in which the $i$th row contains the name of structure $i$, the Bayes estimate for its weight $wiB$ and the weight's associated Dirichlet parameter $\alpha i'$, respectively:

```
Name Weight Dirichlet
Structure_1 0.0028 1.46
Structure_2 0.0014 0.716
Structure_3 0.0018 0.908
Structure 4 0.0013 0.643
...
Structure N 0.0032 2.21
```

As discussed in the Introduction, the posterior distribution for the weights of each conformation, $g(\vec{w}\,|\,\vec{\alpha},S)$, can be obtained from the vector of Dirichlet parameters, $\vec{\alpha}$. Thus, the "ensemble_summary.txt" file contains the full description of the conformational ensemble $\{S,\vec{w}\}$. In addition to the ensemble description, the "ensemble_summary.txt" file enumerates chemical shift offsets and RDC scaling factors for the data provided, as well as reporting the ensemble average radius of gyration and the covariance matrix for the weights (shown truncated):

```
Scaling Factors: 0.792
Chemical Shift Offsets: 0 0 0.118 -2.38
Ensemble Average Radius of Gyration [mean, standard deviation]: 35.124 0.403
Covariance matrix:
5.51930367044e-06-7.72399810807e-09-9.7952378242e-09 -6.93649550766e-09 -1.37003877056e-08 (...)
```

The "ensemble_fit_statistics.dat" file contains summary statistics describing the degree of agreement between the obtained conformational ensemble and the experimental data provided, including root mean squared errors (RMSEs), Pearson and Spearman Correlation coefficients (*see* **Note 2**). The RMSEs can be used to compute measures of goodness-of-fit, such as reduced chi-squared statistics. In addition, the file contains the uncertainty parameter described in Eq. 5, $\sigma_{\vec{w}^B}$, which takes a value between 0 and 1, where a value of 0 corresponds to a situation of total certainty in the Bayes estimate for the weights $\vec{w}^B$ for this particular given set of structures $S$ (i.e., there is only one way to weight these structures to fit the experimental data), while a value of 1 corresponds to total uncertainty in $\vec{w}^B$ and is indicative of a very high level of degeneracy in the experimental data with respect to the set of set $S$. These data are presented in the following form:

```
type RMSE Correlation Spearman
HA 1.002 0.986 0.977
H 1.20 0.965 0.952
N 1.735 0.946 0.871
CA 1.118 0.982 0.943
CB 1.807 0.993 0.896
C 1.201 0.982 0.958
RDC 0.884 0.988 0.988
J 4.32 0.952 0.937
Uncertainty parameter: 0.46816
```

The ensemble average of any quantity $D$ in the ensemble can simply be computed according to

$$\langle D \rangle = \sum_{i=1}^{N} w_i^B D_i$$

where $D_i$ represents the quantity $D$ for structure $i$. The covariance between the weights of conformations $i$ and $j$ can also be calculated analytically from the Dirichlet distribution from

$$\operatorname{cov}(w_i, w_j) = \frac{\alpha_i' \alpha_0' \delta_{ij} - \alpha_i' \alpha_j'}{\alpha_0'^2 (\alpha_0' + 1)}$$

where $\delta_{ij}$ is the Kronecker delta function. This allows us to also compute the variance of any quantity $D$ in our ensemble according to:

$$\operatorname{var}(D) = \sum_i \sum_j D_i D_j \operatorname{cov}(w_i, w_j)$$

One can then construct confidence intervals for our ensemble averages $\langle D \rangle$ using a Gaussian approximation. For example, a 95% confidence interval can be constructed from $CI_{95\%} = 1.54 \times 1.96 \times (\text{var}(D))^{1/2}$, where 1.54 is an empirical factor relating the variational approximation of the posterior distribution to the true posterior distribution under the complete, non-variational BW formalism, and 1.96 specifies the number of standard deviations in a Gaussian distribution to capture 95% of the probability distribution. This allows the user to report the uncertainty in their ensemble average estimates by placing error bounds on predicted quantities of interest from the resulting ensemble, as well as perform statistical hypothesis tests. For example, if one were reporting the ensemble average fraction of helical content, *H*, for an ensemble, one would first measure the helical contents of each conformation in the ensemble, and compute the weighted average of these quantities:

$$\langle H \rangle = \sum_{i=1}^{N} w_i^B H_i$$

The ensemble variance in helical content can then be calculated according to:

$$\text{var}(H) = \sum_i \sum_j H_i H_j \,\text{cov}(w_i, w_j)$$

Having obtained the variance, one can compute a 95% confidence interval from the aforementioned Gaussian approximation, $\varepsilon = 1.54 \times 1.96 \times (\text{var}(H))^{1/2}$. The user can then report the ensemble average helical content as $\langle H \rangle \pm \varepsilon$.

## 4 Notes

1. Any new measurement type that an investigator wishes to include in the procedure will have to be accounted for in the CalcMeas.py script, by adding a new method that calculates predicted measurements from a PDB file in a similar manner to the other such methods. In addition, the format of the experimental data file has to be adjusted to include the new measurements, and equivalent changes have to be made to CreateVBWFiles.py and CalcMeas.py. Extending the VBW algorithm to include additional experimental measurement types is straightforward when these quantities can be predicted from a PDB file with ease, such as PREs or NOEs.
2. The VBW algorithm takes experimental error into account when computing the posterior distribution for the weights. Estimates of these errors are included for chemical shifts, RDCs, and J-couplings in the CreateVBWFiles.py script that gets called by RunVBW.py, and errors of new experimental measurement types that are included in the procedure can be added there.

## References


1. Uversky VN, Oldfield CJ, Dunker AK (2008) Intrinsically disordered proteins in human diseases: introducing the D2 concept. Annu Rev Biophys 37(1):215–246
2. Fisher Charles K, Ullman O, Stultz Collin M (2013) Comparative studies of disordered proteins with similar sequences: application to Aβ40 and Aβ42. Biophys J 104(7):1546–1555
3. Salmon L et al (2010) NMR characterization of long-range order in intrinsically disordered proteins. J Am Chem Soc 132(24):8407–8418
4. Mittag T et al (2010) Structure/function implications in a dynamic complex of the intrinsically disordered Sic1 with the Cdc4 subunit of an SCF ubiquitin ligase. Structure 18(4): 494–506
5. Fisher CK, Huang A, Stultz CM (2010) Modeling intrinsically disordered proteins with Bayesian statistics. J Am Chem Soc 132(42): 14919–14927
6. Fisher CK, Ullman O, Stultz CM (2012) Efficient construction of disordered protein ensembles in a Bayesian framework with optimal selection of conformations. Pac Symp Biocomput 17:82–93
7. Ulrich EL et al (2008) BioMagResBank. Nucleic Acids Res 36(suppl 1):D402–D408
8. Jha AK, Colubri A, Freed KF, Sosnick TR (2005) Statistical coil model of the unfolded state: resolving the reconciliation problem. Proc Natl Acad Sci U S A 102(37):13099–13104
9. Ozenne V et al (2012) Flexible-meccano: a tool for the generation of explicit ensemble descriptions of intrinsically disordered proteins and their associated experimental observables. Bioinformatics 28(11):1463–1470
10. Brooks BR et al (2009) CHARMM: the biomolecular simulation program. J Comput Chem 30(10):1545–1614
11. Fisher CK, Stultz CM (2011) Constructing ensembles for intrinsically disordered proteins. Curr Opin Struct Biol 21(3):426–431
12. Vitalis A, Lyle N, Pappu RV (2009) Thermodynamics of β-sheet formation in polyglutamine. Biophys J 97(1):303–311
13. Svd W, Colbert SC, Varoquaux G (2011) The NumPy array: a structure for efficient numerical computation. Comput Sci Eng 13(2): 22–30
14. Neal S, Nip A, Zhang H, Wishart D (2003) Rapid and accurate calculation of protein 1H, 13C and 15N chemical shifts. J Biomol NMR 26(3):215–240
15. Zweckstetter M, Bax A (2000) Prediction of sterically induced alignment in a dilute liquid crystalline phase: aid to protein structure determination by NMR. J Am Chem Soc 122(15): 3791–3792
16. Han B, Liu Y, Ginzinger S, Wishart D (2011) SHIFTX2: significantly improved protein chemical shift prediction. J Biomol NMR 50(1):43–57

# Part VI

## Toxicity and Pathology

# Chapter 18

# In Vitro Studies of Membrane Permeability Induced by Amyloidogenic Polypeptides Using Large Unilamellar Vesicles

**Ping Cao and Daniel P. Raleigh**

## Abstract

The process of amyloid formation is cytotoxic and contributes to a wide range of human diseases, but the mechanisms of amyloid-induced cytotoxicity are not well understood. It has been proposed that amyloidogenic peptides exert their toxic effects by damaging membranes. Membrane disruption is clearly not the only mechanism of toxicity, but the literature suggests that loss of membrane integrity may be a contributing factor. In this chapter we describe the measurement of in vitro membrane leakage induced by amyloidogenic proteins via the use of model vesicles. We use islet amyloid polypeptide (IAPP, amylin) as an example, but the methods are general.

**Key words:** Amyloid, Membrane disruption, Cytotoxicity, Islet amyloid polypeptide, Membrane leakage

## 1 Introduction

"Amyloidoses" are protein-misfolding diseases that are caused by the transformation of normally soluble proteins or polypeptides into partially ordered insoluble amyloid fibrils. More than 30 different proteins or polypeptides form amyloid deposits that are associated with human disorders, including neurodegenerative diseases such as Alzheimer's disease and Parkinson's disease, and metabolic diseases such as type 2 diabetes [1–4]. The mechanisms of cytotoxicity are not well understood and multiple mechanisms are likely operative in vivo, but the literature suggests that the loss of membrane integrity may contribute to toxicity [5–8]. This has motivated studies of membrane disruption by amyloidogenic proteins, work which builds on the broader literature on membrane active peptides and proteins. This chapter describes the methodology used to characterize membrane leakage induced by amyloid formation in vitro. Islet amyloid polypeptide (IAPP, also known as amylin) is used as an

David Eliezer (ed.), *Protein Amyloid Aggregation: Methods and Protocols*, Methods in Molecular Biology, vol. 1345, DOI 10.1007/978-1-4939-2978-8_18, © Springer Science+Business Media New York 2016

example, but the methods are general. IAPP is a neuropancreatic hormone that plays a role in regulating energy metabolism. The 37-residue polypeptide is stored in the β-cell secretory granules, and is secreted with insulin [9–11]. The polypeptide aggregates by an unknown mechanism in type 2 diabetes and is responsible for pancreatic islet amyloid in the disease. Amyloid formation by human IAPP (hIAPP) is toxic to islet β-cells, induces β-cell dysfunction in type 2 diabetes, and plays a significant role in the failure of islet transplants [12–14]. A wide range of mechanisms of hIAPP-induced cytotoxicity have been proposed, including receptor-mediated mechanisms, permeabilization of the plasma and mitochondria membranes, ER stress, defects in the unfolded protein response, and defects in autophagy [4, 15].

hIAPP is a hydrophobic polypeptide and is cationic at physiological pH and, as expected, interacts with anionic membranes. Interactions of hIAPP with model membranes containing a significant portion of anionic lipids such as phosphatidylglycerol (PG) or phosphatidylserine (PS) have been widely studied. Anionic lipid vesicles, supported bilayers and monolayers accelerate amyloid formation by hIAPP, with larger effects being observed for higher percentages of anionic lipids. hIAPP promotes membrane leakage in these systems [16–19]. However it is important to note that non-cytotoxic variants of IAPP can also efficiently promote leakage of model membranes, making the connection between reductionist in vitro studies with simplified model membranes and the situation in vivo ambiguous [20]. Here we describe methods employed to examine the ability of hIAPP to induce leakage of anionic model membrane systems consisting of large unilamellar vesicles (LUVs) made of mixtures of PG with the zwitterionic lipid phosphocholine (PC). The mole percent of anionic lipid typically ranges from 50 to 20 % in the most common model membrane systems used for studies of IAPP membrane interactions [16, 17]. In the described protocol, we use a 25 mol% anionic model membrane system as an example, but the methods are not limited to a specific composition and can be applied to other stable vesicles.

## 2 Materials

Deionized water and analytical grade reagents are used. Appropriate waste disposal regulations should be followed when disposing of waste materials and appropriate personal protective equipments (including goggles) should be worn and all MSDS data sheets should be carefully checked before using any reagents or solvents.

1. hIAPP is typically prepared by solid-phase peptide synthesis since the peptide is toxic to many cell lines, prone to aggregate and the C-terminus is amidated. Ongoing efforts in a number of laboratories are aimed at developing improved expression

systems for IAPP. The molecule can be synthesized using either (*tert*-Butyl carbamate) t-Boc or 9-fluorenylmethoxycarbonyl (Fmoc) chemistry (*see* **Note 1**). The Alzheimer's Aβ peptide is also often prepared by solid-phase peptide synthesis, although recombinant methodologies are also used. hIAPP is purified by reverse-phase HPLC using a C18 preparative column (*see* **Note 2**). The identity of the pure peptide should be confirmed by mass spectrometry. IAPP and some other amyloidogenic peptides can undergo spontaneous deamidation in which Asn residues are transformed into mixtures of L-Asp, D-Asp, L-sio-Asp, and D-iso-Asp [21]. Thus, it is important to check the integrity of the polypeptide before commencing experiments. The peptide is best stored as a dry powder at −20 °C.

2. Lipid stocks: For the example described here, 1,2-dioleoyl-sn-glycero-3-phosphocholine (DOPC), and 1,2-dioleoyl–sn–glycerol-3phospho-(1'-rac-glycerol) (DOPG) were obtained from Avanti Polar Lipids and used without further purification. Stock solutions of lipids are prepared in chloroform and stored at −80 °C (*see* **Notes 3** and **4**).
3. Sample buffer: 20 mM Tris–HCl, pH 7.4, 100 mM NaCl.
4. Carboxyfluorescein buffer: 70 mM carboxyfluorescein, 20 mM Tris–HCl, pH 7.4, 100 mM NaCl.
5. PD-10 columns (GE Healthcare Life Sciences).

## 3 Methods

All procedures should be carried out at room temperature unless otherwise specified.

### 3.1 Protein Sample Preparation

Handling amyloidogenic proteins can be a challenge and conflicting reports in the literature on the biophysical and cytotoxic properties of these molecules result, in part, from differences in the protocols used to solubilize the protein of interest. A range of methods have been developed to prepare amyloidogenic polypeptides and proteins in initially monomeric states. Note that in many cases, low-order oligomers are detected essentially as soon as the polypeptide is dissolved; thus it can be difficult to be certain that one is starting an experiment from a monomeric state. The details of the methods used for preparing the samples are specific to the protein of interest. The example described below is appropriate for IAPP and the reader is referred to the literature for protocols employed for other proteins.

1. Dried hIAPP is dissolved in 100 % hexafluoroisopropanol (HFIP) to prepare stock solutions and incubated for at least 12 h. The stock solution is filtered through a 0.22 μm filter (*see* **Note 5**).

2. Aliquots of hIAPP stock solutions are freeze dried using a lyophilizer to remove organic solvents. The strongest possible vacuum should be employed and samples should be dried at least overnight since trace amounts of residual organic solvents can influence the properties of the peptide.
3. After overnight lyophilization, the peptide is redissolved in sample buffer at the desired concentration (*see* **Note 6**). Other buffers may be used, but the process of amyloid formation by IAPP is strongly pH dependent and is significantly faster when the side chain of His-18 and the N-terminus are neutral. The rate of hIAPP amyloid formation is also strongly dependent on ionic strength and the nature of the anion.

### 3.2 Preparation of LUVs

The model membrane system used in this example contains 25 % anionic lipids by mole percent (*see* **Note 7**).

1. Lipid stock solutions of DOPC and DOPG are transferred into a round-bottom glass flask at a 3:1 molar ratio (*see* **Note 8**). The organic solvent is evaporated first using a stream of nitrogen gas to form a film at the bottom of the flask (it is recommended to use the highest purity nitrogen and to employ an oil free regulator), and then further dried under a vacuum overnight in order to completely remove residual organic solvent.
2. The resulting lipid film is dissolved in sample buffer and agitated for one hour (stirring or mild shaking) (*see* **Note 9**).
3. After hydration, the lipid suspension is subjected to 10 freeze-thaw cycles and then extruded 15 times through 100 nm pore size filters (Whatman, GE) (*see* **Notes 10** and **11**).
4. The phospholipid concentration of the resulting LUVs can be determined using the method of Stewart [22].
5. Fluorescent vesicles are used for the membrane disruption assays. Fluorescent LUVs incorporating the dye carboxyfluorescein are made using the same protocol described above, except that the dried lipid film is rehydrated with carboxyfluorescein buffer (*see* **Note 12**). Carboxyfluorescein is a relatively small molecule and other larger probes have been developed. Comparative studies can be performed using a range of different sized probes if desired.
6. Nonencapsulated carboxyfluorescein needs to be removed from the carboxyfluorescein-filled vesicles and can be done so using size-exclusion chromatography with a PD-10 column and elution with sample buffer.

### 3.3 Membrane Permeability Measurements

A fluorescence spectrophotometer is used for the membrane leakage assay in the example presented here. A plate reader can also be used; however care must be used to avoid plates that can either disrupt model membranes or can bind IAPP.

1. The peptide solution is added to the concentrated carboxyfluorescein-filled LUVs to a final desired peptide to lipid ratio. The cuvette should be gently shaken for 3 s immediately after mixing. Fluorescence is measured using an excitation wavelength of 492 nm and an emission wavelength of 517 nm. A typical slit width used on the specific instrument described in this example is 1.5 nm. Time dependent studies can be performed in which the leakage is monitored by recording the carboxyfluorescein fluorescence as a function of time after addition of the peptide to the vesicles. The leakage assays should be repeated to obtain reliable estimates of the uncertainty. At least three repeats are recommended, preferably using different peptide stock solutions, to obtain mean values and apparent standard deviations.
2. For each experiment, the baseline fluorescence (Fbaseline) of the carboxyfluorescein-filled LUVs should be measured. The maximum leakage induced by total disruption of the lipid vesicles (Fmax) is determined by the addition of Triton X-100 to a final concentration of 0.2 %. The percent leakage of the dye is calculated as

$$\text{Percentage leakage} = 100 \times (\text{Ft} - \text{Fbaseline})/(\text{Fmax} - \text{Fbaseline})$$

where Ft is the measured carboxyfluorescein fluorescence.

## 4 Notes

1. The IAPP samples used in this example are prepared using Fmoc chemistry and Fmoc-protected pseudoproline dipeptide derivatives are incorporated to facilitate the synthesis and prevent on-resin aggregation. The disulfide bond in IAPP between residues Cys-2 and Cys-7 is formed via oxidation by DMSO in the present example [23].
2. HCl should be used as the ion-pairing agent instead of TFA during HPLC purification of IAPP since TFA can cause problems with cell toxicity assays and it has been shown that TFA influences the aggregation kinetics of some IAPP-derived peptides [24].
3. To prevent potential decomposition of the lipids, they should be stored in the dry form at –80 °C in case long-term storage is required.
4. Glassware, not plasticware, should be used when handling organic solvents such as chloroform.
5. Filtration is required to remove any IAPP pre-fibrillar materials. This is an important step for preparing IAPP samples.

6. The concentration of the peptide buffer solution should be determined to check for any loss during filtration. The concentration can be estimated by measuring the UV absorbance at 280 nm. hIAPP contains one Tyr and 2 Phe residues, but no Trp; thus the extinction coefficient at 280 nm is dominated by the absorbance of the Tyr. A precise extinction coefficient has not been reported for hIAPP, but there will be only a small uncertainty induced by using the standard value for a single Tyr. Peptide concentration can also be determined by quantitative amino acid analysis or by using the Bradford assay [25, 26].
7. The lipid composition can be altered from what described here, but the general preparation method can be used for other symmetric lipid vesicles. Here we use a 25 % anionic membrane system as an example in order to describe the procedures. More complicated lipid mixtures, including ones containing cholesterol, can be prepared and methods are emerging for the preparation of asymmetric LUVs to better mimic the plasma membrane [27].
8. The lipids must be mixed thoroughly to obtain a homogeneous solution when preparing membranes with a mixed lipid composition.
9. There is a gel-liquid crystal transition temperature (Tc or Tm) for each lipid. For the hydration step, the lipid suspension needs to be maintained at a temperature above the highest Tc of any of the mixed lipids.
10. Extrusion is normally used to form large unilamellar vesicles, while small unilamellar vesicles with diameters between 15 and 50 nm are usually prepared by sonication. The pore size of the filter used depends on the desired size of the lipid vesicles (typically in the range of 200–1000 nm for LUVs). Extrusion needs to be performed at a temperature above the Tc of the mixed lipids.
11. LUVs can be stable for up to several days after preparation, however, it is recommended to use freshly prepared vesicles. The uniformity of the lipid vesicles can be checked by light scattering and/or by cryo electron microscopy.
12. Carboxyfluorescein is a fluorescent dye with an excitation at 492 nm and emission of 517 nm. It is commonly used as a probe for membrane permeability. The fluorescence of the dye is self-quenched when it is encapsulated owing to the high local concentration in the interior of the vesicle. Permeabilization of the membrane allows the dye to escape, thereby lowering the concentration and relieving self-quenching. The pH of the buffer may be decreased after dissolving the compound, thus one should be sure to readjust the buffer pH to 7.4 if needed.

## Acknowledgements

This work was supported by NIH grant GM078114 (D.P.R.).

### References


1. Selkoe DJ (2004) Cell biology of protein misfolding: the examples of Alzheimer's and Parkinson's diseases. Nat Cell Biol 6: 1054–1061
2. Chiti F, Dobson CM (2006) Protein misfolding, functional amyloid, and human disease. Annu Rev Biochem 75:333–366
3. Sipe JD (1994) Amyloidosis. Crit Rev Clin Lab Sci 31:325–354
4. Westermark P, Andersson A, Westermark GT (2011) Islet amyloid polypeptide, islet amyloid, and diabetes mellitus. Physiol Rev 91: 795–826
5. Janson J, Ashley RH, Harrison D, McIntyre S, Butler PC (1999) The mechanism of islet amyloid polypeptide toxicity is membrane disruption by intermediate-sized toxic amyloid particles. Diabetes 48:491–498
6. Brender JR, Salamekh S, Ramamoorthy A (2012) Membrane disruption and early events in the aggregation of the diabetes related peptide IAPP from a molecular perspective. Acc Chem Res 45:454–462
7. de Planque MRR, Raussens V, Contera SA, Rijkers DTS, Liskamp RMJ, Ruysschaert JM, Ryan JF, Separovic F, Watts A (2007) Beta-sheet structured beta-amyloid(1-40) perturbs phosphatidylcholine model membranes. J Mol Biol 368:982–997
8. Beyer K (2007) Mechanistic aspects of Parkinson's disease: alpha-synuclein and the biomembrane. Cell Biochem Biophys 47:285–299
9. Lukinius A, Wilander E, Westermark GT, Engstrom U, Westermark P (1989) Co-localization of islet amyloid polypeptide and insulin in the b-cell secretory granules of the human pancreatic-islets. Diabetologia 32: 240–244
10. Kahn SE, Dalessio DA, Schwartz MW, Fujimoto WY, Ensinck JW, Taborsky GJ, Porte D (1990) Evidence of cosecretion of islet amyloid polypeptide and insulin by beta-cells. Diabetes 39:634–638
11. Stridsberg M, Sandler S, Wilander E (1993) Cosecretion of islet amyloid polypeptide (IAPP) and insulin from isolated rat pancreatic-islets following stimulation or inhibition of beta-cell Function. Regul Pept 45:363–370
12. Clark A, Wells CA, Buley ID, Cruickshank JK, Vanhegan RI, Matthews DR, Cooper GJS, Holman RR, Turner RC (1988) Islet amyloid, increased alpha-cells, reduced beta-cells and exocrine fibrosis—quantitative changes in the pancreas in type-2 diabetes. Diabetes Res Clin Ex 9:151–159
13. Lorenzo A, Razzaboni B, Weir GC, Yankner BA (1994) Pancreatic-islet cell toxicity of amylin associated with type-2 diabetes-mellitus. Nature 368:756–760
14. Konarkowska B, Aitken JF, Kistler J, Zhang S, Cooper GJ (2006) The aggregation potential of human amylin determines its cytotoxicity towards islet beta-cells. FEBS J 273: 3614–3624
15. Cao P, Marek P, Noor H, Patsalo V, Tu LH, Wang H, Abedini A, Raleigh DP (2013) Islet amyloid: from fundamental biophysics to mechanisms of cytotoxicity. FEBS Lett 587: 1106–1118
16. Knight JD, Hebda JA, Miranker AD (2006) Conserved and cooperative assembly of membrane-bound alpha-helical states of islet amyloid polypeptide. Biochemistry 45: 9496–9508
17. Hebda JA, Miranker AD (2009) The interplay of catalysis and toxicity by amyloid intermediates on lipid bilayers: insights from type II diabetes. Annu Rev Biophys 38:125–152
18. Jayasinghe SA, Langen R (2007) Membrane interaction of islet amyloid polypeptide. Biochim Biophys Acta Biomembr 1768: 2002–2009
19. Engel MFM (2009) Membrane permeabilization by islet amyloid polypeptide. Chem Phys Lipids 160:1–10
20. Cao P, Abedini A, Wang H, Tu LH, Zhang XX, Schmidt AM, Raleigh DP (2013) Islet amyloid polypeptide toxicity and membrane interactions. Proc Natl Acad Sci U S A 110: 19279–19284
21. Dunkelberger EB, Buchanan LE, Marek P, Cao P, Raleigh DP, Zanni MT (2012) Deamidation accelerates amyloid formation and alters amylin fiber structure. J Am Chem Soc 134:12658–12667
22. Stewart JCM (1980) Colorimetric determination of phospholipids with ammonium ferrothiocyanate. Anal Biochem 104:10–14
23. Abedini A, Singh G, Raleigh DP (2006) Recovery and purification of highly aggregation-prone disulfide-containing peptides: application

to islet amyloid polypeptide. Anal Biochem 351: 181–186

24. Nilsson MR, Raleigh DP (1999) Analysis of amylin cleavage products provides new insights into the amyloidogenic region of human amylin. J Mol Biol 294:1375–1385
25. Bradford MM (1976) A rapid and sensitive method for the quantitation of microgram quantities of protein utilizing the principle of protein-dye binding. Anal Biochem 72: 248–254
26. Zor T, Selinger Z (1996) Linearization of the Bradford protein assay increases its sensitivity: theoretical and experimental studies. Anal Biochem 236:302–308
27. Lin Q, London E (2014) Preparation of artificial plasma membrane mimicking vesicles with lipid asymmetry. PLoS One 9:e87903

# Chapter 19

# Cell Models to Study Cell-to-Cell Transmission of α-Synuclein


**Eun-Jin Bae, He-Jin Lee, and Seung-Jae Lee**



## Abstract

The cell-to-cell transmission of protein aggregates has been implicated in the progression of pathological phenotypes in neurodegenerative disorders such as Alzheimer's disease, Parkinson's disease, and amyotrophic lateral sclerosis. In recent years, several experimental model systems have been developed to study the mechanisms of cell-to-cell transmission. Herein, we describe cell culture models with which cell-to-cell transmission of α-synuclein can be quantitatively analyzed. The principle underlying these models could be applied to developing model systems for transmission of other protein aggregates, such as tau and TDP-43.



**Key words** Neurodegenerative diseases, Parkinson's disease, α-Synuclein, Protein aggregation, Transcellular transmission


## 1 Introduction

Deposition of specific protein aggregates is a common feature of the major neurodegenerative disorders, such as Alzheimer's disease (AD) and Parkinson's disease (PD) [1]. Protein aggregate pathologies spread progressively from small regions of brain to larger brain areas, as disease progresses. The patterns of protein aggregate spreading appear to partly, if not perfectly, explain the symptomatic progression patterns of specific diseases. Recent studies suggested that cell-to-cell transmission of protein aggregates might be the underlying mechanism for pathological propagation and hence disease progression [2, 3].

PD is characterized by the presence of α-synuclein aggregates in Lewy bodies and Lewy neurites. α-synuclein is a cytosolic protein, highly expressed at the presynaptic terminals in the central nervous system (CNS). α-Synuclein is composed of 140 amino acid including three regions: amphipathic N-terminus, central hydrophobic region, known as a non-amyloid β


David Eliezer (ed.), *Protein Amyloid Aggregation: Methods and Protocols*, Methods in Molecular Biology, vol. 1345, DOI 10.1007/978-1-4939-2978-8_19, © Springer Science+Business Media New York 2016

component (NAC), and C-terminus acidic tail. Aggregated forms of α-synuclein are the major components of hallmark pathological inclusions in a group of diseases, known as "synucleinopathies," including PD, dementia with Lewy bodies (DLB), and multiple system atrophy (MSA) [4, 5]. Braak and colleagues suggested that in PD, α-synuclein aggregates first appear in the lower brainstem regions and olfactory bulb in the CNS and sequentially spread to other regions as disease progresses [6]. We and others showed that α-synuclein aggregates are transmitted between neuronal cells [7–10], and the transmission event involves exocytosis and the subsequent endocytosis of the aggregates [11, 12].

Understanding the mechanism of cell-to-cell transmission of α-synuclein might unveil therapeutic targets for stopping or slowing the progression of PD and other synucleinopathies. Herein, we describe cell-based methods to quantitatively analyze cell-to-cell the transmission of α-synuclein aggregates.

## 2 Materials

### 2.1 Cell-to-Cell Transfer of α-Synuclein in SH-SY5Y Co-culture System

#### 2.1.1 Materials for Cell Culture

1. DMEM/High glucose (Hyclone, Logan, UT).
2. Characterized Fetal Bovine Serum, Canadian Origin (Hyclone).
3. 1000X Penicillin-Streptomycin 10,000 U/mL (Invitrogen, Carlsbad, CA).
4. All-*trans*-retinoic acid (100 mM dissolved in DMSO) (Sigma Aldrich, St Louis, MO).
5. Growth media: DMEM/high glucose containing 10 % characterized fetal bovine serum and 1X penicillin-streptomycin.
6. Growth media without antibiotic: DMEM/high glucose containing 10 % characterized fetal bovine serum.
7. Growth media with retinoic acid: 50 μM all-*trans*-retinoic acid in growth media with antibiotic:
8. 12-Well tissue culture plates (Nunc, Rochester, NY).
9. Glass cover slips (Fisher Scientific, Rochester, NY).
10. Poly-L-lysine: 0.1 mg/mL in distilled water (DW) (Sigma Aldrich).
11. Qtracker Cell Labeling Kit (Invitrogen Q25011MP).
12. Qtracker labeling solution: Premix 1 μL each of Qtracker Kit component A and component B in a 1.5 mL microcentrifuge tube. Incubate for 5 min at room temperature protected from light.

#### 2.1.2 Buffers and Reagents for Immunofluorescence Staining

1. 4 % Paraformaldehyde (Sigma) solution: Dissolved in PBS.
2. Permeabilization solution: 0.1 % triton X-100 in PBS.

3. Blocking solution: 5 % Bovine Serum Albumin (Sigma) prepared in PBS.
4. Diluted Topro-3 iodide: Dilute TOPRO-3 iodide (Invitrogen) in PBS (1:1,000 dilution).
5. Prolong gold anti-fade reagent (Invitrogen).

*2.1.3 Antibodies for Immunofluorescence Staining*

1. α-Synuclein monoclonal antibody Ab274 [13].
2. α-Synuclein polyclonal antibody (Cell Signaling Technology; #2642, Beverly, MA).
3. GFP N-terminus polyclonal antibody (Cell Signal Technology; #2555).
4. GFP C-terminus polyclonal antibody (Santa Cruz Biotechnology; sc-5384, Santa Cruz, CA).
5. (Cy2, Rhodamine Red X, etc.)-fluorescence-conjugated secondary antibodies (Jackson Immunoresearch Laboratories, Inc, West Grove, PA).

*2.1.4 Materials for Adenovirus Preparation*

1. pacAd5 CMV K-N pA shuttle vector (CELL BIOLABS, INC; VPK-252, San Diego, CA).
2. pacAd5 9.2-100 Ad backbone vector (CELL BIOLABS, INC; VPK-252, San Diego, CA).
3. 293 AD cell line (CELL BIOLABS, INC; AD-100).
4. Pacl (New England Biolabs; #R0547L, Beverly, MA).
5. Lipofectamine 2000 (Life Technologies; #11668019).
6. PCR purification kit (QIAquick).

## 2.2 Cell-to-Cell Transmission of α-Synuclein in the Dual-Cell BiFC System

*2.2.1 BiFC Cell Lines*

Figure 1 shows the constructs of BiFC complement system.

1. V1S: SH-SY5Y stably expressing Venus N-terminal fragment conjugated human α-synuclein.
2. SV2: SH-SY5Y stably expressing Venus C-terminal fragment conjugated human α-synuclein.

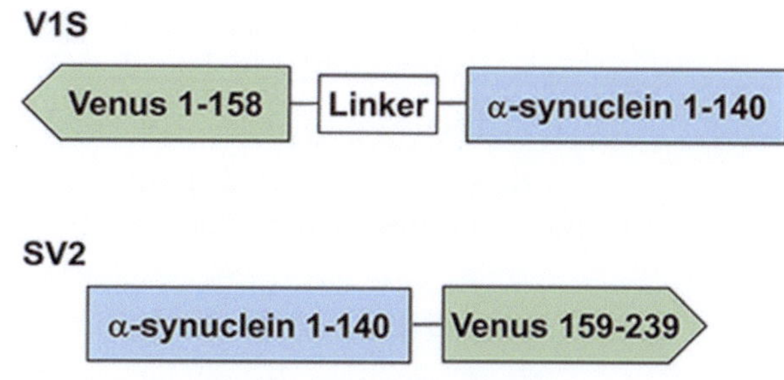


**Fig. 1** Constructs used in the dual-cell BiFC system

*2.2.2 Materials for Cell Culture*

1. Geneticin Selective antibiotic (G418 Sulfate) from Gibco.
2. G418 media: Growth media with 300 μg/mL G418 Sulfate.
3. 12-Well culture plates.
4. Flamed glass cover slips (Fisher Scientific).
5. Poly-L-lysine (0.1 mg/mL in DW) from Sigma Aldrich.

*2.2.3 Buffers and Reagents for Immunofluorescence Staining*

1. Paraformaldehyde (4 % in PBS).
2. Blocking solution (5 % BSA/3 % goat serum in PBS).
3. Topro-3 iodide (Invitrogen).
4. Prolong gold anti-fade reagent (Invitrogen).

*2.2.4 Antibodies for Immunofluorescence Staining*

1. GFP N-terminus polyclonal antibody (Cell Signal Technology; #2555).
2. GFP C-terminus polyclonal antibody (Santa Cruz Biotechnology; sc-5384).
3. Fluorescence-conjugated secondary antibody (Jackson Immunoresearch Laboratories, Inc).

## 3 Methods

### 3.1 Cell-to-Cell Transfer of α-Synuclein Aggregates in SH-SY5Y Co-culture System

*3.1.1 Adenovirus Preparation by Using RAPAd CMV Adenoviral Bicistronic Expression System for α-Synuclein Overexpression*

1. After cloning human α-synuclein into pacAd5 CMV K-N pA shuttle vector, digest a sufficient amount (20 μg) of the shuttle vector and pacAd5 9.2-100 Ad backbone vector with PacI to linearize the plasmids.
2. To confirm the complete digestion by PacI, run 0.5 μg of each digested DNA and undigested DNA on a 0.8 % agarose gel.
3. Cleanup digested DNA using PCR purification kit (we use the QIAquick kit from Qiagen).
4. Seed $4 \times 10^6$ cells of 293 AD cell line in a 100 mm culture dish using growth media without antibiotic a day before transfection.
5. After 24 h, transfect shuttle vector and backbone vector to 293 cells when the confluency becomes 70–80 % using Lipofectamine 2000 (*see* **Note 1**).
6. The next day, aspirate the medium containing transfection reagent and add 10 mL of fresh growth media.
7. Incubate the cells (for 7–15 days) until plaques are visualized.
8. When the plate is ready for harvest (>50 % of cells lifted), collect the crude viral lysate.
9. Release viruses from cells by three freezing and thawing cycles (30 min each in −80 °C and quick thawing in 37 °C water bath).
10. Centrifuge the cell lysate at $3{,}000 \times g$ for 15 min to remove the cell debris.
11. Aliquot and store the crude viral lysate at −80 °C.

*3.1.2 Differentiation of SH-SY5Y Cells*

1. Seed $1 \times 10^6$ cells of SH-SY5Y onto 100 mm culture dish using growth media a day before differentiation.
2. Next day, change media using growth media with retinoic acid (day 0), then change media using fresh growth media with retinoic acid every 2–3 days. On the day 5 after starting differentiation, the cells are ready for adenoviral transduction. Reserve some differentiated cells without transduction to serve as recipient cells.

*3.1.3 Adenoviral Transduction of SH-SY5Y Cells for Overexpression of α-Synuclein*

1. To calculate the amount of virus required for infection, determine the optimal multiplicity of infectious unit (MOI).
2. After the optimal MOI has been determined, calculate the volume of virus required using the following equation: Required volume of virus = (No. of cells) × (MOI)/Concentration of virus (titer) (*see* **Note 2**).
3. After calculating the required volume of virus, aspirate the growth media and add adenovirus (MOI 30) diluted in 1/2 total volume of fresh growth media to cells (for example, if cells were cultivated with 10 mL of media in 100 mm culture dish, after aspiration of culture media, virus is diluted with 5 mL of fresh media and added to cells).
4. After incubation for 1.5 h at 37 °C, add the remaining 1/2 volume using growth media with retinoic acid. Transfected cells will served as donor cells.

*3.1.4 Co-culture of SH-SY5Y Cells*

1. On day 6 of differentiation label recipient cells with Qtracker 585. First prepare 1 nM Qtracker labeling solution in 1.5 mL microcentrifuge tube (*see* **Note 3**).
2. Add 1 mL of fresh growth media to the tube and vortex vigorously for 30 s.
3. Add this 1 mL of Qtracker labeled media solution to the cells on a cover slip.
4. Incubate the cells for 1 h at 37 °C.
5. After the incubation, wash the cells five times with fresh DMEM.
6. To co-culture the recipient cells with donor cells, begin by trypsinizing the donor cells.
7. Add $6 \times 10^4$ donor cells to $6 \times 10^4$ recipient cells on the cover slip.
8. Incubate cells for 1–3 days.
9. To fix the cells for immunofluorescence cell staining, wash the cover slips twice with ice-cold PBS. Add 1 mL of 4 % paraformaldehyde in PBS to cells. Incubate for 30 min at room temperature. After the incubation, wash cells three times with ice-cold PBS.

10. Permeabilize the cells with ice-cold permeabilization solution for 5 min at room temperature.
11. Rinse the cells with ice-cold PBS three times. Add 1 mL of blocking solution to each well, and then incubate for 30 min at room temperature with shaking.
12. After the blocking add appropriate primary antibodies in blocking solution. And incubate the dish for 30 min at room temperature with shaking.
13. Wash with ice-cold PBS for 20 min each for three times.
14. Add fluorescent dye-conjugated goat anti-mouse or rabbit antibody in blocking solution.
15. Incubate for 30 min at room temperature with shaking.
16. Wash with ice-cold PBS four times, 30 min each.
17. Incubate with diluted Topro-3 iodide for 10 min at room temperature with shaking.
18. Wash with ice-cold PBS three times, 5 min each.
19. Put a drop of prolong gold anti-fade reagent on slide glass and cover with the cover slip. After drying for several hours at RT, seal the cover slips with clear nail polish.
20. Store at 4 °C.
21. Obtain the images with confocal laser scanning microscope, and quantify the fluorescence intensity using image analysis software.
22. Quantification of transmission efficiency: Calculate the transmission efficiency using the following equation: Transmission efficiency = (Number of α-synuclein (+) and Q tracker (+) cells/Number of Q tracker (+) cells) × 100.
23. Quantify the level of transmitted α-synuclein using the following equation: Level of transmitted α-synuclein = (Integrated density of α-synuclein in α-synuclein (+) and Q tracker (+) cells) − (Area of selected cell × mean fluorescence of background readings).

### 3.2 Cell-to-Cell Transmission of α-Synuclein Aggregates in the Dual-Cell BiFC System

#### 3.2.1 Establishment of BiFC Co-culture System

1. Generate the two BiFC DNA constructs containing α-synuclein fused with either N-terminus (V1S) or C-terminus (SV2) fragment of Venus fluorescence protein (Fig. 2).
2. Transfect either V1S or SV2 construct to SH-SY5Y cells using electroporation.
3. Select transfected cells on G418 media for 2–3 weeks until colonies emerged.

#### 3.2.2 Co-culture of V1S and SV2 BiFC Cells

1. Seed the $1 \times 10^6$ cells of V1S and SV2 cells each to a 12-well plate using G418 media the day before co-culture.

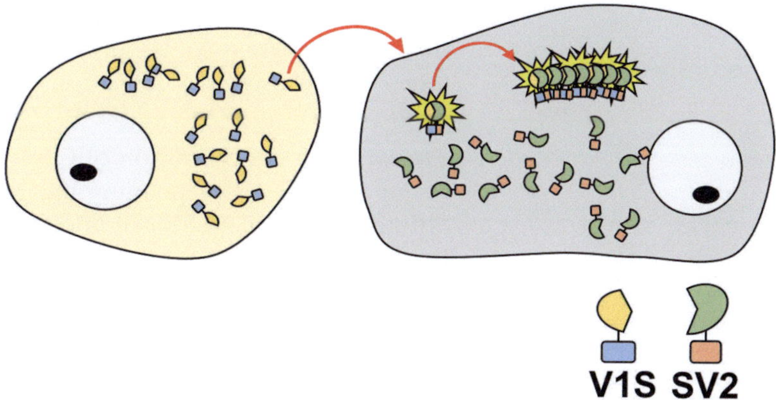


**Fig. 2** A scheme of cell-to-cell transmission of $\alpha$-synuclein in the dual-cell BiFC co-culture system

2. Coat the cover slip with 0.1 mg/mL of poly L-Lysine in DW for 30 min at room temperature and wash the cover slips with sterile water twice.
3. Trypsinize V1S and SV2 cells. Pipette $7.5 \times 10^6$ cells of V1S and SV2 to new 15 mL conical tube.
4. After centrifugation at $100 \times g$ for 5 min, individually resuspend the V1S and SV2 cells with 0.5 mL of growth media.
5. Mix 0.5 mL of V1S and SV2 cells. Seed the V1S and SV2 mixture onto the coated cover slip (*see* **Note 4**).
6. Incubate for 3 days at 37 °C.
7. After 3-day incubation, fix and stain the cells with protection from light.
8. Obtain the images with a confocal laser scanning microscope, and quantify the fluorescence intensity using image analysis software.
9. Quantification of transmission efficiency: Calculate the percentage of cells with aggregate transmission using the equation: (BiFC(+) cell number/Total cell number) × 100.
10. Quantify the levels of "seeded" α-synuclein aggregation using the equation: Integrated fluorescence intensity in BiFC (+) cell − (Area of selected cell x mean fluorescence of background readings).

## 4 Notes

1. For example, 20 μg of pacAd5 CMV K-N pA shuttle vector and 4 μg of pac Ad5 9.2-100 Ad backbone vector are mixed with 60 μL of lipofectamine 2000 reagent according to the manufacturer's recommendation. The mixture is added drop wise directly into the growth medium.

2. For example, when the $1 \times 10^7$ cells are infected with adenoviruses whose titer is $6 \times 10^9$ pfu/mL calculate the required volume of virus using the following equation: $(1 \times 10^7 \text{ cells}) \times 30$ MOI / $6 \times 10^9$ pfu/mL = 0.05 mL (50 μL of virus).

3. The working concentration of Qtracker label reagent is in the range of 2–15 nM depending on the cell type. In the case of differentiated SH-SY5Y cell line, the final working concentration is 2 nM.

4. For controls to exclude nonspecific fluorescence or autofluorescence in V1S and SV2 culture, seed $7.5 \times 10^6$ cells of V1S and SV2 cells separately to different cover slips.

## References


1. Ross CA, Poirier MA (2004) Protein aggregation and neurodegenerative disease. Nat Med 10(Suppl):S10–S17
2. Polymenidou M, Cleveland DW (2012) Prion-like spread of protein aggregates in neurodegeneration. J Exp Med 209:889–893
3. Lee SJ, Desplats P, Sigurdson C, Tsigelny I, Masliah E (2010) Cell-to-cell transmission of non-prion protein aggregates. Nat Rev Neurol 6:702–706
4. Spillantini MG, Schmidt ML, Lee VM, Trojanowski JQ, Jakes R, Goedert M (1997) Alpha-synuclein in Lewy bodies. Nature 388:839–840
5. Spillantini MG, Crowther RA, Jakes R, Hasegawa M, Goedert M (1998) alpha-Synuclein in filamentous inclusions of Lewy bodies from Parkinson's disease and dementia with lewy bodies. Proc Natl Acad Sci U S A 95:6469–6473
6. Braak H, Ghebremedhin E, Rub U, Bratzke H, Del Tredici K (2004) Stages in the development of Parkinson's disease-related pathology. Cell Tissue Res 318:121–134
7. Desplats P, Lee HJ, Bae EJ, Patrick C, Rockenstein E, Crews L, Spencer B, Masliah E, Lee SJ (2009) Inclusion formation and neuronal cell death through neuron-to-neuron transmission of alpha-synuclein. Proc Natl Acad Sci U S A 106:13010–13015
8. Kordower JH, Chu Y, Hauser RA, Freeman TB, Olanow CW (2008) Lewy body-like pathology in long-term embryonic nigral transplants in Parkinson's disease. Nat Med 14:504–506
9. Li JY, Englund E, Holton JL, Soulet D, Hagell P, Lees AJ, Lashley T, Quinn NP, Rehncrona S, Bjorklund A, Widner H, Revesz T, Lindvall O, Brundin P (2008) Lewy bodies in grafted neurons in subjects with Parkinson's disease suggest host-to-graft disease propagation. Nat Med 14:501–503
10. Mendez I, Vinuela A, Astradsson A, Mukhida K, Hallett P, Robertson H, Tierney T, Holness R, Dagher A, Trojanowski JQ, Isacson O (2008) Dopamine neurons implanted into people with Parkinson's disease survive without pathology for 14 years. Nat Med 14:507–509
11. Lee HJ, Suk JE, Bae EJ, Lee JH, Paik SR, Lee SJ (2008) Assembly-dependent endocytosis and clearance of extracellular alpha-synuclein. Int J Biochem Cell Biol 40:1835–1849
12. Lee HJ, Patel S, Lee SJ (2005) Intravesicular localization and exocytosis of alpha-synuclein and its aggregates. J Neurosci 25:6016–6024
13. Lee HJ, Bae EJ, Jang A, Ho DH, Cho ED, Suk JE, Yun YM, Lee SJ (2011) Enzyme-linked immunosorbent assays for alpha-synuclein with species and multimeric state specificities. J Neurosci Methods 199:249–257

# Chapter 20

# Preparation of Amyloid Fibrils Seeded from Brain and Meninges

**Kathryn P. Scherpelz, Jun-Xia Lu, Robert Tycko, and Stephen C. Meredith**

## Abstract

Seeding of amyloid fibrils into fresh solutions of the same peptide or protein in disaggregated form leads to the formation of *replicate fibrils* [1], with close structural similarity or identity to the original fibrillar seeds. Here we describe procedures for isolating fibrils composed mainly of β-amyloid (Aβ) from human brain and from leptomeninges, a source of cerebral blood vessels, for investigating Alzheimer's disease and cerebral amyloid angiopathy. We also describe methods for seeding isotopically labeled, disaggregated Aβ peptide solutions for study using solid-state NMR and other techniques. These methods should be applicable to other types of amyloid fibrils, to Aβ fibrils from mice or other species, tissues other than brain, and to some non-fibrillar aggregates. These procedures allow for the examination of authentic amyloid fibrils and other protein aggregates from biological tissues without the need for labeling the tissue.

**Key words** β-Amyloid, Alzheimer's disease, Solid-state NMR, Fibril structure, Fibril polymorphism, Protein aggregation, Protein aggregation diseases, Neuritic plaques, Cerebral amyloid angiopathy

## 1 Introduction

Amyloid fibrils and pre-amyloid protein and peptide aggregates are believed to be pathogenic in Alzheimer's disease and many other diseases within and outside of the central nervous system. The study of the structure of such aggregates remains an important objective both in understanding the disease processes, and in designing diagnostic and therapeutic agents.

A major advance in our understanding of the structure of these protein aggregates has come from the use of solid-state NMR spectroscopy, which allows the interrogation of solid but noncrystalline materials. Solid-state NMR spectroscopy has shown that most amyloid fibrils contain a core of parallel, in-register β-sheets [2–4]. More recent studies have elucidated additional structural details, including the supramolecular organization of peptides within the fibril.

David Eliezer (ed.), *Protein Amyloid Aggregation: Methods and Protocols*, Methods in Molecular Biology, vol. 1345, DOI 10.1007/978-1-4939-2978-8_20, 

The formation of amyloid fibrils and pre-amyloid protein/peptide aggregates represents a failure of protein folding. Aβ, for example, is unstructured in solution, and never adopts a unique three-dimensional fold. In 2005, Petkova et al. found documented polymorphism in Aβ amyloid fibrils [1], based on differences in fibrillization conditions, i.e., whether otherwise identical solutions of Aβ peptide were allowed to form fibrils under "quiescent" or "agitated" conditions (gentle swirling). These results clearly demonstrated that for Aβ amyloid fibrils, structure was not uniquely determined by amino acid sequence. In addition to structural differences, observed at the level of electron microscopy and solid-state NMR spectroscopy, these authors observed differences in cytotoxicity of these two fibril types in vitro. Subsequent studies elucidated detailed structural models for these two types of Aβ fibrils [5–7]. Similar types of polymorphism have been observed with other fibrils of other proteins or peptides (e.g., β2-microglobulin [8] among others). Polymorphism is now believed to be a general property of amyloid fibrils. The variation in fibril structure is also highly reminiscent of the strain phenomenon found in yeast and mammalian prions [9–11].

One of the critical findings from the above studies is that differences in fibril structure can be propagated to "progeny" fibrils through seeding [1]. Thus, Aβ fibrils formed under "agitated" conditions can be used as seeds to form fibrils from fresh, disaggregated Aβ solutions, and these will have strong resemblance to the "parent" fibrils structurally—and these structural properties will be relatively independent of fibrillization conditions. The seeds, in other words, "trump" the fibrillization conditions, presumably because most or all of the fibril polymorphism arises at the level of nucleation.

These findings present an excellent opportunity for examining the structures of fibrils from biological material, especially where the usual labeling procedures used in NMR spectroscopy cannot be readily applied. In particular, although the above-cited studies documented polymorphism of Aβ fibrils, it was not known which of these structures—or both or neither—would be found in the brains of individuals with Alzheimer's disease and the related condition of cerebral amyloid angiopathy. Reasoning that one can make isotopically labeled Aβ fibrils that are replicates of authentic brain Aβ fibrils from patients dying with Alzheimer's disease, we used such material to isolate amyloid fibrils. We then used these fibrils as seeds for generating isotopically labeled replicate Aβ fibrils for study by solid-state NMR. We have shown that amyloid fibrils can be retrieved from brains at autopsy, and then used to generate fibrils from fresh solutions of synthetic Aβ peptides. The synthetic peptides can be labeled by standard procedures. The brain amyloid seeds fibril formation in solutions of synthetic, disaggregated Aβ peptides, and these fibrils have been studied by solid-state NMR, electron microscopy, X-ray diffraction, and other biophysical techniques. Using these procedures, we have interrogated the structure of amyloid fibrils occurring in the brains of patients dying with or of Alzheimer's disease and its variants [12, 13]. NMR spectra of the replicate fibrils show surprisingly sharp peaks, and in general, one or two structural populations in each brain. We have also shown that sampling from multiple regions with a single brain yields replicate fibrils of identical structure, i.e., without variation depending on anatomic location within a single brain. Importantly, we have also observed differences among patients in the structures of their brain-derived amyloid fibrils.

Here, we present procedures for isolating Aβ fibrils from human brains obtained at autopsy from patients dying with or of Alzheimer's disease. The procedure is depicted schematically in Fig. 1. We have recently also isolated Aβ fibrils from leptomeninges, as a source of cerebrovascular amyloid, and present these procedures below. We also describe procedures for using such biochemically isolated amyloid as a seed, to generate replicate progeny fibrils.

We focus on Aβ fibrils. These procedures, however, should be adaptable to diverse types of samples, in particular (1) fibrils composed of other peptides and proteins; (2) Aβ and other fibrils from non-human species; (3) tissues other than brain; and (4) some non-fibrillar aggregates of Aβ, including soluble oligomeric species.

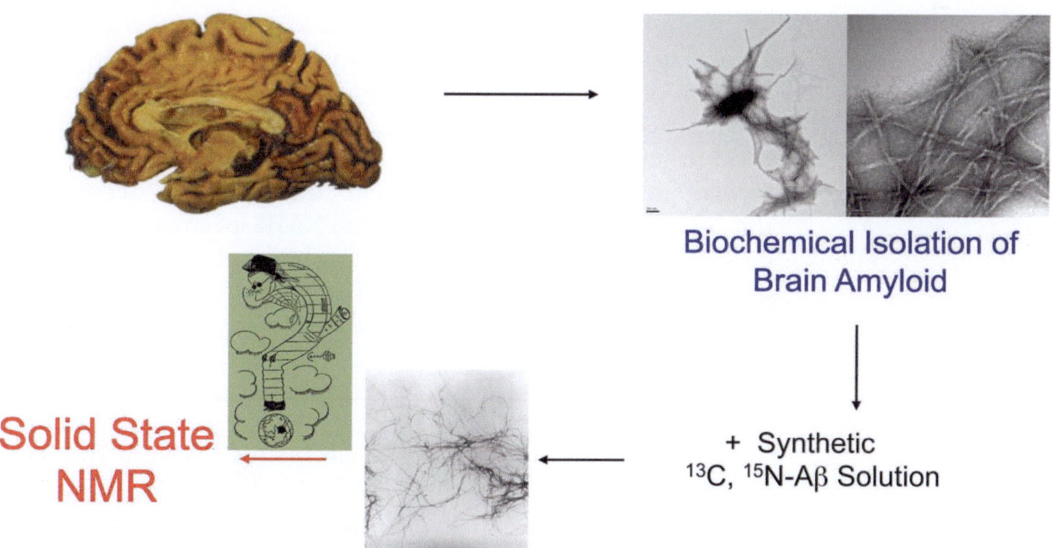


**Fig. 1** Schematic of procedure for forming amyloid fibrils seeded from material obtained from brain or leptomeninges. Amyloid fibrils are isolated by biochemical procedures, and are then added to synthetic (or expressed), isotopically labeled Aβ40 or Aβ42. The resulting replicate fibrils can then be interrogated using solid-state NMR spectroscopy

The following are the main issues and questions about procedures for isolating amyloid fibrils from human brain or other tissues.

*How harshly to treat the tissues, i.e., how rigorously to purify the amyloid.* Early procedures for isolating amyloid fibrils from brain took advantage of the fact that fibrils are resistant to denaturants, including SDS and acidic conditions, and are protease resistant [14, 15]. Prolonged treatment of amyloid fibrils under "harsh" conditions will eventually lead to some loss and degradation of amyloid fibrils, however. Amyloid fibrils constitute a minute fraction of the brain even in patients with advanced Alzheimer's disease; thus it is necessary to eliminate most of the material in brains in order to obtain amyloid fibrils that can effectively seed Aβ solutions. Isolation procedures inevitably involve the use of organic solvents, denaturants and proteases. Here we describe procedures to "navigate" the trade-off between obtaining greater purity in the isolated brain amyloid, versus loss of seeding material and potential alteration of the material from harsh isolation procedures.

*Which fractions to retain from brain parenchyma or meninges.* As discussed below, there are slight differences (e.g., fractions from ultracentrifugation steps) between procedures for isolating amyloid from brain parenchyma or leptomeninges. These differences were based solely on the empirical assessment of the ability of various fractions to seed amyloid formation in solutions of Aβ40. Amyloid represents a fairly small fraction of the total mass of these tissues, and differences presumably reflect difference in the overall composition of the tissues.

*Conditions for seeding, particular with respect to the number of "generations" of progeny fibrils.* This issue follows directly from the previous one. In earlier studies, more rigorous purification of brain amyloid reduced the overall yield of fibrils. This, in turn, necessitated re-seeding procedures, i.e., the use of first-generation brain-seeded fibrils to form sufficient quantities of second- and third-generation fibrils for solid-state NMR spectroscopy. The procedures presented below allow for the generation of sufficient quantities of labeled fibrils for solid-state NMR from as little as one gram of starting human brain material.

## 2 Materials

1. Brains: Brain tissue is obtained at the time of autopsy and is either used immediately or frozen at −80 °C (*see* **Note 1**).
2. Homogenization buffer: 10 mM Tris, 3 mM EDTA, 0.25 M sucrose, 50 μg/mL gentamicin sulfate, 0.25 μg/mL amphotericin B, and protease inhibitor (one Roche complete protease inhibitor tablet is added to ~100 mL of homogenization buffer before use).

3. Tenbroek homogenizer
4. First ultracentrifugation buffer: Identical to homogenization buffer, except that solid sucrose is added to a final concentration of 1.9 M.
5. Wash buffer: 50 mM Tris-HCl, pH 8.00
6. Digestion buffer: 50 mM Tris pH 8.00, 2 mM CaCl, 0.2 mg/mL collagenase CLS3 (Worthington), and 0.01 mg/mL DNase I. Between 1 and 10 mL collagenase solution were used, based on the size of the sample.
7. SDS-second ultracentrifugation buffer: 50 mM Tris pH 8.00, also containing 1.3 M sucrose and 1 % (w/v) SDS.
8. Seeding buffer: 10 mM sodium phosphate pH 7.40, also containing 0.01 % (w/v) $NaN_3$.
9. Phosphate buffer (PB): 10 mM sodium phosphate, pH 7.40, also containing 0.01 % (w/v) $NaN_3$.
10. Sonifier: We have used two particular sonifiers: Branson S-250A Sonifier with a tapered 1/8th inch micro-tip horn, and Branson 450 Sonifier equipped with a 1/8th inch micro-tip.
11. CNBr: Crystalline CNBr should be stored at –20 °C under N2 atmosphere, with the bottle inverted so that the CNBr is continuously re-sublimed. The last quarter or third of the bottle should be discarded.
12. 88 % Formic acid: This is concentrated formic acid as available from vendors. The actual formic acid varies slightly by lot.
13. Solvent A: Filtered, deionized water with 0.1 % (v/v) TFA.
14. Solvent B: Acetonitrile with 0.1 % (v/v) TFA.

## 3 Methods

### 3.1 Amyloid Extraction

1. Brain or leptomeningeal is obtained at autopsy and is placed into a disposable plastic container *without fixative* (*see* **Note 2**).
2. The tissue is grossly dissected to separate meninges and visible blood vessels.
3. In brain samples, gray and white matter are separated from one another with a scalpel or razor blade.
4. At this point, it can be used immediately, or stored frozen (–20 °C).
5. The tissue is then homogenized in 20 volumes of ice-cold homogenization buffer in a Tenbroek homogenizer. The slurry is then stirred overnight at 4 °C.
6. Solid sucrose is then added to the sample to 1.20 M. The sample is centrifuged for 45 minutes at 25,000×*g* at 4 °C (*see* **Note 3**). The supernatant and upper layer are discarded and the pellet is retained.

7. The pellet is resuspended in first ultracentrifugation buffer. It is ultracentrifuged for 30 min at 125,000 × *g* at 15 °C (*see* **Note 4**). For brain parenchymal samples, the upper layer is reserved; for meningeal samples, the pellet is reserved.
8. The reserved portion is washed twice in wash buffer. Approximately 10 volumes of wash buffer is added to the reserved portion, which is then centrifuged for five minutes at 13,400 × *g* at 4 °C using a tabletop centrifuge (*see* **Note 5**).
9. The pellet is resuspended by adding approximately 10 volumes of digestion buffer and vortexing for a few seconds. The samples are incubated overnight (approximately 16 h) at 37 °C. Tubes are placed horizontally and swirled or shaken vigorously (approximately 200 rpm).
10. The samples are washed twice as in **step 8**, above, and then resuspended in SDS-second ultracentrifugation buffer.
11. The samples are then ultracentrifuged for 30 minutes at 200,000 × *g* at 15 °C.
12. Typically, the sample size of the ultracentrifuge tube is 12 mL. The upper 2 mL of the slurry is removed and discarded, and this volume, containing mainly lipids, is replaced with 2 mL of distilled $H_2O$. The mixture is gently and repeatedly (~20 times) pipetted to disperse the solid material. The slurry is then recentrifuged for 30 min at 200,000 × *g* and at 15 °C. Any additional lipid-rich material is discarded.
13. The supernatant is then discarded and the pellet gently washed once with distilled $H_2O$, and then twice additionally with PB. In these washes, approximately 10 volumes of liquid is added to the solid; the mixture is then briefly vortexed and then centrifuged for 5 minutes at 13,400 × *g*.
14. This isolate is then sonicated at low power for ~ 20 s. in an ice bath for brain parenchyma, or ~30 s. for leptomeningeal samples. This step serves to disperse the isolated brain or meningeal material for assays to quantify Aβ. More vigorous sonication is needed to disperse amyloid fibrils for seeding Aβ solutions.
15. Total protein is measured using any convenient protein assay, such as the BCA assay (Pierce). Aβ content is quantified after digestion with CNBr by LC-MS (unpublished, *see* Subheading 3.3 below; a similar procedure has been published by Kuo et al. [16]. In addition, the isolate is examined by transmission electron microscopy (TEM), which generally shows some collagen fibrils and amorphous material of unknown composition in addition to amyloid fibrils.
16. The final pellet contains approximately 6- to 12-fold enrichment of amyloid seeding activity. Of the brain extract, in most samples Aβ represents ≤1 % of the protein mass.

17. The isolate (*see* **Note 6**) can be frozen and stored at −20 °C, or used for seeding. Depending on the quantity of starting material, it may be convenient to make aliquots for freezing. Each aliquot should contain sufficient material to give total protein concentrations of approximately 5 mg/mL when suspended in seeding buffer (1.8 mL, as described below).

### 3.2 Seeded Growth of Amyloid Fibrils

1. An aliquot of the above material is thawed and suspended in 1.8 ml of seeding buffer such that the slurry contains approximately 5 mg/ml of insoluble material.
2. The suspension is vigorously sonicated (*see* **Note 7**). The material is then frozen in liquid nitrogen and thawed to room temperature.
3. A solution of synthetic or expressed Aβ is then made. In most of our experiments, this is synthetic Aβ40 (see, for example, [12, 13, 17]). Comparable results can be obtained using soluble, synthetic Aβ42, and cross-seeding can occur between Aβ40 and Aβ42, albeit at slower rates than self-seeding (Fig. 2). Peptide is dissolved in neat dimethyl sulfoxide (DMSO) to a peptide concentration of 6 mM. This is then added to the sonicated brain extract so that the final peptide concentration is 100 μM (i.e., 0.8 mg of Aβ40).
4. Typically, we used 1 g of frozen brain tissue and 1 mg of synthetic peptide for seeding experiments.
5. Fibril growth is monitored by TEM and/or ThT fluorescence [18].
6. Successful seeding is shown by the appearance of long fibrils in TEM images and/or a rise and plateau in signal in ThT fluorescence (*see* **Note 8**) within a few hours.
7. Fibril growth is allowed to continue for approximately 24 h (or less, *see* **Note 9**), after which fibrils are pelleted by ultracentrifugation and lyophilized for solid-state NMR measurements (*see* **Note 10**).

### 3.3 Quantitation of Aβ40 and Aβ42 in tissue isolates by CNBr Cleavage and LC-MS

1. An aliquot of brain isolate containing 5–10 μg of protein is dispersed in approximately 25 μL of PB in a conical glass tube.
2. A fresh solution of CNBr at 5 mg/mL in 88 % formic acid is then prepared in a fume hood.
3. 160 μL of the CNBr solution is added to each sample. A sample of 25 μL PB and 160 μL of the CNBr solution is used as a blank.
4. Samples are covered with parafilm and incubated at 37 °C overnight; tubes are shaken (200 rpm).
5. Next, 1.4 mL of distilled or Milli-Q $H_2O$ is added to each tube, and the material is lyophilized. Since CNBr is volatile, most of the residual CNBr is removed by lyophilization.

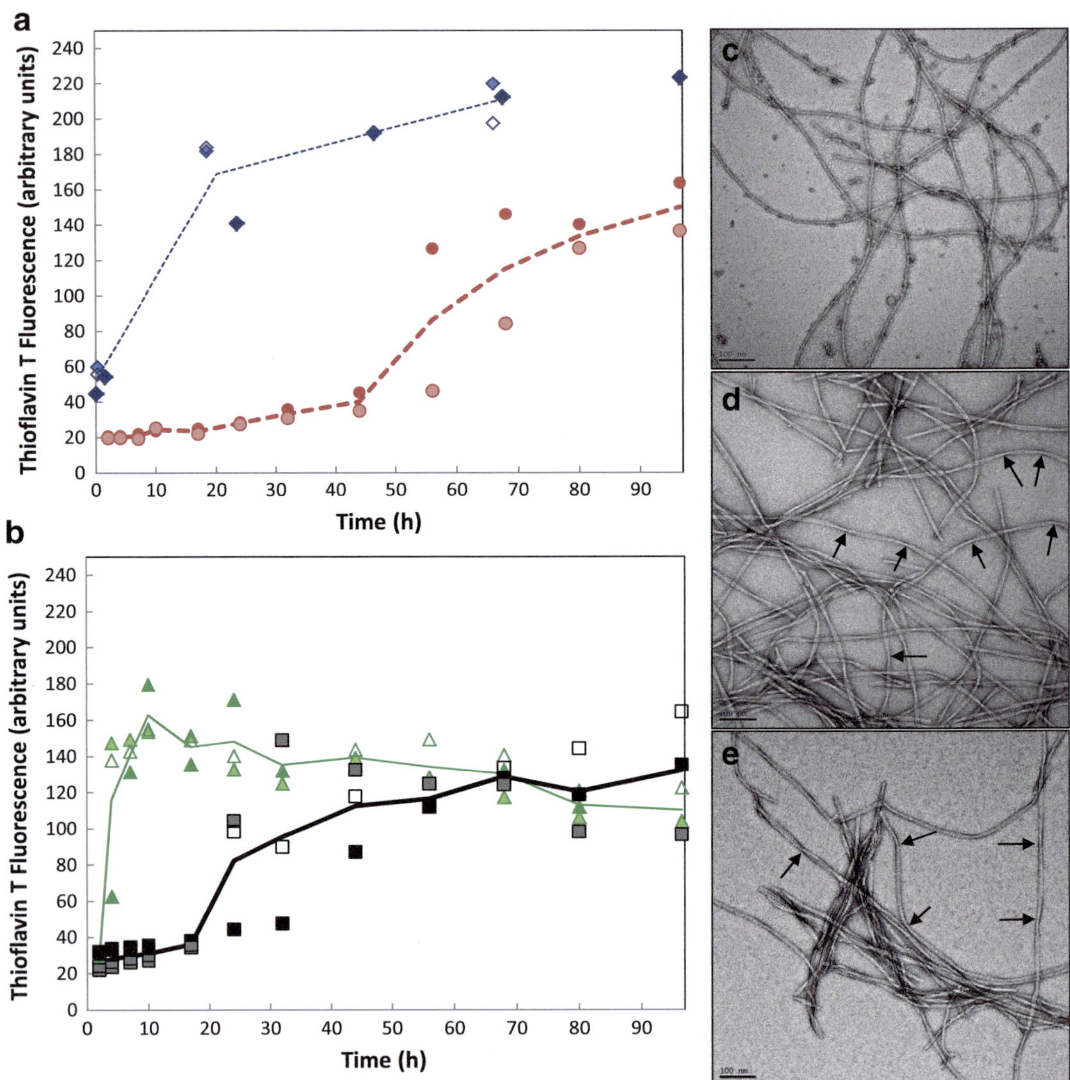


**Fig. 2** Synthetic Aβ40 or Aβ42 fibrils used for seeding fibrillization of soluble Aβ40. The resulting fibrils contain mainly Aβ40 peptide. (**a**) Unseeded growth of 50 μM Aβ42 (*blue diamonds*) and 100 μM Aβ40 (*red circles*). Note that Aβ42 reaches higher maximum values for ThT fluorescence (arbitrary units), even at half the concentration of Aβ40. (**b**) Growth of Aβ40 fibrils after seeding by 10 % preformed Aβ40 (*green triangles*) or Aβ42 (*black squares*). Assays were performed in duplicate or triplicate; lines demarcate the averages. Different fillings of the symbols (squares, triangles, circles of a single color) represent replicate seeding reactions done at the same time, using a single fibril sample for seeding. (**c**) TEM images of Aβ42 fibrils. (**d**) TEM images of Aβ40 fibrils. (**e**) TEM images of fibrils formed using Aβ42 seeds and soluble Aβ40. Arrows point at fibril twists. Scale bar is 100 nm in each image

6. After lyophilization, soluble material is dissolved by adding 40 μL of distilled or Milli-Q $H_2O$. This step is repeated once or twice (*see* **Note 11**). The liquid is transferred to an Eppendorf tube.
7. This procedure usually leads to the transfer of some solid material particles, these being observed in every tube. For this reason, this material is centrifuged for 5 min at 13,400 × *g* and the supernatant is transferred to appropriate vials.

8. LC-MS should be performed with the next day or so.
9. It is beyond the scope of this chapter to consider all of the possibilities for performing LC-MS analyses. We have used the following procedure. We use an Agilent 1290 Infinity UHPLC and 6460 Triple Quad LC/MS instrument, with a Zorbax SB-C18 column with 1.8 μm pores and dimensions of 2.1 × 50 mm, and a 2.1 × 5 mm guard column with the same immobile phase. The column is heated to 40 °C and a flow rate of 0.3 mL/min is used. Peptides are eluted using the following gradient: 5–20 % solvent B over 4 min, followed by 20–43 % over 3 min, followed by 43–65 % over 1.4 min, followed by a 6-min wash of 100 % solvent B, followed by a 9-min equilibration of the column in 5 % solvent B.
10. We quantify Aβ(36–40) and Aβ(36–42). Other peptides are present in much lower quantities, but in principle the technique can be scaled up to detect other C-terminal fragments derived from Aβ. Dynamic multiple reaction monitoring was employed to identify and quantitate peaks of these two peptides.
11. The sequences of these peptides in the underivatized state are VGGVV and VGGVVIA, respectively. Standards are synthesized using FMOC chemistry, and purified by routine reverse-phase HPLC techniques. Standard peptides are quantified using amino acids analysis [19, 20].
12. Synthetic Aβ40 and Aβ42 fibrils can also be used as controls for the efficiency of CNBr cleavage, recovery of C-terminal peptides, the lyophilization and transfer steps, and LC-MS (*see* **Note 12**).
13. Figure 4 shows a sample of LC-MS data obtained on two brain samples. In this figure, panel A represents standards, i.e., two synthetic peptides standards, VGGVV and VGGVVIA, which were co-injected. Panels B and C represent analyses of brain amyloid from two different patients, with different ratios of Aβ40:Aβ42. Ratios of Aβ40:Aβ42 are calculated using standard curves based on multiple injections of the synthetic peptides on the day of the experiment (*see* **Note 13**).

## 4 Notes

1. All tissues should be treated with proper biosafety procedures. Although in our usual practice brain tissue is stored in a –80 °C freezer, we have observed no difference in quality of NMR spectra of brain-derived material stored at –20 °C.
2. It is critical that these brains not be put into fixatives such as formalin, which can modify and crosslink proteins. Since fixation is routine for brains at autopsy, and brains in particular may be routinely placed into fixative solutions immediately

after removal from the skull, it is often necessary to specify that this tissue is to remain unfixed.

3. For example, Sorvall RC 5C Plus centrifuge.
4. For example, Beckman XL-80 ultracentrifuge.
5. For example, Eppendorf Centrifuge 5415D, for which $13{,}400 \times g$ is equivalent to 10,000 rpm.
6. This procedure is designed to be fairly mild. Compared to our earlier procedure [12], the more current protocol [13] decreases the concentration and time of exposure to denaturants (SDS), reduces the time of exposure to collagenase, and eliminates other proteolytic digestion steps. Although collagenase shows significant substrate specificity, this specificity is not absolute. The goal is not to purify the fibrils rigorously, but rather to enrich the tissue isolate in amyloid fibrils. At the concentrations of collagenase and SDS used in this procedure, neither reagent has a significant effect on seeding by synthetic Aβ40 fibrils (*see* Fig. 3). The addition of collagen (from rat tail tendon, mainly type I collagen) did not have a significant effect on seeding.
7. Sonifiers differ considerably. In one typical procedure, a Branson S-250A sonifier with a tapered 1/8$^{th}$ inch micro-tip horn is used at lowest power, 10 % duty factor, for 10 min. In another procedure, a Branson 450 Sonifier probe sonicator equipped with a 1/8$^{th}$ inch micro-tip, was used at output 7, 80 % duty factor, for 8 min. The sample is kept in an ice bath. In addition, we typically include a one min break from sonication between the fourth and fifth minutes in order to avoid overheating the small volume of material.
8. Fibrils should appear in TEM images within approximately four hours. It is important to compare seeded fibril formation with unseeded control samples. In unseeded fibril growth, few or no fibrils should be apparent by either TEM or ThT fluorescence in the first 24 h.
9. Longer incubation times, up to 48 h, have been used as well. Ideally, [12], fibrils can be prepared in quantities sufficient for solid-state NMR measurements by a single seeding step and an incubation time of ≤48 h. In contrast to longer incubation times or multiple rounds of seeding, a single and short incubation step tends to minimizes the possibility of selective amplification of or suppression of fibril types, and interconversion of fibril types. Whether this can be achieved depends on many factors, especially the quantity of Aβ fibrils in any given brain sample and the quantity of brain material available.
10. For some solid-state NMR experiments, large quantities of material are needed, and therefore multiple rounds of seeding may be necessary. In such cases, it is important to confirm the

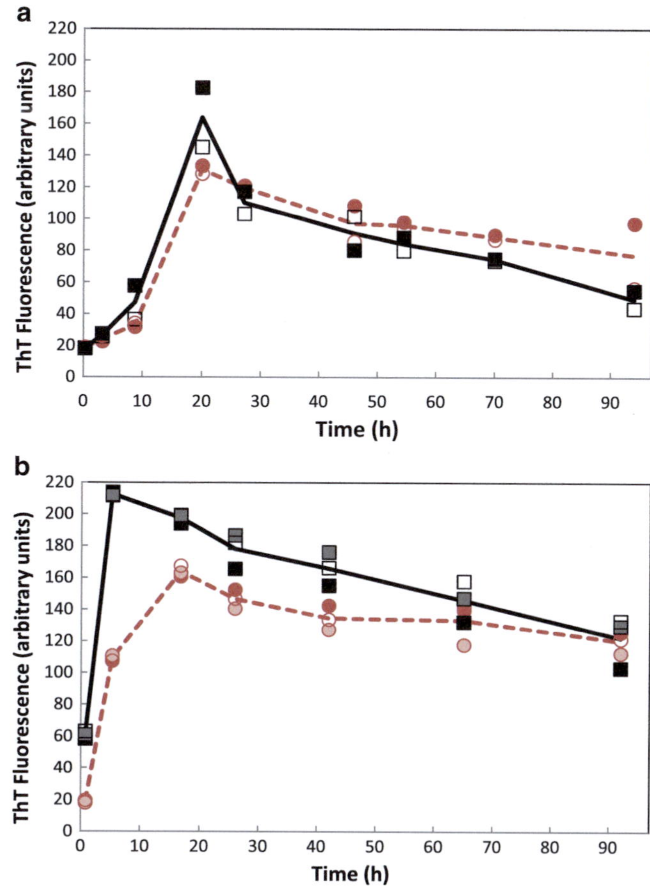


**Fig. 3** Collagenase and SDS treatments do not affect the seeding by synthetic Aβ40 fibrils. (**a**) Seeding of solutions of synthetic Aβ40 by 0.5 % (w/v) of synthetic Aβ40 fibrils, treated (*black filled square*) or not treated previously (*red filled circle*) with an overnight collagenase digestion. (**b**) Seeding of solutions of synthetic Aβ40 by 2 % (w/v) of synthetic Aβ fibrils, treated (*black filled square*) or not treated previously (*red filled circle*) with an incubation in 2 % (w/v) SDS. Assays were performed in duplicate or triplicate; lines demarcate the averages with (*continuous line*) and without (*dashed line*) treatments

reproducibility of fibril morphology by TEM and, when possible, chemical shift measurements from round to round.

11. The C-terminal peptides of Aβ (i.e., Aβ36-n, where $n = 39–42$) are fully solubilized by this procedure, owing in part to the residual acid in the lyophilized powder.
12. Typical efficiencies of the CNBr cleavage are 40–50 % for synthetic fibrils. These efficiency values almost certainly represent maximal values. Efficiency of cleavage of the CNBr cleavage of any given lot of synthetic Aβ fibrils declined over time, to approximately 25–30 % over the course of several months, presumably due to oxidation of Met35 of Aβ in the fibrils.

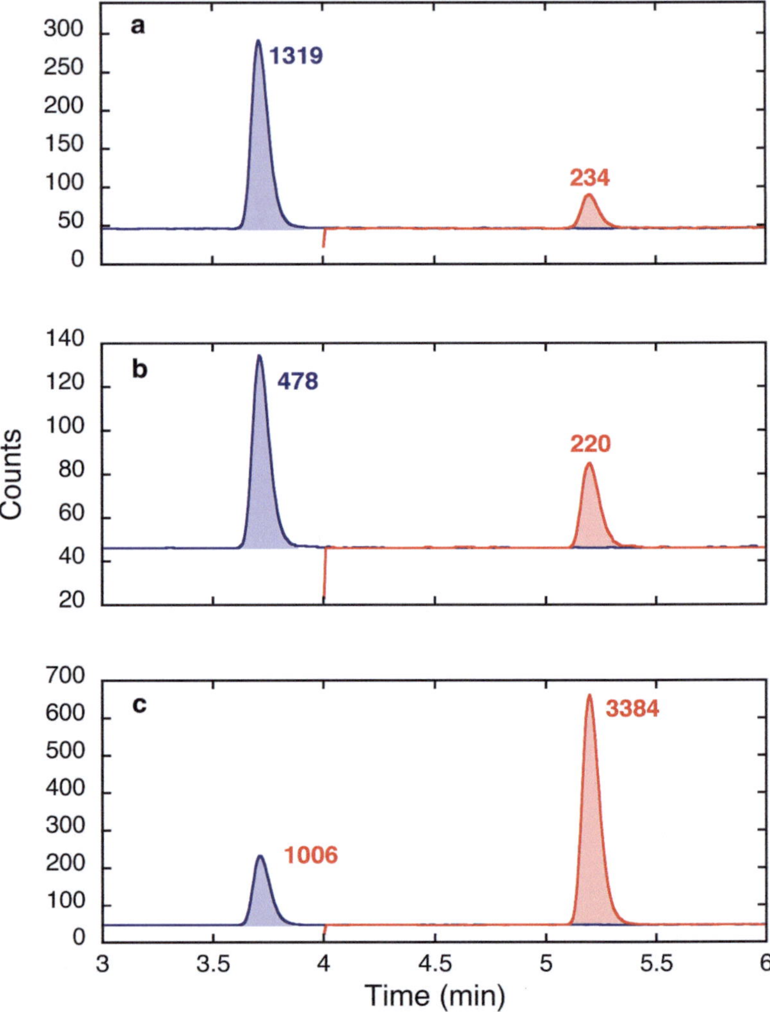


**Fig. 4** LC-MS analysis of peptide standards (**a**) and two brain samples (**b** and **c**) treated as described in the text. The panels are arranged as "chromatographs" in which the y-axis represents ion flow. In each case, the mass transition 430.3 → 72.1, including the mass of VGGVV (a measure of Aβ40), is shown in *blue*, and the mass transition 614.4 → 313.2, including the mass of VGGVVIA (a measure of Aβ42), is shown in *red*. The samples are analyzed by LC-MS in triplicate; the data shown in the figure represent a single determination for each sample

13. Standard curves are constructed from pure, synthetic peptides on the day of analyses. For the two peptides, VGGVV and VGGVVIA (indicative of Aβ40 and Aβ42), one typical set of results was as follows. A solution of VGGVV at $8.45 \times 10^{-4}$ M (0.363 μg/ml) in 0.1 % TFA/$H_2O$ was diluted with $H_2O$, and 4 μL of the diluted solutions was injected for LC-MS. These solutions gave a curve with the equation $y = 2793.7x + 39.35$ ($x$ in pmol injected; $R = 0.999$). A solution of VGGVVIA at

$2.38 \times 10^{-4}$ M (0.146 μg/ml) in 0.1 % TFA/$H_2O$ was diluted with $H_2O$, and 4 μL of the diluted solutions was injected for LC-MS. These solutions gave a curve with the equation $y = 1501.3x + 1.42$ ($x$ in pmol injected; $R = 0.999$). Thus, for the sample shown in Panel B of Fig. 4, with peak size readings of 478 and 220, the Aβ40:Aβ42 ratio is 0.157:0.146 = 1:0.93. For the sample shown in Panel C of Fig. 4, with peak size readings of 1006 and 3384, the Aβ40:Aβ42 ratio is 0.346:2.253 = 1:6.51.

## Acknowledgments

This work was supported in part by the Intramural Research Program of the National Institute of Diabetes and Digestive and Kidney Diseases, of the National Institutes of Health and by N.I.H. grant R01 NS042852 (to S.C.M.).

### References

1. Petkova AT, Leapman RD, Guo ZH, Yau WM, Mattson MP, Tycko R (2005) Self-propagating, molecular-level polymorphism in Alzheimer's β-amyloid fibrils. Science 307:262–265
2. Benzinger TLS, Gregory DM, Burkoth TS, Miller-Auer H, Lynn DG, Botto RE, Meredith SC (1998) Core structure of Alzheimer's β-amyloid fibrils determined by solid-state NMR. Proc Natl Acad Sci U S A 95:13407–134012
3. Tycko R (2014) Physical and structural basis for polymorphism in amyloid fibrils. Protein Sci 23:1528–1539
4. Tycko R (2011) Solid-state NMR studies of amyloid fibril structure. Annu Rev Phys Chem 62:279–299
5. Petkova AT, Yau WM, Tycko R (2006) Experimental constraints on quaternary structure in Alzheimer's β-amyloid fibrils. Biochemistry 45:498–512
6. Petkova AT, Ishii Y, Balbach JJ, Antzutkin ON, Leapman RD, Delaglio F, Tycko R (2002) A structural model for Alzheimer's β-amyloid fibrils based on experimental constraints from solid state NMR. Proc Natl Acad Sci U S A 99:16742–16747
7. Paravastu AK, Leapman RD, Yau WM, Tycko R (2008) Molecular structural basis for polymorphism in Alzheimer's β-amyloid fibrils. Proc Natl Acad Sci U S A 105: 18349–18354
8. Yamaguchi K, Takahashi S, Kawai T, Naiki H, Goto Y (2005) Seeding-dependent propagation and maturation of amyloid fibril conformation. J Mol Biol 352:952–960
9. Prusiner SB (2013) Biology and genetics of prions causing neurodegeneration. Annu Rev Genet 47:601–623
10. Toyama BH, Weissman JS (2011) Amyloid structure: conformational diversity and consequences. Annu Rev Biochem 80:557–585
11. Tycko R, Wickner RB (2013) Molecular structures of amyloid and prion fibrils: consensus versus controversy. Acc Chem Res 46:1487–1496
12. Paravastu AK, Qahwash I, Leapman RD, Meredith SC, Tycko R (2009) Seeded growth of β-amyloid fibrils from Alzheimer's brain-derived fibrils produces a distinct fibril structure. Proc Natl Acad Sci U S A 106:7443–7448
13. Lu JX, Qiang W, Yau WM, Schwieters CD, Meredith SC, Tycko R (2013) Molecular structure of β-amyloid fibrils in Alzheimer's disease brain tissue. Cell 154:1257–1268
14. Roher AE, Kuo YM (1999) Isolation of amyloid deposits from brain. Methods Enzymol 309:58–67
15. Roher AE, Lowenson JD, Clarke S, Wolkow C, Wang R, Cotter RJ, Reardon IM, Zurcherneely HA, Heinrikson RL, Ball MJ, Greenberg BD (1993) Structural alterations in the peptide backbone of β-amyloid core protein may account for its deposition and stability in Alzheimers-disease. J Biol Chem 268:3072–3083
16. Kuo YM, Kokjohn TA, Beach TG, Sue LI, Brune D, Lopez JC, Kalback WM, Abramowski D, Sturchler-Pierrat C, Staufenbiel M, Roher AE (2001) Comparative analysis of amyloid-beta chemical structure and amyloid plaque morphology of transgenic mouse and Alzheimer's disease brains. J Biol Chem 276:12991–12998

17. Sciarretta KL, Gordon DJ, Petkova AT, Tycko R, Meredith SC (2005) Aβ40-lactam(D23/K28) models a conformation highly favorable for nucleation of amyloid. Biochemistry 44:6003–6014
18. LeVine H 3rd (1999) Quantification of β-sheet amyloid fibril structures with thioflavin T. Methods Enzymol 309:274–284
19. Heinrikson RL, Meredith SC (1984) Amino acid analysis by reverse-phase high-performance liquid chromatography: precolumn derivatization with phenylisothiocyanate. Anal Biochem 136:65–74
20. Mora R, Berndt KD, Tsai H, Meredith SC (1988) Quantitation of aspartate and glutamate in HPLC analysis of phenylthiocarbamyl amino acids. Anal Biochem 172:368–376
21. Powers, JM (1995) Practice guidelines for autopsy pathology: autopsy procedures for brain, spinal cord, and neuromuscular system. Arch Path Lab Man 119:777–783

# Index



David Eliezer (ed.), *Protein Amyloid Aggregation: Methods and Protocols*, Methods in Molecular Biology, vol. 1345, DOI 10.1007/978-1-4939-2978-8, © Springer Science+Business Media New York 2016

MIX
Papier aus verantwortungsvollen Quellen
Paper from responsible sources
FSC® C105338

If you have any concerns about our products,
you can contact us on
**ProductSafety@springernature.com**

In case Publisher is established outside the EU,
the EU authorized representative is:
**Springer Nature Customer Service Center GmbH**
**Europaplatz 3, 69115 Heidelberg, Germany**

Printed by Libri Plureos GmbH
in Hamburg, Germany